CAMBRIDGE LIBRARY COLLECTION

Books of enduring scholarly value

Botany and Horticulture

Until the nineteenth century, the investigation of natural phenomena, plants and animals was considered either the preserve of elite scholars or a pastime for the leisured upper classes. As increasing academic rigour and systematisation was brought to the study of 'natural history', its subdisciplines were adopted into university curricula, and learned societies (such as the Royal Horticultural Society, founded in 1804) were established to support research in these areas. A related development was strong enthusiasm for exotic garden plants, which resulted in plant collecting expeditions to every corner of the globe, sometimes with tragic consequences. This series includes accounts of some of those expeditions, detailed reference works on the flora of different regions, and practical advice for amateur and professional gardeners.

The Trees of Great Britain and Ireland

Although without formal scientific training, Henry John Elwes (1846–1922) devoted his life to natural history. He had studied birds, butterflies and moths, but later turned his attention to collecting and growing plants. Embarking on his most ambitious project in 1903, he recruited the Irish dendrologist Augustine Henry (1857–1930) to collaborate with him on this well-illustrated work. Privately printed in seven volumes between 1906 and 1913, it covers the varieties, distribution, history and cultivation of tree species in the British Isles. The strictly botanical parts were written by Henry, while Elwes drew on his extensive knowledge of native and non-native species to give details of where remarkable examples could be found. Each volume contains photographic plates as well as drawings of leaves and buds to aid identification. The species covered in Volume 3 (1908) include cedar, hornbeam, southern beech, hickory, maple and redwood.

Cambridge University Press has long been a pioneer in the reissuing of out-of-print titles from its own backlist, producing digital reprints of books that are still sought after by scholars and students but could not be reprinted economically using traditional technology. The Cambridge Library Collection extends this activity to a wider range of books which are still of importance to researchers and professionals, either for the source material they contain, or as landmarks in the history of their academic discipline.

Drawing from the world-renowned collections in the Cambridge University Library and other partner libraries, and guided by the advice of experts in each subject area, Cambridge University Press is using state-of-the-art scanning machines in its own Printing House to capture the content of each book selected for inclusion. The files are processed to give a consistently clear, crisp image, and the books finished to the high quality standard for which the Press is recognised around the world. The latest print-on-demand technology ensures that the books will remain available indefinitely, and that orders for single or multiple copies can quickly be supplied.

The Cambridge Library Collection brings back to life books of enduring scholarly value (including out-of-copyright works originally issued by other publishers) across a wide range of disciplines in the humanities and social sciences and in science and technology.

The Trees
of Great Britain
and Ireland

VOLUME 3

HENRY JOHN ELWES
AUGUSTINE HENRY

CAMBRIDGE
UNIVERSITY PRESS

CAMBRIDGE
UNIVERSITY PRESS

University Printing House, Cambridge, CB2 8BS, United Kingdom

Published in the United States of America by Cambridge University Press, New York

Cambridge University Press is part of the University of Cambridge.
It furthers the University's mission by disseminating knowledge in the pursuit of
education, learning and research at the highest international levels of excellence.

www.cambridge.org
Information on this title: www.cambridge.org/9781108069342

© in this compilation Cambridge University Press 2014

This edition first published 1908
This digitally printed version 2014

ISBN 978-1-108-06934-2 Paperback

This book reproduces the text of the original edition. The content and language reflect
the beliefs, practices and terminology of their time, and have not been updated.

Cambridge University Press wishes to make clear that the book, unless originally published
by Cambridge, is not being republished by, in association or collaboration with, or
with the endorsement or approval of, the original publisher or its successors in title.

The original edition of this book contains a number of colour plates,
which have been reproduced in black and white. Colour versions of these
images can be found online at www.cambridge.org/9781108069342

THE TREES OF GREAT BRITAIN AND IRELAND

LEBANON CEDAR AT HIGHCLERE CASTLE

From a Drawing by Charlotte Lady Phillimore.

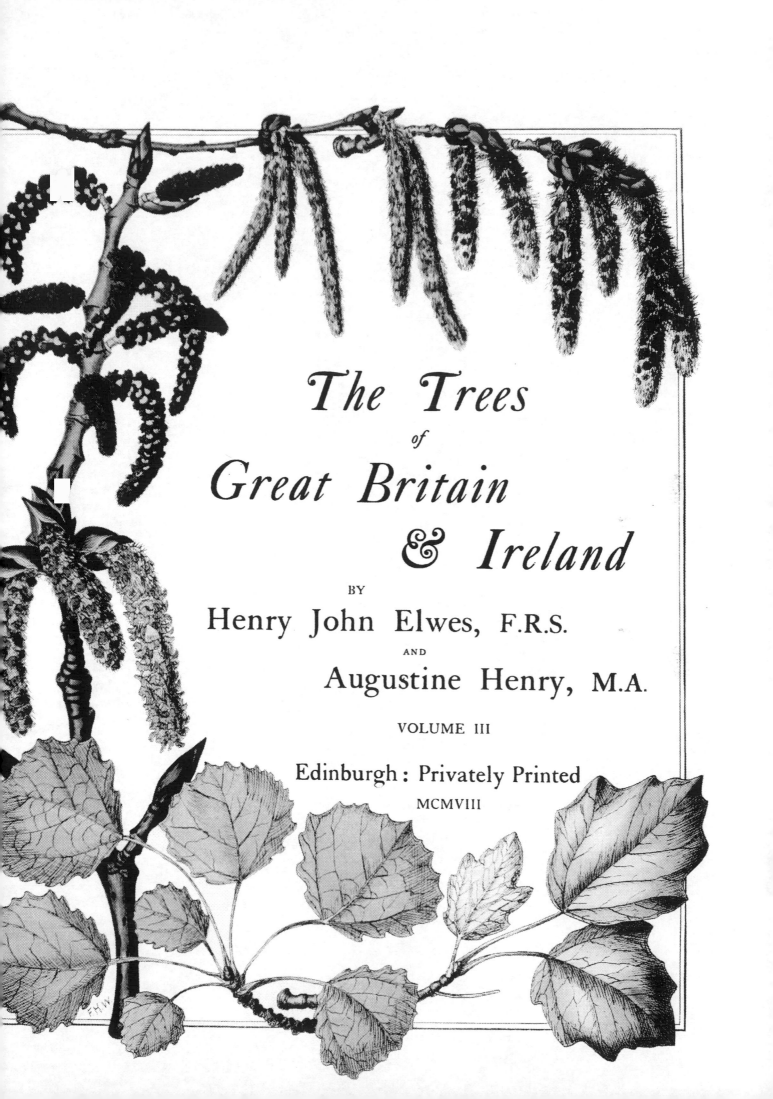

The Trees

of

Great Britain

& Ireland

BY

Henry John Elwes, F.R.S.

AND

Augustine Henry, M.A.

VOLUME III

Edinburgh: Privately Printed

MCMVIII

CONTENTS

iii

ILLUSTRATIONS

CEDRUS

Cedrus, Lawson, **Agric. Man.** 379 (1836); Loudon, **Arb. et Frut. Brit.** iv. 2402 (1838); Bentham et Hooker, **Gen. Pl.** iii. 439 (1880); Masters, **Journ. Linn. Soc. (Bot.)** xxx. 30 (1893).
Larix, Miller, **Dict.** No. 3 (1724) (*ex parte*).
Pinus, Linnæus, **Sp. Pl.** 1001 (1753) (*ex parte*).
Pinus, section *Cedrus*, Parlatore in DC. **Prod.** xvi. 2, p. 407 (1864).
Abies, Poiret in Lamarck, **Dict.** vi. 510 (1804) (*ex parte*).

TREES belonging to the tribe Abietineæ of the order Coniferæ, with evergreen foliage, borne, for the most part, in tufted flat masses on the ramifications of the branches, which arise irregularly and not in whorls from the stem.   Bark dark grey and smooth on young stems and branches; ultimately on old trunks thick and fissuring into irregular longitudinal plates, roughened externally by small scales.

Branchlets of two kinds : long shoots bearing in spiral order solitary leaves, and short shoots or spurs with leaves in pseudo-verticels.   Buds minute ovoid, with a few brown scales, which persist after the opening of the bud, either sheathing the base of the long shoots or surrounding the annual rings of the short shoots.   Long shoot with a solitary terminal bud, prolonging the growth of the branchlet in the following year; and with a few lateral buds solitary in the axils of some of the leaves and usually developing into short shoots.   Short shoot with a terminal bud only, which, in the following year, either lengthens slightly the spur and adds to it a whorl of leaves with or without flowers, or occasionally develops into a long shoot.   Long shoots, slightly furrowed, between the slightly raised decurrent bases of the pulvini, the free ends of which project and bear leaves, and on older branchlets, from which the leaves have fallen, remain persistent as slight prominences.

Leaves, deciduous in the third to the sixth year, variable in length, the shortest on the spurs, articulated at the base, acicular, rigid, sharply pointed, more or less triangular in section, stomatic on all sides; fibro-vascular bundle undivided, hypoderm thick, with two resin canals close to the epidermis on the lower surface.

Flowers, monœcious, terminal, solitary on the older leaf-bearing short shoots. Male flowers, erect, catkin-like, cylindrical, about 2 inches long; anthers numerous, spirally crowded, bi-locular, dehiscing longitudinally; connective prolonged into an ovate denticulate crest; pollen grains globose, without wings, borne to the female flowers by the wind.   Female flowers appearing as small purplish cones, about $\frac{1}{2}$ inch in length; composed of numerous spirally arranged, closely appressed, irregularly dentate, sub-orbicular scales, each subtended by a short, included, obovate, denticulate bract; ovules, two on each scale, inverted.

Ripe cones, solitary, erect, on short stout peduncles, dull-brown, resinous, ellipsoid or cylindrical; rounded, flattened or depressed at the apex. Bracts obsolete or minute and ragged. Scales numerous, closely imbricated, woody, fan-shaped; upper expanded part thin, transversely oblong, with denticulate rounded or sloping wings, brown-tomentose in greater part beneath, almost glabrous above; claw thickened, obcuneate, with a raised ridge between the depressions for the seeds on the upper surface, the lower surface being slightly hollowed by the pressure of the seeds of the adjoining scale. Seeds, two on each scale, $\frac{1}{4}$ to $\frac{1}{2}$ inch long, with resin-vesicles on both surfaces, brown, irregularly triangular; surmounted by a membranous brownish wing, broadly triangular or hatchet-shaped, about twice as long as the body of the seed. Cotyledons, nine or ten.

The flowers appear in July or August, the pollen being shed profusely in October. During winter the cones remain small, and only begin to grow in the following April, attaining about half or two-thirds their full size in October of the second year. They are fully ripe in October or November of the third year, *i.e.* about twenty-six months after the first appearance of the flowers. In their native forests the dissemination of the seed is caused by the autumnal rains, the cones not disarticulating in dry weather. After being soaked with rain, the scales and seeds separate from the axis of the cone (which remains persistent on the branch) and fall to the ground, the seeds with their light wings being blown, when there is a breeze, to a little distance from the parent tree. In England, irregularities occur in the period when the cones disarticulate, dependent, probably, on the absence of heavy rains in the autumn in certain seasons.

Seedling.—Plants raised from seed gathered on Mount Lebanon in 1904, and sown at Monreith in April 1905, averaged 9 inches high in the following September,[1] and showed the following characters:—Tap-root, about 9 inches long, slender, with a few lateral fibres. Caulicle, 2 inches long, slightly furrowed, glabrous. Cotyledons, ten, sessile, $1\frac{3}{4}$ inch long, curved, tapering to a sharp point, triangular in section, the upper two sides stomatiferous, the lower side green and narrow. Young stem glabrous, bearing in a whorl, just above the cotyledons, the first seven leaves, $\frac{5}{8}$ inch long, linear, flattened, sharp-pointed, stomatiferous on both surfaces, deeply grooved below, slightly convex above, sharply serrate in margin. Above the whorl, leaves, gradually increasing in size to $1\frac{1}{8}$ inch long, arise in spiral order, similarly serrate and stomatiferous, but almost rounded in section; in addition, the stem gives off at irregular intervals five or six small branchlets.

With regard to the different forms of the cedar, which inhabit four distinct and isolated areas, opinions are much at variance as to their rank. They differ more or less in the length of their leaves, and in the size and shape of the cones, cone scales, and seeds, and in the young stage they differ in habit; but in their native forests they all assume, when old, the flattened form which is sometimes erroneously considered to be peculiar to the Lebanon cedar. This is caused by the inflection of the leading shoot, which is followed by a diminution in the rate of vertical growth, the lateral branches at the same time thickening and growing out horizontally. An

---

[1] This growth is quite exceptional in my experience.—(H. J. E.)

important difference is the height attained in the wild state, the deodar becoming very tall, the Cyprus cedar remaining short, with the Lebanon and Algerian cedars intermediate in size. They differ in their period of vegetation. At Kew the deodar is the first to put forth young leaves in spring; the Lebanon usually follows a fortnight later; and the Algerian generally comes out last, after an interval of a few days. They may be correctly considered geographical races of the same species; but for arboricultural purposes it is most convenient to rank them as distinct species.

CEDRUS LIBANI, Lebanon Cedar

Cedrus Libani, Barrelier, *Plantæ, Icon.* 499 (1714); Loudon, *Arb. et Frut. Brit.* iv. 2402 (1838); Ravenscroft, *Pinet. Brit.* iii. 247 (1884); Kent, Veitch's *Man. Coniferæ*, 415 (1900).
Cedrus patula, Koch, *Dendrol.* ii. 268 (1873).
Pinus Cedrus, Linnæus, *Sp. Pl.* 1001 (1753).
Larix Cedrus, Miller, *Gard. Dict.* ed. viii. No. 3 (1768).
Larix patula, Salisbury, *Trans. Linn. Soc.* viii. 314 (1807).
Abies Cedrus, Poiret in Lamarck, *Dict.* vi. 510 (1804).

Leading shoot of young trees erect or slightly bent, not pendulous. Branchlets not pendulous, glabrous or with slight short pubescence. Leaves up to $1\frac{1}{4}$ inch in length, broader than thick. Cones large and broad, ellipsoid, 3 to $4\frac{1}{2}$ inches long, $1\frac{3}{4}$ to $2\frac{1}{2}$ inches wide; scales 2 inches or more in width, with the claw inflected almost at a right angle.

VARIETIES

1. Var. *argentea*, Antoine et Kotschy, *Iter. Cilic.* No. 417. Trees with glaucous foliage, growing wild in the Cilician Taurus, intermingled with the ordinary form. This variety appears in cultivation, but is rarer than the glaucous form of *C. atlantica*.

2. Var. *decidua*, Carrière, *Conif.* 372 (1867). Leaves deciduous. A tree of this kind, slow in growth and bushy in habit, was obtained by Sénéclauze in 1851. Kent mentions one growing at Westgate near Chichester.[1] Webster reports[2] another, 65 feet high, growing on Lord Derby's property in Kent, and said to be in perfect health, though from its bare appearance in winter it has often been supposed to be dying.

3. Var. *tortuosa*. On the lawn of a private house at Dulwich, belonging to the Dulwich College estate, there is a remarkable cedar, a photograph of which was sent to Kew in 1903. The stem and all the branches are spirally twisted.

The Lebanon cedar is variable in habit, and numerous supposed varieties are mentioned by Beissner, as *nana*,[3] a dwarf form; *stricta*, narrowly pyramidal in habit; *pendula*, with pendulous branches and branchlets; and *viridis*, with bright green shining foliage.

---

[1] But on writing to Captain Norman, who was the authority for this, he tells me that the tree is now dead, and that in his opinion the deciduous habit, which was regular and unfailing, was due to constitutional weakness, caused by uncongenial surroundings, in proof of which he states that another tree at the same place raised from a seed taken from the same cone, was much more robust and showed no abnormal tendency.—(H. J. E.) [2] *Hardy Coniferous Trees*, 27 (1896).

[3] A specimen of the dwarf cedar, only 4 feet high and of considerable age, is growing in grounds adjoining one of the oldest parks at Hemel Hempstead. The branches are flattened, horizontal, and very close together, giving the plant a dense, stiff appearance.—*Gard. Chron.* xix. 563 (1896).

Supposed hybrids between the Lebanon and Atlantic cedars have been recorded,[1] but on insufficient evidence.   (A. H.)

DISTRIBUTION

The best account of the Cedar of Lebanon known to me is the classical paper by Sir Joseph Hooker published in the *Natural History Review*, vol. ii. p. 11 (1862), and as this gives a careful summary of the facts bearing on the specific identity of the forms of cedar, I summarise it as follows :—In the autumn of 1860 Sir J. Hooker went to Syria in company with Captain Washington, Hydrographer of the Navy, and Captain Mansell, R.N., and arrived at Beyrout on 25th September.   The party proceeded to the Lebanon, where Captain Mansell made a detailed survey of the basin where the cedars grow, at the head of the Kedisha valley, 15 miles from the sea in a straight line.   At that time the other groves were apparently unknown, though Professor Ehrenberg informed Sir Joseph Hooker that he found many trees in forests of oak on the road from Bsharri to Bshinnate.   The Kedisha valley at 6000 feet elevation terminates in broad, flat, shallow basins, and is two or three miles across and as much long.   It is three or four miles south of the summit of Lebanon, which is about 10,200 feet in height, the chapel in the cedar grove being about 6200 feet.   The cedars grow on a portion of the moraine which borders a stream, and nowhere else ; they form one grove about 400 yards in diameter, and appear as a black speck in the great area of the corrie and its moraines, which contain no other arboreous vegetation, nor any shrubs but a few small barberry and rose bushes.   The number of the trees is about 400, and they are disposed in nine groups, corresponding with as many hummocks of the range of moraines ; they are of various dimensions, from 18 inches to upwards of 40 feet in girth ; but the most remarkable and significant fact connected with their size, and consequently with the age of the grove, is that there is no tree of less than 18 inches girth, and that no young trees, seedlings, nor even bushes of a second year's growth were found.   Calculating from the rings in a branch of one of the older trees, now in the Kew Museum, the younger trees would average 100 years old, the oldest 2500, both estimates no doubt being widely far from the mark.   Sir Joseph goes on to say, that the word cedar as used in the Bible applies to other trees, and he doubts whether the cedar of Lebanon is the one which supplied the timber used in building Solomon's temple.   He thinks that the cypress or the tall fragrant juniper of the Lebanon (*Juniperus excelsa*) would have been not only much easier to procure, but far more prized on every account.[2]   Between individuals from the Lebanon and the common Asia Minor form there is said to be no appreciable difference by those who have examined both, but there are two distinct forms or varieties in Asia Minor, one having shorter, stiffer, and more silvery foliage than the other ; this is the silver cedar, *C. argentea*, of our gardens.   Northern Syria and Asia Minor form one botanical province, so that the Lebanon groves,

---

[1] Beissner, *Nadelholzkunde*, 301, 302 (1891).

[2] But at a later period Sir J. Hooker changed his opinion on this subject, and believed that the wood used by Solomon and by Nebuchadnezzar in buildings was the Lebanon cedar.

though so widely disconnected from the Taurus forest, can be regarded in no other light than as an outlying member of the latter. After speaking of the Algerian cedar and the deodar, Sir Joseph says that it is evident that the distinctions between them are so trivial, and so far within the proved limits of variation in coniferous plants, that it may reasonably be assumed that all originally sprang from one. There are no other distinctions whatever between them of bark, wood, leaves, male cones, anthers, or in their mode of germination, growth, or hardiness (but this has not been confirmed during the severe winters of a later date in England). Though the difference in the shape of the scales and seeds of *Deodara* and *Libani* are very marked, they vary much, many forms of each overlap, and further transitions between the most dissimilar may be established by the inter-calation of seeds and scales from *C. atlantica.* Sir Joseph accounts for the difference in the habit of the three forms in a great measure by the climate of the three localities : the most sparse, weeping, long-leaved cedar is from the most humid region, the Himalaya; whilst *atlantica*, the form of most rigid habit, corresponds with the climate of the country under the influence of the great Sahara desert. No course remains, then, but to regard all as species, or all as varieties, or *Deodara* and *atlantica* as varieties of one species, and *Libani* as another. The hitherto adopted and only alternative of regarding *Libani* and *atlantica* as varieties of one species and *Deodara* as another species must be given up.

Ravenscroft, in *Pinetum Britannicum*, gives a very full account of the cedars of Lebanon from various sources, with four good illustrations from photographs taken by F. M. Good of Winchfield, and there are many points in his account worth referring to.

Mr. Ridgway of Fairlawn, who visited them in 1862, says[1] that there is a young tree 50 yards west of the chapel, of exactly the same form and habit as a deodar in his park near Tonbridge. It has the same graceful drooping habit, the same light silvery green, and none of the usual rigid horizontal form of the cedar. He says the remainder of the race of trees vary from 20 to 25 feet in girth; some are as tall and straight as poplars, some not above 20 feet high, and gnarled and stunted. Ravenscroft gives in a table the facts relating to the number of trees found in the accounts of various authors who have written on the Cedars of Lebanon, commencing with Belon in 1550 and ending with Canon Tristram in 1864. Of the older ones there were 28 in Belon's time, which are now reduced to about half that number. There is a gap of some centuries—Ravenscroft says probably more than 1000 years—between the cedars of the second size and the older ones, and again a very long interval of growth between all the young trees, which are now about 400. I do not find any reliable information, taken from an actual count of the number of rings in any of the old Cedars of Lebanon, as to what their possible age may be. Ravenscroft has gone very carefully into the question of the age of the Cedars of Lebanon, which, he says, may be from 4000 to 5000 years old; and he further gives a table based on 200 measurements of cedars of all ages in England, which shows that the average growth in height in England is about 1 foot per annum for trees up to sixty years

[1] *Gard. Chron.* 1862, p. 572.

old, and from 6 to 9 inches in trees of 100 to 200 years. He gives the average breadth of the annual rings per annum in trees of from twenty to fifty years as from $3\frac{1}{2}$ to $9\frac{1}{4}$ lines, and in trees from 60 to 200 years as $3\frac{1}{4}$ lines.

In *Gardeners' Chronicle*, xii. 204 (1879), S. R. Oliver writes :—"And now about the cedars themselves. The guardian told us that there are exactly 385 trees, large and small, but the smallest must be at least from 50 to 80 years old, and no younger trees are springing up—a fact to which it would be well to draw the attention of the public. At this time of year (28th September) innumerable seeds, which are scattered everywhere beneath the trees from the fallen and expanded cones, are germinating, scattered by the wind ; these germinating seeds extend far beyond the actual area covered by the remaining trees ; and if it were not that they are trodden under foot, or, what is still more destructive, eaten up by the goats, a few decades of years would soon see a fair sprinkling of healthy young cedars enlarging the borders of the grove. In 100 years the grove would be increased into a wood, and five centuries hence the wood would have become a forest. At present, for want of proper protection against the goats and thoughtless tourists, the present grove is dwindling away, and another generation will exclaim against our supineness in thus allowing a relic of the past to die out prematurely. For a small sum of money a stone wall might be built, enclosing the area of the cedar grove sufficiently well against goats. Future travellers ought to be warned by the guardian to confine their steps to certain paths, so as not to injure the young trees ; and stringent precautions should be taken against the disfigurement of the trees now existent, by the cutting of names, tearing down of branches for the cones, etc. It would be easy to build such a wall so as not to be an eyesore or disfigurement, by taking advantage of the sinuosities of the numerous small valleys which permeate the vicinity. I am sure that many travellers would contribute small donations should a subscription list be opened for such a purpose.[1] As the property of the cedars belongs to the Patriarch of the Maronites, by name Butross Massaad, who resides by the Dog river, not far from Beyrout, it would be necessary to obtain his co-operation, and I hope, through the aid of the Consul-General for Syria, to have an interview with him on the subject before I leave the neighbourhood. Most of the single trees of antique growth average 20 to 30 feet in girth at about 6 feet from the ground, but the enormous fathers of the forest are in reality a congeries of two, three, or even more trees which have grown so closely together as to coalesce and actually form a single trunk. Among the younger specimens twin and triplet trees are rather the rule than the exception, and this will explain such a girth as Dr. Wartabet measured round the largest tree on the slope north of the Maronite chapel overlooking the ravine, viz. 48 feet. This tree is by no means one of the oldest, but is at its full growth of maturity, and in vigorous health. The hoar, gaunt, and withered trunks of greatest antiquity are around the usual camping ground at the S.E. corner of the group."

Dr. A. E. Day wrote to me as follows on their actual condition more recently, in a letter dated Beirut, Syria, 9th November 1903 :—

---

[1] Rustem Pasha informed Sir W. Thiselton Dyer that he had built a wall to protect the young cedars from grazing, but at a later period this was broken down.

"To the best of my knowledge there are five groves of cedars in Lebanon. The best known one, and that containing the oldest trees, is one in northern Lebanon above Bsharri. [Plate 127, from a photograph by Dr. Van Dyck, shews one of these trees.] The condition of that has, I think, not changed much in thirty years. I am sure that no new trees have grown up in it. A few of the oldest ones have lost branches, or have entirely perished. The grove is a favourite resort in summer for Syrians and for foreigners. A few hours south and west of Bsharri is the village of Hadeth-el-Jubbeh, or Hadeth, as it is often called, though there are a number of Hadeths in Lebanon. Within a half-hour to the south of Hadeth is a fine grove of young trees which, I think I have been informed, was started and has been preserved by a Greek or Maronite bishop. The remaining three groves are near each other, on the western slope of the main ridge of Lebanon, the most northern one being a few miles south of the Beirut-Damascus road as it crosses the ridge. The most northern of the three is above the village of Ain-Zahalta, the next is above Bârûk, and the third is above Maâsir, each being known by the name of the village near it, being also the property of that village. The smallest grove, but that containing the oldest trees, is that of Maâsir. The Bârûk grove is the most extensive of all the five in Lebanon, and contains many young trees in all stages of growth. Most of the trees are upon a very steep slope, but in the upper part of the grove there are various knolls and hollows, affording a few charming spots for camping. I am sorry to say that this fine grove suffers much from being cut. The people of Bârûk obtain from it roof-beams and wood for fuel, and I am informed that they are discussing selling a large part of it to be felled for pitch. I have failed to find a single large tree in the Bârûk grove which has not been cut off, with the result that several branches have taken the place of the principal stem. The ordinary Arabic name of the cedar is 'Arz,' but the natives of the villages near the three southern groves call the tree 'Ubhul.'"

The cedar is also found in the Taurus and Anti-Taurus ranges in Asia Minor, extending from the province of Caria[1] in the west to near the frontier of Armenia in the east. It forms a considerable part of the coniferous forest, which, in a few scattered localities, covers the mountains between 4000 and 7000 feet. It is usually associated with *Abies cilicica, Juniperus excelsa*, and *J. fœtidissima*; and is occasionally mixed with *Pinus Laricio*. In Lycia, dense woods of cedar were observed by Luschan[2] in the Baba Dagh and between Zumuru and the Bulanik Dagh. The tree, however, appears to attain its maximum development in the Cilician Taurus, where there are fine forests of great extent in the Bulgar Dagh, which have been visited by Tchihatcheff,[3] Kotschy,[4] and W. Siehe.[5] The latter states that the climate in which the cedar grows is a severe one, the snow lying several feet deep on the ground for about five months of the year. He describes

---

[1] Collected in Caria by Pinard, according to Boissier, *Flora Orientalis*, v. 699 (1881). Dr. Stapf informs us that Luschan also saw the cedar in this province.

[2] Cf. Stapf, *Beiträge Flora Lycien, Carien, u. Mesopotamien*, 2 (1885).

[3] *Asie Mineure*, ii. 496 (1860).

[4] *Reise Cilicischen Taurus*, 58, 370 (1858).

[5] *Gartenflora*, 1897, pp. 182, 206. Siehe has sent seed from the Cilician Taurus to various places, and I have two vigorous young trees raised from them.

the forest as an open one, the trees standing isolated and attaining about 130 feet in height and 10 feet in girth, and none of a larger size were noticed. Haussknecht found the cedar in the Berytdagh in Cataonia; Heldreich collected it in the Davros Dagh in Pisidia; and two or three other localities, where the tree is apparently neither abundant nor remarkable for size, are mentioned by Boissier.

I saw the cedar in the Ak Dagh, on the road between Makri and Cassaba, in 1874, where the trees were growing in open woods at about 5000 to 6000 feet elevation, and were not anything like as large as those in the Lebanon.

INTRODUCTION

We have no certain evidence as to the earliest introduction of the cedar into England; but Loudon, p. 2412, considers that Evelyn was most probably the introducer of the tree, as he states in the third edition of the *Sylva* (p. 125), published in 1679, that he had received seeds from Mount Libanus.

It has been supposed that Dr. Uvedale got the seeds which were planted by him at Enfield between 1665 and 1670 from Evelyn, who, however, does not mention this in the second edition of the *Sylva*, published in 1670; and until this tree is dead or cut down we shall not know its age for certain.[1]

The oldest specimen[2] of cedar in the British Museum is in a volume of *Herb. Sloane*, ix. fol. 90, the title-page of which bears the following inscription:—" Plants gathered about London about the year 1682 for my own (*i.e.* Sir Hans Sloane's) collection."

Sir Stephen Fox was credited by his descendant, Lord Holland, with having introduced and planted the cedar at Farley, near Salisbury (cf. Loudon, p. 2413), which was cut down in 1813, when it weighed over 13 tons. Quenby Hall, Leicestershire, is also mentioned as having the oldest cedar in England, but this rests on family tradition only, and the tree at Quenby in 1837 was only 47½ feet by 7 feet 9 inches in girth.

In *Country Life*, May 2, 1903, the late Mr. C. J. Cornish gives an account of a cedar at Childrey Rectory, near Wantage, which, "according to unbroken tradition," was planted by Dr. Edward Pocock, who was chaplain to the Turkey Company at Aleppo in 1629, and afterwards chaplain to the Embassy at Constantinople. He returned home in 1641 and was appointed to the living of Childrey in 1642. In 1903 it was still growing vigorously and increasing rapidly in size, and measured 25 feet in girth at five feet from the ground, and covered an area of about 1600 square yards. Though it has suffered much from the loss of branches broken by the weight of snow about twenty years ago it now presents a very handsome appearance as shown by the photograph which is given on p. 567 of *Country Life*, No. 330.

Lord Savile informs me that a cedar, which he remembers as being the tallest that he ever saw, grew at Rufford Abbey. This is believed to have been planted

---

[1] Boulger, in his biographical sketch of Uvedale in *Journ. Bot.* xxix. 13 (1891), gives some details of the Enfield cedar, but has not been able to verify the statement that it dates from 1670. The Enfield cedar is figured in *Gard. Chron.* xxxii. 31, f. 12 (1902). Cf. also *Gard. Chron.* viii. 505 (1890).

[2] The statement in *Gard. Chron.* ii. 194 (1887), that there is mention in Belon's works, which were published in 1553 and 1558, that the cedar of Lebanon existed in France before 1558 is erroneous; and it is probable that the tree was not introduced into France till 1734. Cf. Loudon, p. 2414.

by Charles II., who used to visit and stay at Rufford, where his rooms are now known as "the King's rooms." Its stump is now surrounded by iron railings and labelled "Cedrus Libani, planted by King Charles II."

Loudon considered that the cedars at Chelsea[1] mentioned by Sir Hans Sloane in 1685 as then existing (*Ray's Letters*, p. 176), but now dead, and those at Chiswick House, which are still flourishing, were the oldest in England. But I am informed by Mr. Challis, gardener to the Earl of Pembroke at Wilton House, that "in the year 1874 a very large cedar was cut down there, whose stem up to about 18 feet from the ground was nearly uniform in size, and then divided into twelve distinct branches, each nearly equal in size to a good-sized tree, some of them extending horizontally 70 feet from the trunk. The circumference of the bole five feet from the ground was 36 feet, a transverse section measured when down 11 feet 9 inches, and the number of concentric rings, after several careful counts, some of the rings being somewhat indistinct, was 236. A section of this stem was sent to the South Kensington Museum."

If this is correct, and it seems to me that the exact statement of so experienced a gardener as Mr. Challis cannot be questioned, the tree must have been introduced in 1638, before Evelyn's time, and was not only the oldest but also the largest cedar on record in England. I have taken great pains to verify this statement by seeing the section mentioned; but though careful search has been made in the Records of the British Museum (Natural History), as well as at the Victoria and Albert Museum at South Kensington, and in the letter books of the Royal Horticultural Society, this wonderful specimen cannot now be traced or discovered.

CULTIVATION

The seeds of the cedar, whether imported or home grown, should be sown under glass in the spring; for though they will germinate in the open air, their growth is so slow for the first three or four years that much time and loss will be saved by protecting them with a frame. If sown in pots they should be planted out in a frame at a year old, as the roots soon become cramped and pot-bound, and the young plants do not make good roots for some time if they have once been so checked. At two or three years old they may be planted in rich soil about the beginning of May when the buds are starting, and will require some years more in the nursery before finally planting them out.

The Lebanon cedar requires a warm, deep, well-drained soil to bring it to perfection, and does not grow so well in the colder and moister parts of England. When once established it will endure our most severe winters without much injury, though it often suffers from heavy snowstorms, which break the branches. The seedlings vary considerably in habit, in vigour, and in colour, and as they do not bear pruning well when the branches become large, it is best to cut off the lower ones when quite small, so as to encourage an upright growth.

---

[1] The last of the cedars in the Physic Garden at Chelsea, which had been dead for some years, was removed in 1904. In 1882 it was 60 feet high and 13 feet 9 inches in girth at 3 feet from the ground. *Gard. Chron.* xxxv. pp. 185, 224 (1904). Cf. also *ibid.* xxvi. 336, f. 70 (1886), where a figure of the tree is given.

As most people prefer the spreading forms of cedar for lawns or parks, the Lebanon cedar is probably the best for such places; but when surrounded by other trees it may be drawn up to a great height with few side branches, though I should prefer the Algerian cedar for planting in such situations.

Generally it may be said that the Lebanon cedar is the best for the hotter and drier parts of England, and the deodar for the moister and milder districts. The Algerian cedar seems to be hardier, and according to Sargent this is also the case in the United States; but none of the cedars succeed in New England, though near Philadelphia, Washington, and at Biltmore, North Carolina, there are fine specimens of the Algerian form.

The transplantation of large cedars is rarely desirable, but has been sometimes effected with success. A case is recorded[1] in which a cedar at Southsea 30 feet high, with a spread of 36 feet, was transplanted at a cost of about £100.

Mr. J. W. Odell, gardener at the Grove, Stanmore, in a communication to the Royal Horticultural Society on 14th February 1899, states that during a recent gale a large branch was broken off a cedar there, which showed that a great mass of adventitious roots had started from the seat of a previous injury and grown downwards towards the base of the tree, between the splintered portions of the wood. I observed a precisely similar occurrence in a cedar which was partly blown down at Stoke Hall, Notts, in October 1903. The roots were bright reddish-brown in colour, and the thicker ones, an inch in diameter, were covered with rough pustules. Some of these were sent to the Museum at Kew.

REMARKABLE TREES

Among the existing trees in England it is difficult to say which is the finest. If height and girth combined are taken there is none to equal the splendid tree at Pains Hill, near Cobham, now the property of C. Combe, Esq., of Cobham Park. An account of this place, published in *Country Life* for March 19, 1904, states that these cedars were probably planted between 1750 and 1760 by the Hon. Charles Hamilton. In 1781 Sir Joseph Banks visited Pains Hill with the younger Linnæus, who said that he saw there a greater variety of fir trees than he had seen anywhere else. Curiously enough, Loudon, though he often mentions Pains Hill, gives no measurements, and neither Strutt, Lambert, nor Lawson alludes to the cedars there; but when I saw them in 1904 I measured the largest (Plate 128) to be from 115 to 120 feet high by 26 feet 5 inches in girth. It grows on sandy soil near the lake and divides into several tall, straight stems, which form a spreading crown, and seems to be in perfect health.

The next finest of this type that I have seen is perhaps a tree standing in Goodwood Park, near the kitchen garden, which, when I measured it in 1906, was about 95 feet high, though its flat top makes the exact height difficult to ascertain, and 26½ feet in girth, the branches spreading over an area of 133 paces in circumference (Plate 129). Goodwood[2] is perhaps more celebrated for its cedars than any other

---

[1] *Gard. Chron.* xxv. 42 (1899).

[2] Cf. *Gard. Chron.* xxvii. 124 (1900), where the finest cedar at Goodwood was reported to be 29½ feet in girth in 1900.

place in England, as in 1761 many hundreds were planted by Peter Collinson for the Duke of Richmond. Loudon, on page 2414, quotes a MS. memorandum of Collinson's as follows:—"I paid John Clark, a butcher of Barnes, who was very successful in raising cedars, for 1000 plants of Cedar of Lebanon, 8th June 1761, £79 : 6s., on behalf of the Duke of Richmond. These 1000 cedars were planted at five years old, in my sixty-seventh year, in March and April 1761 ; in September 1761 I was at Goodwood and saw these cedars in a thriving state. This day, 20th October 1762, I paid Mr. Clark for another large portion of cedars for the Duke of Richmond. The duke's father was a great planter, but the young duke much exceeds him, for he intends to clothe all the naked hills above him with evergreen woods." Of the cedars at Goodwood, Loudon goes on to say that 139 remained in 1837. According to Kent (*op. cit.* 419, note *), eleven fine cedars were uprooted in Goodwood Park by the fierce gale of 3rd March 1897.

There are some splendid cedars at Wilton of which Lambert[1] writes as follows:—"I am indebted to the Hon. and Rev. W. Herbert (author of the celebrated work on *Amaryllidaceæ*) for the following interesting particulars respecting the cedars at Highclere, the seat of the Earl of Carnarvon : 'The two oldest cedars at Highclere were raised in 1739 from a cone brought from Lebanon by Dr. Pococke[2] in 1738. They were stunted plants for some time, and removed to their present situation in 1767. The largest of the two measured, in 1829, 9 feet in circumference, having grown only an inch in the last two years, the chalk being unfavourable to its growth. The largest cedar at Highclere, though much younger, measured in 1830, at three feet from the ground, 10 feet 1 inch in circumference ; it was reared from a cone, which came from the Wilton cedars in 1772, and was about 48 years old before it bore. It was known to the late Earl of Carnarvon that the cedars at Wilton were kept by his grandmother, the Countess of Pembroke, in pots at her window, till growing too large, they were planted upon the lawn, between the house and the water, a situation very favourable to their growth. Supposing them to have been 48 years old, when the cone was gathered from them in 1772, they must have been raised as early as 1724. It is most probable they were between 1710 and 1720 ; for the Countess of Pembroke who cultivated them died before her husband, who married again after her death, and died in 1733. The oldest cedars at Highclere are, therefore, now (in 1831) 92 years old ; those at Wilton at least 106, probably between 110 and 120. Dr. Pococke found the circumference of the largest cedar with a round or single stem to be 20 feet ; but he does not state how near the ground he measured it.'" I saw these trees in 1903 and measured them carefully ; the best was then about 108 feet high and 21 in girth, with a spread of 109 feet. This tree has lost a large limb, the hollow caused by which has been carefully filled with cement.

At Strathfieldsaye there are also some splendid cedars, a group of which on strongish clay soil have the same upright, small-branched character as the Windsor trees. The best of these is 110 feet high by 11 feet 9 inches in girth, with a clean

---

[1] *Genus Pinus*, ii. 91 (1832).

[2] This is not confirmed by Mr. Challis's statement on p. 459 ; and probably all the Wilton cedars were not of the same age. Dr. Richard Pococke travelled in the East during 1737 to 1742.

bole of about 40 feet (Plate 130).   When mentioned by Loudon it was considered the tallest in England, being then 108 feet high by 9 feet in girth.   At Combe Bank, near Sevenoaks, there is a magnificent cedar, which Henry in 1904 measured about 105 feet high with a girth of 20 feet.   There are no really large cedars at Syon, Kew, or Woburn.   There is a magnificent tree at Blenheim, 28 feet in girth, but of no great height, and having the spreading habit which we usually see in this tree in England (Plate 132).

Probably the tallest cedar in England is one in the pleasure grounds of Petworth Park, which I measured carefully in 1905, and could not make it less than about 125 feet high by 14½ feet in girth.   It is remarkable for having a trunk clear of branches for no less than 56 feet, where a small limb comes off, but, with this exception, it is clean up to about 80 feet.   Probably this habit is due to its growing in a sheltered position, more or less shut in by other trees, on a deep bed of sandy loam. Owing to its position this tree is very difficult to photograph (Plate 131).

The next finest tree of this type which I have seen is one in the Belvedere Plantation at Windsor.   This, according to Menzies, who says that the ground is marked in a map of 1750 as open, cannot be more than 150 years old; and it is at least 115 feet high with a girth of 16 feet.   It is without any large branch until it reaches a height of 60 feet or more, and carries nearly the same girth to this elevation; so that a plank 60 feet long and 3 or 4 feet wide at the top end could probably be cut from it.   Menzies figures this tree [1] and gives the dimensions in 1864 as only 75 feet by 12 feet 10 inches, which was probably less than its actual size at that time.   There are several other fine trees in the same drive, but none equal to this; and a young one close by, which was planted by Mr. Simmonds, Deputy Surveyor of Windsor Forest, about thirty-five years ago, is now about 40 feet high, and has the same straight-growing upright habit which cedars seem to develop best on deep sandy soil.

In Hertfordshire there are many fine cedars, of which the most notable are growing on a lawn at Bayfordbury.[2]   Mr. Clinton Baker tells me that they were raised from seeds of the Enfield Cedar, and planted in 1765.   They have been measured at various intervals by his grandfather, father, and himself, as follows :—

	No. 1 at 3 ft.		No. 1 at 5 ft.		No. 3 at 3 ft.[2]		No. 3 at 5 ft.		No. 7 at 1 ft.	
	Ft.	In.	Ft.	In.	Ft.	In.	Ft.	In.	Ft.	In.
1822 . . .	10	0	10	5	10	10	10	8	14	5
1837 . . .	12	2	...		...		13	0	17	6
1865 . . .	15	0	16	0	15	11	16	5	22	1
1880 . . .	16	0	17	3	17	3	18	0	...	
1895 . . .	16	7	...		18	9	...		26	0
1900 . . .	16	9	18	0	19	5	19	5	27	1
1904 . . .	17	0	18	2	19	5	19	9	27	3

At Langleybury, Herts, a large cedar[3] was growing in the grounds of

[1] *History of Windsor Great Park and Windsor Forest*, Plate 14.
[2] Figured in *Gard. Chron.* xxvi. 521, f. 102, and 553, f. 109 (1886).
[3] *Gard. Chron.* xiv. 392 (1880).

E. H. Loyd, Esq., in 1880, which at 4 feet from the ground measured 22 feet 4 inches in girth, with a height of 107 feet. At Chart Park, Deepdene, Surrey, a tree 95 feet high is 19 feet 3 inches in girth, and divides at 12 feet up into ten upright stems.

At Chorleywood Cedars, near Rickmansworth, there are seven very fine cedars, standing on high ground, which form a landmark in the country, and are said to measure about 23 feet in girth. Another at the same place was recently struck by lightning, and cut down.

At Beechwood, near Dunstable, the seat of Sir Edgar Sebright, Bt., there are some very fine old cedars, of which the largest, as I am informed by Miss F. Woolward, measures 100 feet by 28 feet 4 inches, with a spread of branches 46 feet across. Another, 90 feet by 23 feet, has branches from 50 to 60 feet long.

At Chiswick House there are a number of very fine cedars still surviving, though not so many as when the late Mr. Barron, Superintendent of the Royal Horticultural Society's Gardens, measured them in 1882. The two largest trees at that time were 16 feet and 18 feet in girth, and when I saw them in 1904 the two largest were 16 feet 5 inches and 18 feet 5 inches. These are supposed to have been planted about 1720, but are nothing like so fine as many trees at a greater distance from London.

One of the most remarkable cedars in England, on account of its habit, stands in what was probably a dense grove of tall silver firs near the site of the old house at Stratton Strawless, the home of Robert Marsham, who planted it when $1\frac{1}{2}$ feet high, in 1747. When described by Grigor[1] in 1841 its stem was 44 feet high, free from branches, and 12 feet 2 inches in girth at 6 feet. His plate shows that it has changed but little now. When Mr. Birkbeck showed it to me in April 1907 it was about 80 feet high and $16\frac{1}{2}$ feet in girth, and though some branches in the crown had been broken off, it looks remarkably vigorous (Plate 133).

A fine tree of the same type, but not equal to the last mentioned, is in a sheltered part of the grounds at Gosfield Hall, Essex, the property of Mrs. Lowe. It is 80 to 90 feet high, by $14\frac{1}{2}$ feet in girth, with a clean stem up to about 60 feet, and a flat, spreading crown of branches at the top.

A cedar which is growing at Birchanger Place, near Bishop-Stortford, for a photograph of which I am indebted to the owner, T. Harrison, Esq. (Plate 134), is strikingly different in habit, and of its type is one of the most beautiful and perfectly shaped in England. It is about 60 feet high and 17 feet in girth, the branches covering an area at least 100 yards in circumference. Another tree of the same type, but not so symmetrical, grows at Billing Hall, the seat of Valentine Cary Elwes, Esq., near Northampton, and measures about 60 feet by 19 feet 5 inches. The branches, which spread over an area about 100 paces round, are supported by a great number of wooden props.

In the west of England this tree does not attain the same size and beauty as in the drier counties, the largest I have seen in Devonshire being at Bicton, which is about $21\frac{1}{2}$ feet in girth. At Castlehill, in the same county, there is a tree about 80 feet by 14 feet 9 inches; and at Sherborne Castle, in Dorsetshire, there are a

[1] *Eastern Arboretum,* p. 84, plate opposite p. 104 (1841).

number of fine trees, the largest of which I found to be about 105 feet by 16½ feet, dividing at about 15 feet up into five or six tall, straight stems.

In Wales I have seen none remarkable for size except a tree at Maesleugh Castle which is about 100 feet by 16½ feet, with a clean stem about 20 feet high.

In Lincolnshire and Yorkshire, even where the soil is good, the cedar does not attain the same dimensions as in the south of England, but it ripens seed at least as far north as Syston Park, where there are some trees near the house in an exposed position at an elevation of about 500 feet above sea-level, which show remarkable variation in colour. When I saw them on 16th June one was only just opening its buds, and looked quite black in comparison with others whose new leaves were well out and of a very glaucous colour. This colour is reproduced by their seeds, for two young trees raised from them, which were kindly given me by Sir John Thorold, are so glaucous that every one who has seen them in my nursery has mistaken them for *C. atlantica glauca*, while two seedlings of *C. atlantica* from Cooper's Hill are not distinguishable from *C. Libani*.

In Cumberland there are two splendid cedars at Eden Hall, the seat of Sir R. Musgrave, Bart., which, according to a paper[1] by Mr. Clark of Carlisle, were supposed to be 270 years old, and one of them measured 86 feet by 22½ feet, the other 86 feet by 21 feet, with a spread of 101 feet in diameter. At Alnwick Castle, Northumberland, there is a tree in the wood near the Duchess bridge, measuring 69 feet by 7 feet 3 inches.

The finest avenue of cedars I know in England is that at Dropmore, of which I give an illustration taken from a photograph made in 1903 (Plate 135). This avenue is said[2] to be composed of Lebanon cedars planted probably about 1844, and if really so young as this, is a very remarkable instance of the rapid growth of the cedar in this country. There is, however, some doubt as to whether they are Cedars of Lebanon or Algerian cedars, and though I have made inquiries from Mr. Fortescue I cannot ascertain with certainty their origin.

The best account I know of the Cedar of Lebanon in Scotland is given in the *Transactions of the Horticultural Society*, vi. 429, in 1826, by Mr. J. Smith, then gardener to the Earl of Hopetoun, and as this shows the rate of growth of the cedar to be, even in that latitude, greater than that of any other tree, I quote it as follows :—

"The extensive pleasure grounds at this place were laid out about the year 1740, and in that and the subsequent years a great variety of curious ornamental trees was planted, which are now of considerable size, and in great beauty and perfection : among these are three cedars, which were planted in the year 1748. The two largest are growing in a favourable deep soil, which although not wet inclines to be moist ; the third is on a gravelly soil, beside a rill of water. Their situation is well sheltered, and about 100 feet above the level of the sea. In the year 1797 the third tree was the largest, and Dr. Walker,[3] who noted its size at that date, ascribes its superiority to the wetness of its situation. He has

---

[1] *Trans. Eng. Arb. Soc.* 1887, p. 135.   [2] *Gard. Chron.* xxv. 138, fig. 52 (1899).
[3] *Essays on Natural History*, 69 (1808).

stated that it was 5 feet and 1 inch in circumference, but omitted to mention at what height from the ground this measurement was taken. In 1801 the dimensions of these trees, as well as of other kinds planted at the same period, were taken; the observations were repeated in 1820, and I am now enabled to add the present size of those which had been before noticed, as well as some others of different kinds but of the same age, which were not before attended to. The circumference of the trunks is taken in all cases at three feet above the ground, and it will be seen by comparing the different measures how much the cedars have exceeded all the other trees :—

	1801.		1820.		1825.	
	Ft.	In.	Ft.	In.	Ft.	In.
First cedar	10	0	13	1½	14	0
Second cedar . . .	8	6	10	9½	11	4
Third cedar . . .	7	10	9	9½	10	8
Sweet chestnut . . .	10	1	11	7	12	0
Beech	9	4	9	11	10	3
Sycamore	8	11	9	7½	9	11 "

I visited Hopetoun, the seat of the Marquess of Linlithgow, in April 1904, and found that two of these cedars still survive in good condition, the larger being about 80 feet high and 23 feet 8 inches in girth, the other about 88 feet by 13 feet.

There is a fine cedar at Biel, East Lothian, the seat of Mrs. N. Hamilton Ogilvy, which is said to have been planted in 1707 by Lord Belhaven, to commemorate the Union of England and Scotland. According to Mr. S. Ross[1] it was, in 1883, 75 feet high by 17½ feet in girth; but I am informed by Mr. T. Muir that it is now 85 feet high by 19 feet 9 inches at 1½ feet from the ground, with a spread of 101 feet.

At Moncreiffe House near Perth, the seat of Sir R. Moncreiffe, there is a well-shaped tree, which Hunter[2] mentioned as bearing many cones and measuring 66 feet by 11 feet. In 1907, when I saw it, it was about 80 feet by 14½ feet at 3 feet from the ground. At Dupplin Castle, the seat of the Earl of Kinnoull, there are two cedars of which the best shaped measures 86 feet by 16 feet 10 inches, and the other is 18 feet 8 inches in girth at 3 feet. At Murthly there are two good trees, which, though probably not much over seventy years old, measure 74 feet by 9 feet 3 inches and 70 feet by 10 feet 6 inches respectively.

The best I have heard of in the west of Scotland are one at Mount Stuart in Bute, which Mr. Renwick tells me is 64 feet by 8 feet 3 inches, and another at Erskine House, near Renfrew, which is 62 feet by 10 feet at 1½ feet from the ground.

In the N.E. of Scotland it also grows well; there are two very fine trees at Beaufort Castle. According to the measurements given me by Mr. G. Brown the largest of these is 73 feet by 22 feet 8 inches at 3 feet from the ground, dividing at five feet into four large stems, which measure from 9 to 11 feet in girth. The other is the same height and 16 feet in girth. At Brahan Castle there are also some fine cedars.

---

[1] *Woods and Forests*, Dec. 26, 1883, p. 59. [2] *Woods, Forests, and Estates of Perthshire*, p. 135 (1883).

In Ireland the Lebanon cedar has been rarely planted in comparison with its frequency in England; and Henry has not seen any large trees except one at Carton, which in 1903 was 93 feet by 14 feet 9 inches, and is said to have been the first planted in Ireland; and six fine trees [1] at Anneville near Dundrum, Co. Dublin, the largest of which was 14½ feet in girth in 1904.

There is an excellent article on cedars by Dr. Masters in the *Gardeners' Chronicle* for Oct. 17, 1903, giving an illustration of the historic tree in the Jardin des Plantes at Paris, about which many incorrect statements have been published. Carrière [2] gives 1736 as the date at which it was planted, from seed brought from England by Bernard de Jussieu in 1735. From this seed was also derived the cedar at Montigny (Seine et Oise), and the one at Beaulieu, near Geneva. Carrière states that the cedars at Geneva produce seeds so freely that but for the scythe of the mower it would form forests on the shores of the Lake. In a letter from M. Maurice de Vilmorin I learn that the Montigny cedar [3] is now probably the best in France. About 1855 it was 7 metres in girth at two metres from the ground, and it is now 7.90 metres at the same height. There is another tree at Vrigny, the residence of M. Duhamel de Monceau, near Pithiviers, Loiret. His notes of 1874 state that this tree, planted in 1744, had suffered much from the frost of 1870-71, when two-thirds of its branches were frozen. It measured about 8 metres in girth.

I saw a very fine cedar in the grounds of M. Philippe de Vilmorin at Verrières, near Paris, in May 1905, which measured 87 feet by 13 feet; and also visited the tree in the grounds of Madame Chauvet at Beaulieu, near Geneva, which is now considered to be the finest on the Continent, though not equal to several English trees. It is a well-shaped spreading tree about 100 feet high, though difficult to measure exactly, and 16 feet in girth, with a spread of 102 feet.

TIMBER

What is called cedar in commerce is usually the wood of *Cedrela odorata*, a tree found in the West Indies and Central America. The wood of the so-called pencil cedar, *Juniperus virginiana*, is also often known as cedar, [4] and this can be distinguished at once by its colour and smell from the true cedar. A case was recently tried in London with regard to the quality of the cedar used in panelling a room at Packington Hall, in which it was stated in evidence by a so-called expert that there were three kinds of cedar known in the trade, "English grown, pencil cedar, and Californian cedar," "the latter used for inferior work." This is a not unusual instance of the gross ignorance which prevails in England among users of timbers as to their names and native countries, and this ignorance has led to many costly lawsuits. The Lebanon cedar grows so fast in England under favourable circumstances that the wood is of a much softer character than it is in Syria, but it may be used for

---

[1] These are said by Loudon, *Arb. et Frut. Brit.* i. 114 (1838), to have been brought direct from the Lebanon by an ancestor of Lord Tremblestown, and to be the oldest in Ireland.

[2] *Traité Conif.* 78 (1867).

[3] An account of it in *Revue Horticole*, 1907, p. 465, gives the dimensions as 105 feet high by 24 feet in girth at one metre from the ground.

[4] In the Eastern States it is known as red cedar, but this term is applied to *Thuya plicata* in the Pacific States.

many purposes of internal decoration; and the best instance of such use that I know is at Broom House, Fulham, the residence of Miss Sulivan. This lady, having a cedar blown down on her lawn, had it cut into boards, of which there were sufficient to floor and panel the whole of a good-sized drawing-room. When the wood is carefully selected, its pale pink colour and handsome figure make it very ornamental. Its value in commerce is, however, low, because neither the supply nor the demand is regular; and the cost of removing and sawing up large cedar trees is so great, that I was offered a tree containing over 300 cubic feet for nothing if I could get it away; and the Earl of Powis told me that some large trees which were blown down at Walcot were unsaleable, and were eventually used as a cheap material for the kennels of the United Foxhounds. (H. J. E.)

CEDRUS BREVIFOLIA, CYPRUS CEDAR

Cedrus brevifolia.
Cedrus Libani, Barrelier, var. *brevifolia*, Hooker, *Journ. Linn. Soc.* (*Bot.*), xvii. 517 (1879); Beissner, *Nadelholzkunde*, 300, fig. 75 [1] (1891).

Resembling *C. Libani* in characters of leading shoot and branchlets, but with very short leaves, not exceeding $\frac{1}{2}$ inch in length. Cones smaller than those of *C. atlantica*, which they resemble in other respects.

The cedar was discovered in the mountains of Cyprus in 1879 by Sir Samuel Baker, whose specimens were described in the same year as *Cedrus Libani*, var. *brevifolia*, by Sir J. D. Hooker, who considered this form to agree more closely with the Algerian than the other cedars, resembling it in the small size of the cones and in the general characters of the foliage.

The best account [2] of this cedar forest is by Sir Robert Biddulph, who wrote as follows in 1884 to the Director of Kew:—

"The cedar forest occupies a ridge on the principal watershed of the southern range, and about 15 miles west of Mount Troodos. The length of the forest is about 3 miles, its breadth very much less. A few outlying cedar trees were visible on neighbouring hills, but on the ridge they were quite thick, and probably many thousands in number. I took the height above the sea by an aneroid barometer, and found it to be 4300 feet. The trees are very handsome and in good condition, but comparatively young. The smallest seemed to be from ten to fifteen years old; the largest, I am told by the principal forest officer, are probably not over sixty or seventy years. The worst feature is that there were no seedlings or young trees under ten years; and indeed this is the same with regard to the pine forests. It would seem as if the great influx of goats has been comparatively recent. I made a tour through the heart of the forest last August. I started from a point on the west coast, and from thence ascended to the main watershed, and kept along the top till I reached Mount Troodos, taking three days to do it. The country through which we passed on the first day was perfectly uninhabited, and a mass of hills and forest, chiefly *Pinus hale-*

[1] Beissner's figure represents a ripe cone, collected on Mt. Troodos by Herr v. St. Paul.
[2] Published in *Nature*, xxix. 597 (1884). Cf. also *Proc. R. Geog. Soc.* xi. 709 (1889).

pensis and the Ilex. The trees were in very great number, but there was a scarcity of young trees, and most of the old ones had been tapped for resin. On the second day we passed through the cedar forest, and the same sort of country as before, the *Pinus Laricio* beginning at an altitude of 4000 feet. We got as far as the monastery of Kykou that day, and the next day I continued along the watershed to the camp at Troodos. Our road as far as Kykou was a mere track on the side of the hill, in some parts rather dangerous, and we had to lead our ponies on foot, in many parts very steep. The difficulty on the road is the want of water at that elevation. We halted the first night at a beautiful spring, but we had to carry with us food for man and beast for the whole party, muleteers, etc. The scenery was wild and romantic. This spot is the centre of the *moufflon* ground; three of them were at the spring when we approached it. It gave me a clearer idea of the forests of Cyprus than I ever had before."

Madon, who wrote for the Government in 1881 a report[1] on the forests of Cyprus, states that none of the trees were then apparently over eighty years old; but that all were in a vigorous state of vegetation, with numerous young trees of every age covering the soil. In addition to the main forest, three outlying clumps were seen by Madon,—one on the other side of the Ogostina valley, a group of forty-four very young trees at the Kykou monastery, and a third group much lower down. He considered that the cedars formerly covered the whole of the mountain heights from Machera to Livrami, being limited below by the zone of the olive tree. The timber can be recognised in the houses at Campo and in the carvings of the Kykou monastery, showing that the tree was formerly felled for building purposes. Madon noticed what has been confirmed by other observers, that the foliage varied in tint, most of the trees being glaucous.

Hartmann, who has recently visited Cyprus, reports[2] that the trees are remarkable for their broad, umbrella-like crowns, and average about 40 feet in height, 6 feet in girth, and 100 years in age. (A. H.)

I am informed by Mr. C. D. Cobham, Acting Chief-Secretary to Government, in a recent letter, that the Cyprian cedars now occupy an area of about 500 acres in the centre of the Papho Forest, of which the summit, Tripylos, is 4640 feet above the sea. The cedars are mixed with pines and Ilex. There are also a few young trees at Kykou Monastery, a few in the vineyards at Chakistra, and one good specimen tree at Pedoullas. This last was purchased by the Government to preserve it from being cut for building material. There are a number of seedlings in the cedar forest, but these do not seem to have been affected by the exclusion of goats, as animals avoid the cedar when they can find other food. The largest tree in the forest is in Argakis Irkas Teratsa, near the Kykou goatfold. It stands about 60 feet high, and measures 11 feet 6 inches in girth at five feet from the ground. A photograph of this tree is so precisely like a Lebanon cedar standing on my own lawn, which I see as I write, that I need not reproduce it. I may add that some cones sent from Cyprus in February 1905 were smaller than the cones from Syria

---

[1] *Parly. Paper: Encl. 2 in Cyprus, No. 366, of* 1881, p. 28.
[2] *Mitt. Deut. Dendr. Ges.* 1905, p. 181.

or those grown in England. Though at the time I did not think they were mature, yet the seeds contained in them have germinated and produced young plants, which in July were just putting forth their second whorl of leaves, but by the following May had increased very little in size, being much smaller than those of the same age from Swiss and English seed.

Plants were raised at Kew from seed received in 1881; and two, now growing in the cedar collection at Kew, have attained only 6 feet in height, and are remarkable for their singularly short leaves and stunted bushy appearance. A number of them were killed by the winter, having been planted out when too young, which seems to show that this variety is more tender than the Lebanon tree. (H. J. E.)

CEDRUS ATLANTICA, ATLAS OR ALGERIAN CEDAR

Cedrus atlantica, Manetti,[1] *Cat. Plant. Hort. prope Modiciam, Suppl. Secundum*, 9 (1845); Ravenscroft, *Pinet. Brit.* iii. 217 (1884); Kent, Veitch's *Man. Coniferæ*, 409 (1900); Masters, *Gard. Chron.* x. 425, f. 53 (1891).
Cedrus africana, Knight, *Syn. Conif.* 42 (1850).
Cedrus Libani, Mathieu, *Flore Forestière*, 564 (1897).
Pinus atlantica, Endlicher, *Syn. Conif.* 137 (1847).
Pinus Cedrus, Linnæus, var. *atlantica*, Parlatore, DC. *Prod.* xvi. 2, p. 108 (1864).
Abies atlantica, Lindley and Gordon, *Journ. Hort. Soc.* v. 214 (1850).

Young trees stiffer in habit than the Lebanon cedar, and with an erect leader. Branchlets not pendulous, covered with short dense pubescence. Leaves up to an inch (occasionally in cultivated specimens 1¼ inch) long, usually as thick as or thicker than broad. Cones shorter and more cylindrical than in *C. Libani*; scales 1½ inches in width, claw inflected.

VARIETIES

Var. *glauca*.—In the cedar forests of Algeria a certain proportion of the trees have glaucous foliage, the leaves being marked above with conspicuous white stomatic bands; but there is no other difference, and no foundation exists for the opinion, first mooted by Jamin,[2] that the glaucous variety constitutes a distinct species.[3] The glaucous tint is an essentially unstable character,[4] trees occurring in the wild state in which glaucous leaves appear only on some of the branches. This variety often arises in cultivation.

Beissner[5] mentions several varieties, which have been obtained in cultivation, as *pyramidalis*, *columnaris*, and *fastigiata*,[6] characterised by peculiarities of habit; and a variegated form in which the foliage of the young shoots is yellowish,[7] but so far

---

[1] Manetti gives the name only without any description, in the second supplement to his catalogue (1845), and not in the first supplement (1844) as usually stated. Endlicher first described the Atlas cedar from plants 6 inches high, sent in 1847 by Manetti from the Royal Gardens at Monza (Modicia) near Milan.

[2] Decaisne, *Rev. Hort.* ii. 41 (1853). Cf. *Gard. Chron.* 1853, p. 132.

[3] *Cedrus argentea*, Renou, *Ann. Forest.* iii. 2 (1854).

[4] Cf. Fliche in Mathieu, *Flore Forestière*, 564, note 2 (1897). [5] *Nadelholzkunde*, 304 (1891).

[6] Var. *fastigiata*, a pyramidal form, with branches ascending like those of the Lombardy Poplar, originated as a seedling in Lalande's nursery at Nantes. Cf. *Gard. Chron.* vii. 197 (1890).

[7] Var. *aurea*, young foliage of a rich golden colour, which changes to the normal green of the species in the second year. This variety is mentioned by Kent, *loc. cit.*

as we have seen these are not distinguishable as they get older. At Glasnevin there is a remarkable tree about forty years old, of which the stem is erect for about 25 feet, and beyond this bends over almost horizontally, extending laterally outwards for almost 12 feet; and Elwes saw one of very slender and pendulous habit at Angers in France.

DISTRIBUTION

This cedar occurs in Algeria and Morocco. In the latter country its distribution is still scarcely known, though it was in Morocco that the Atlas cedar was first discovered. Philip Barker Webb visited[1] Tangiers and Tetuan in the spring of 1827, and from a native received branches of cedar which had been collected in the impenetrable mountains of the province of El Rif, where there were said to be vast forests. Webb's specimens are preserved in the museum at Florence, where I saw them in December 1906. His discovery was published in an article[2] by De Candolle in 1837. Dr. Trabut[3] states that the tree occurs in the mountains behind Tetuan ; and it is supposed[4] to exist to the south-east of Fez, where the traveller Rohlfs states that he saw larch growing.

In Algeria the cedar[5] forms a considerable number of isolated forests, none of them of great extent, at altitudes between 4000 and 6900 feet. The tree appears to be indifferent to soil, as it grows both on limestone and on sandstone formations. No meteorological observations have been regularly taken in the cedar forests; but in general, where the tree flourishes, snow lies for several months during winter, the temperature descending to 5° Fahr., and frost prevailing until May. In summer the weather is dry with moderate temperatures.

In the following detailed account I have supplemented my own observations by consulting both the special pamphlet[6] concerning the cedar, published by order of Governor-General Cambon, and M. Lefebvre's excellent book[7] on the forests of Algeria.

The chief forests are those in the vicinity of Ouarsenis, Téniet-el-Hâad, and Blida, and in the Djurdjura range in the province of Algiers ; and those on Mt. Babor, in the Mâadid mountains south of Sétif, and in the Aurès and Belezma mountains near Batna.

The forest[8] of Ouarsenis, the most westerly in Algeria, lies in the mountains south of Orléansville. Here the cedar, mostly in mixture with *Quercus Ilex*, only covers an area of 250 acres. The forest near Blida, which is often visited by tourists, as it lies near the railway not far from Algiers, is 1700 acres in extent, and consists of cedars either growing pure or in mixture with the evergreen oak ; and it is, generally speaking, in a poor condition. In the Djurdjura range, extending in an interrupted band on both slopes for nearly 40 miles, are the remains of an ancient forest, most of the trees either growing singly or in small groups on rocks and precipices,

[1] Gay, *Bull. Soc. Bot. France*, iii. 39 (1856).   [2] *Bibliothèque Universelle de Genève*, 1837, pp. 439, 440.
[3] *Les Zones Botaniques de l'Algérie*, 7 (1888).   [4] Lefebvre, *Les Forêts de Cèdre*, 1 (1894).
[5] A fine picture of a forest in Algeria is given in *Garden and Forest*, viii. 335, f. 47 (1895).
[6] *Les Forêts de Cèdre* (Alger-Mustapha, 1894).   [7] *Les Forêts de l'Algérie*, pp. 406-421 (Alger-Mustapha, 1900).
[8] Hutchison, *Trans. R. Scot. Arb. Soc.* xiii. 211, states, but does not give his authority, that cedars were cut here, the diameter of which was so great, that it was necessary to join two saw-blades, each 6½ feet long, in order to fell the trees.

between 4900 and 6500 feet; but on the Häizer peak M. Britsch saw a few trees on the north slope as high as 7100 feet.

The forest on Mt. Babor is of no great extent, but is an interesting one, consisting of a mixture of cedar, *Quercus Mirbeckii*, and *Abies numidica*, and will be described in our account of the last-named species. The brigadier in charge of this forest informed me that he had measured there a cedar 62 feet in girth. In the mountains of Mâadid there are four distinct forests, generally speaking in bad condition, and yielding scarcely any timber, though in one of them, called Ouled Khellouf, there are said to be some very large trees.

The forests which are the most important from every point of view are those in the west near Téniet-el-Hâad, and those in the east in the vicinity of Batna, visited by me last January.

The cedar occurs around Batna, both on the Aurès range and its spur Belezma. The forest of Sgag is 23 miles distant from Batna and covers 1200 acres. Between Batna and Biskra, about 20 miles north of the latter place, the forest of Djebel Lazereg is 1350 acres in area, and is noted for producing a peculiar kind of cedar timber, pink in colour and with a juniper-like odour. A very fine forest of considerable extent, 28,000 acres, lies around Mt. Chélia, the highest point in Algeria, 7500 feet altitude, 43 miles to the south-east of Batna; but it was practically inaccessible in January. In one part of it, the forest of Beni-Oudjana, 44,666 trees have been marked for felling, estimated to contain 3,615,000 cubic feet of timber, which will be offered for sale by the Government in the course of the present year.

I visited the forest of Belezma, which is only 12 miles to the north-west of Batna. The whole wooded area here under government control is 140,000 acres in extent; but of this the cedar occupies only 22,000 acres, ascending the mountain to its summit, 6900 feet, and descending on northern slopes to 3600 feet, and on southern slopes to 4300 feet. The forest was badly treated in former years, whole tracts of the finest trees having been clean cut away and the timber used in building the town of Batna. The drought which prevailed from 1875 to 1881 caused serious damage to the remaining trees, and many died, most of which, except those that have been lately felled, are still standing. Felling is done regularly every year, only dead trees being removed. The sapwood of these has rotted away, but the heartwood remains quite sound and unaltered. This timber is mainly used for railway sleepers, though some has been utilised in house-building and for making wood pavement and furniture. None of it appears ever to have been exported; and it is a great pity to see such excellent wood utilised only for rough purposes. The price obtained for it is as it stands very low, 1d. to 2d. per cubic foot; yet it is fairly accessible, as the haulage to Batna is very cheap, but the rate by railway from there to Philippeville, the nearest seaport, is 15s. per ton.

It snowed very heavily during my stay at the forester's house near the top of the mountain; but so far as I could see, the cedar only grows here in a dense condition in the young stage, there being in the ravines fine stands of cedars 30 feet high, which are slightly mixed, like the rest of the forest, with *Quercus Ilex*, *Juniperus thurifera*, and *Juniperus phœnicea*. These young trees are narrowly pyramidal in form, with erect stiff leaders; but in slightly older trees the leader begins to incline over on one side, and the branches to thicken and elongate, and this process being continued, eventually the tree assumes when old the habit of Lebanon cedars, as we see them in English parks. In other parts of the forest the older trees are more or less scattered with the same admixture of junipers and evergreen oak, the undergrowth being Phillyrea and broom. The cedar appeared to be slow in growth, the annual shoots of young vigorous trees not exceeding three or four inches in length. From observations made in one section of this forest the tree shows at different ages the following dimensions :—

Age.		Diameter.		Height of Market-able Timber.	Total Height.
		Feet.	Inches.	Feet.	Feet.
125 years	. . . .	2	$7\frac{1}{2}$	46	98
160 ,,	. . . .	2	$11\frac{1}{2}$	$52\frac{1}{2}$	105
200 ,,	. . . .	3	3	59	115
255 ,,	. . . .	4	3	59	125
305 ,,	. . . .	4	11	59	125

An official document, which I saw at Batna, gave the total number of sound trees over 40 inches in girth as 265,500, estimated to contain between ten and eleven million cubic feet of timber, the total timber in the forest, young and old trees, cubing 16,000,000 feet. In addition, there is still standing 900,000 cubic feet of dead timber. In a few spots, as in the Chellala-Bordjen section, there are rather dense stands of old trees, which run to 7000 cubic feet per acre; but there are large tracts in parts of the forest which have scarcely 150 cubic feet to the acre.

The tree produces seed abundantly every two or three years; and regeneration is good in favourable situations, as in northerly ravines. The cones[1] disarticulate in November, after the autumnal rains, but if the weather is exceptionally dry, do not open. Seedlings appear under dense cover, but in such situations grow slowly, and do much better in the partially open places between large trees. The wide-spreading branches which the tree ultimately produces show, I think, that in old age it requires a great deal of light, and tends to grow in a more or less isolated condition; but until middle age the trees bear crowding without injury. In the bare parts of the mountain, where the trees were cut away many years ago, artificial planting has been tried on a small scale, and has succeeded on northern slopes when two-year-old seedlings have been planted in autumn. Plants put out in the spring on the southern slopes have died of drought, which is the great enemy to both artificial and natural regeneration.

The forest of Téniet-el-Hâad is about a day's journey from Algiers—four hours by rail and thence seven hours by the coach to the town, which is distant from the cedars about an hour's walk. The mountain-range runs in a N.W.-S.E. direction, the cedars ascending to the summit of the crest, 5900 feet, and descending on the north side to 4250 feet, and on the south side to 4900 feet, there being a zone of

---

[1] Only the central part of the cone contains good seed. In January the basal scales of many cones were still remaining around the central axis, the other scales having fallen much earlier.

Quercus Ilex below, with which the cedar slightly mingles. The cedar forest occupies 2300 acres, four-fifths of this being on the north slope and one-fifth on the south slope, and consists of a mixture in varying proportions of cedar and *Quercus Mirbeckii*, the latter a beautiful tall tree with semi-evergreen foliage, often attaining 12 feet in girth. This mixed forest is nowhere very dense, except where there are young stands, and grows upon sandstone—the undergrowth being chiefly *Rosa* and *Rubus*, with *Juniperus* in the lower zone. The tallest cedar does not, I believe, exceed over 120 feet; and the largest, which I measured and photographed (Plate 136, B), are *La Soltane*, 98 feet high by 24 feet in girth, and *Le Massaoud* (Plate 136, A), 108 feet by 23 feet. Trees of peculiar shape are common; one, 108 feet by 19 feet, dividing into two stems at eight feet up; and another, *Le Cèdre Parasol*, which stands on a rocky promontory, being a low tree with a peculiar broad-shaped umbrella-like crown. Around the forester's house, Le Rond Point, at 4600 feet, there is a plateau of some extent, with many fine old trees having the habit of the Lebanon cedar as we see it in England.[1] No felling is done at present in this forest, which is rapidly improving in value owing to the entirely successful natural regeneration, cedars being present in all stages of growth.

The wood of the cedar, though without resin-canals, contains a quantity of resin, which gives it a peculiar, penetrating, and distinctive odour.[2] At Batna, *libanol*, a kind of resin, is obtained by distillation of the sawdust of old trees. This product is very valuable in the treatment of inflammation of the mucous membranes, and is said to be curative in influenza. Cedar wood contains a large amount of white sapwood, 25 to 50 annual rings, with a brown or brownish-yellow heartwood. The heartwood is homogeneous and fine in the grain, and takes an admirable polish. It lasts indefinitely, trees which were cut down fifty years ago in the forest at Batna remaining still on the ground quite sound, and when not exposed to the air is imperishable. Pieces of cedar wood have been found in tombs which are supposed to belong to the Punic period, and portions of ancient mosques built of cedar are in perfect condition. Placed in water, the heartwood becomes very hard; and vats made of it, which have been buried in sand for thirty years, are not only well preserved, but cannot be cut by an axe. The wood of dead trees can be used at once, but that of living trees requires to be seasoned carefully for six or twelve months. Though the timber is used in building, it is rather heavy for that purpose, and has no great elasticity or resistance to flexion under a heavy weight. It is, however, well suited for the finest kinds of cabinet-making.   (A. H.)

CULTIVATION

The seed ripens in most seasons in England at least, as well as that of the Lebanon cedar, and will sometimes come up naturally near the parent trees, as at Cooper's Hill near Windsor, from whence I transplanted two self-sown seedlings to my own garden.

---

[1] An excellent illustration in *Garden and Forest*, viii. 335 (1895), shows the flat-topped habit of mature trees in their native forest.

[2] The odour disappears after exposure to the air for a few years, and is not noticeable in the cedar furniture which is so common in the houses at Batna and Téniet-el-Hâad. Cf. Lefebvre, *Les Forêts de l'Algérie*, 350 (1900).

When staying at Heythrop Park, Oxfordshire, in March 1901, I went out on a morning when the frost was so hard that the hounds could not hunt till noon, and found seeds which had germinated on the ground beneath a glaucous cedar. The radicles were protruding from the seeds, in some cones which had not fallen; I took them home and planted them, and have now several healthy young trees about a foot high.

I also sowed a quantity of imported seed in the open field, where they germinated well, but the plants were all destroyed by mice, frost, and drought in the first season, though seedlings raised in the nursery stood the winter without protection. As the seed can be procured in quantity at a cheap rate from Messrs. Vilmorin of Paris, I should recommend its being sown in a frame and protected for two or three years, after which it will require two to three years more in the nursery before planting out.

The tree seems to like lime in the soil, and will, in my opinion, prove a valuable timber tree if planted in open woods, in warm, dry soils, sufficiently close together to prevent its branches from developing too much, and possibly if mixed with beech it might thrive better than alone.

As regards the relative rate of growth of the Atlas and Lebanon cedars we have the evidence of M. André Leroy, the well-known nurseryman of Angers, who, in the *Belgique Horticole*, 1867, p. 59, gives the following measurements :—

LEBANON CEDAR				ATLAS CEDAR			
AGE.		Height.		AGE.		Height.	
		Metres.	Centimetres.			Metres.	Centimetres.
1 year	.	0	6-8	1 year	.	0	10-15
2 years	.	0	12-15	2 years	.	0	20-30
3 ,,	.	0	18-25	3 ,,	.	0	40-50
4 ,,	.	0	36	4 ,,	.	1	0
5 ,,	.	0	50	5 ,,	.	1	75
6 ,,	.	0	75	6 ,,	.	2	50
7 ,,	.	1	0	7 ,,	.	3 and upwards.	

After seven years of age, he states that the annual growth was often more than one metre, and mentions a tree only twelve years old, from seed, which was one metre in circumference (I presume at the ground). He also says that it is easier to transplant, and endures exposure and bad soil better than the Lebanon cedar, and believes that it will prove a valuable tree for planting on barren wastes where nothing else will thrive.

These remarks, no doubt, will apply better to the soil of Central France than to England, but I have the highest possible opinion of the hardiness of the tree, and have found it endure the damp, cold, and early and late frosts of the Cotswold hills in a way that few other conifers will do. So far as my experience goes, however, it is not a tree which can be transplanted without some care in a small state, and when it has had its roots cramped in small pots, as is often done by nurserymen for convenience of sale, is rather apt to die. I am not aware that it has ever yet been tried in quantity under forest conditions; but, so far as I have seen, it is not subject to insect or fungoid diseases which attack and kill the deodar.

Many of the grafted trees of the glaucous variety, which are usually sold by nurserymen, are one-sided and unsightly objects, for a good many years after planting at any rate; and though it is claimed by some that grafting, if properly done, does not permanently disfigure the tree, yet I would always prefer seedlings. Even if not quite so glaucous in colour as the best of the others, a certain number of this tint will generally appear among them.

The date at which the Algerian cedar was first introduced to this country is somewhat uncertain; but it must have been subsequent to 1844, and if any older ones exist they cannot be recognised with certainty. Several trees appearing older than this have been supposed to be African, on account of their habit and cones, but there is nothing on record to prove it.

According to Ravenscroft, the oldest of which we have an exact record were raised at Eastnor Castle in 1845, from cones gathered by Lord Somers himself at Téniet-el-Hâad. In December 1860 the tallest of these was $18\frac{1}{2}$ feet; in December 1866, 31 feet. When I measured it in 1906, it was 77 feet by 8 feet 1 inch.

REMARKABLE TREES

The tallest tree that I have measured in England is at Linton Park, Kent, and is a glaucous tree, which, from its shape, seems to be grafted, though there is no evidence of this. It was 80 feet high in 1902.

The largest recorded at the Conifer Conference in 1891 was at Mulgrave Castle, Yorkshire,[1] the seat of the Marquess of Normanby. It was then 66 feet by 5 feet 10 inches. Mr. Corbett informs me that it is now 72 feet by 8 feet 4 inches.

On Ashampstead Common, Berks, there is a handsome and well-grown tree which has grown up in a semi-wild condition among other trees, and which was 63 feet by $6\frac{1}{2}$ feet when I last saw it in 1907 (Plate 137).

At Ashridge there are several fine glaucous trees, raised from seeds, which were brought by Earl Brownlow, in 1862, from Téniet-el-Hâad; the best of them already measures 58 feet by 6 feet. At Merton Hall, Norfolk, there is a very well-shaped tree measuring 60 feet by 6 feet.

At Bicton there is a fine tree measuring 68 feet by 7 feet 6 inches. At Coldrinick, in Cornwall, there is a well-shaped tree which, in 1905, was 64 feet by $5\frac{1}{2}$ feet. At Heanton Satchville, North Devon, I saw a healthy young tree in a shrubbery, which was clear of branches to 20 feet up, and though 48 feet high, was only 2 feet 7 inches in girth, showing the ability of this cedar to thrive without much space, even in a climate so much damper and cooler than that of Algeria.

At Tortworth there is a cedar about 50 feet high with very short leaves, and remarkably fastigiate habit, which seems to belong to the variety named *fastigiata*.

In Scotland I have not seen any so large as in England; but the tree grows

[1] A tree at Grimston, near Tadcaster, Yorkshire, reported in 1900 to be 70 feet high and 13 feet in girth at three feet from the ground, which was said to be sixty-five years old, is probably a Lebanon cedar. Cf. *Gard. Chron.* xxviii. 210 (1900).

well at Murthly and other places.   At Smeaton-Hepburn, a tree,[1] planted in 1847, was, in 1902, 69 feet high and 6½ feet in girth.   At Fordell, in Fifeshire, the property of Lord Buckinghamshire, I am informed by Mr. Sibbald that a number of cedars were planted by Mr. Fowler, then head gardener, 42 years ago on a damp sandy soil and well sheltered by other trees.   The average height of the Algerian cedars in 1906 was 48 feet, with an average girth of 4 feet 4 inches, and of the deodars 33 feet by 3½ feet.   The majority of them are in good health, though the Algerian have made by far the best trees, and as the soil and climate of Fifeshire do not seem to be so favourable to the growth of trees generally as those of Perthshire, Morayshire, or parts of Ross-shire, this seems to prove that the tree may be planted in Scotland with good hopes of success.

The finest Atlas cedar in Ireland is at Fota, and is of the glaucous variety.   It was planted, according to Lord Barrymore, in 1850, and measured in 1904 83 feet high by 7 feet 7 inches in girth (Plate 138).   At Carton, the seat of the Duke of Leinster, a tree, which is, from its habit, apparently an Atlas cedar, was, in 1903, 80 feet high by 9 feet in girth.   At Powerscourt a glaucous specimen was in the same year 50 feet high by 5 feet in girth.

In the south of France and North Italy this tree grows better and faster than in England.   Perhaps the best that I have seen are in the public garden at Aix en Savoie, where there is a grove of splendid trees 90 to 95 feet high, though only planted in 1862.   They average 6 to 7 feet in girth, and there are many self-sown seedlings near them.   On the shores of the Lago Maggiore the tree succeeds perfectly, several fine trees in the grounds of the Villa Barbot near Intra being 90 feet or over, and one 7½ feet in girth.   It seemed to me likely to become a most valuable forest tree in this region.   (H. J. E.)

CEDRUS DEODARA, Deodar

Cedrus Deodara, Lawson, *Agric. Man.* 381 (1836); Loudon, *Arb. et Frut. Brit.* iv. 2428 (1838); Brandis, *Forest Flora*, 516 (1874), and *Indian Trees*, 691 (1906); Ravenscroft, *Pinet. Brit.* iii. 225 (1884); Masters, *Gard. Chron.* x. 423, f. 52 (1891); Kent, Veitch's *Man. Coniferæ*, 411 (1900).

Cedrus Libani, Barrelier, var. *Deodara*, Hooker, *Himal. Journ.* i. 257 (1854), *Nat. Hist. Rev.* ii. 11, tt. 1-3 (1862), and *Fl. Brit. Ind.* v. 653 (1888); Collett, *Flora Simlensis*, 486 (1902); Gamble, *Ind. Timbers*, 710 (1902).

Cedrus indica, Chambray, *Arb. Res. Conif.* 341 (1845).

Pinus Deodara, Roxburgh, *Hort. Beng.* 69 (1814).

Abies Deodara, Lindley, *Penny Cycl.* 9 (1833).

Young trees with pendulous leader.   Branchlets always pendulous, grey and densely pubescent.   Leaves up to 2 inches long, as thick as broad.   Cones large and broad, ellipsoid, 4 to 5 inches long by 3 to 4 inches in diameter, rounded at the apex; scales 2 to 2½ inches wide, with claw not inflected, usually less tomentose than in the other cedars.

[1] Sir A. Buchan-Hepburn in *Proc. Berwick Nat. Club*, xviii. 210 (1904).

Varieties

A considerable number of varieties have arisen in cultivation, ten being mentioned by Beissner.[1]

1. Var. *albo-spica*. Growing shoots during spring and early summer of a milky-white colour. Trees of this kind at Dropmore[2] are pyramidal in habit, and make splendid growth. At Grayswood,[3] Haslemere, a bushy form with this peculiar foliage has been noted.

2. Var. *robusta*. Branchlets stout; leaves longer and thicker than in the ordinary form.

3. Var. *crassifolia*. Branches short and stout; branchlets not pendulous; leaves short and thick.

4. Var. *verticillata*. Branchlets whorled.

5. Var. *fastigiata*. Fastigiate in habit.

6. Varieties with variegated foliage and with bright yellow leaves have also been noted. The glaucous tint has appeared in cultivation, and is met with in the wild state. A very glaucous tree at Castlewellan has been named var. *nivea*.[4] Trees with thin, shining, deep green foliage have been distinguished as var. *viridis*.

(A. H.)

Distribution

The deodar is found in the Western Himalaya; and extends eastwards to the Dauli river in Kumaon, occurring at 4000 to 10,000 feet, most common at 6000 to 8000 feet. It extends westwards through Kashmir to the Peiwar forests in the Kuram valley of Afghanistan.

According to Gamble, from whom I take the most of the following account, it is a gregarious tree, but rarely forms pure forests, though exceptions are met with, generally in the form of sacred groves; usually it is associated with *Picea Morinda* and *Pinus excelsa*, and three species of oak in their various zones. Sometimes the silver fir (*Abies Pindrow*) accompanies it, but more rarely; the cypress (*Cupressus torulosa*) in its favourite localities joins it; the yew is often found under it; and at low elevations it mixes with *Pinus longifolia*.

Among other trees commonly found with it may be mentioned *Betula alnoides*, *Populus ciliata*, *Æsculus indica*, elm, hazel, hornbeam, maples, bird-cherry, holly (*Ilex dipyrena*), *Pieris ovalifolia*, and rhododendron; while among the shrubs commonly found in deodar forests may specially be noted species of *Berberis*, *Indigofera*, *Desmodium*, *Cotoneaster*, *Euonymus*, *Salix*, especially *Salix elegans*, *Viburnum*, *Lonicera*, *Parrotia*, and rose, while *Clematis montana*, *Vitis semicordata*, and ivy, are frequently met with climbing over and festooning its branches.

In the outer ranges the deodar forests chiefly clothe the northern and western slopes of the ridges, while in the interior hills, to which the rainfall of the south-west monsoon still reaches, they are found on all aspects, but less pure. Beyond the region of the south-west monsoon the deodar is still found, but gets

---

[1] *Nadelholzkunde*, 307, 308.
[2] *Gard. Chron.* xxxvii. 44, 76 (1905).
[3] *Gard. Chron.* xxxvii. 59, 105 (1905).
[4] *Ibid.* xxv. 399, fig. 146 (1899).

gradually scarcer, and in such places its companions may be *Pinus Gerardiana* and *Quercus Ilex.*

The deodar can attain a very great size.[1]   Thomson[2] mentions one near Nachar, on the Sutlej, that was 35½ feet in girth.   Dr. Stewart measured one at Kúarsi in the valley of the Ravi that was 44 feet at 2 feet, and 36 feet at 6 feet from the ground; another about 900 years old was 34½ feet in girth.   Minniken records a tree at Punang, in Bashahr, that was 150 feet high and had a girth of over 36 feet, the clean bole being 45 feet long.   Dr. Schlich measured a tree in the Sutlej valley 250 feet high with a girth of 20 feet.

In the Dumrali block in the Tehri-Garhwal leased forests a fallen tree was unearthed 90 feet long and over 7 feet in diameter, which had been dead for at least 100 years, and was, when it fell, probably 550 years old.   When cut up it gave 460 metre-gauge sleepers.   I am indebted to Mr. J. H. Lace for the illustration (Plate 139) representing a group of deodars in the Himalayas.

A great section in the corridor of the forest school at Dehra Dún is 23 feet in girth, with 665 annual rings.   The number of annual rings to the inch varies much according to the elevation and rainfall, but averages about 8 to 12, though in the Kuram valley Bagshawe found an average of about 21.

As an ornamental tree there are few in the world that can compare with the deodar.   From the Lebanon cedar and the Atlas cedar it differs somewhat in appearance, but even to an expert, in the collections of Europe, it is not always easy to recognise to which of the three species a given specimen belongs.   Roughly, however, the deodar is distinguished by means of its drooping branches and its longer needles.   Two well-marked varieties are recognisable in the forests, the one with dark green, the other with silvery foliage.   The latter variety, well known in European collections, is found wild in ravines at a comparatively low level. Gamble saw it in Jaunsar, in the upper Dharagadh, in ravines at from 4000 to 6000 feet, and believes that the variety comes true from seed.

Deodar trees are often lopped for litter, and if the leading shoot is not damaged, the tree grows on well enough; when the leading shoot is cut or damaged, the tree shows a great tendency to form others; and frequently several erect shoots, with the appearance of young trees, may be seen growing up straight from its branches.   The deodar may be almost said to produce coppice shoots, for, as Brandis remarks, if only a small branch be left to a stump, it will send out shoots and grow well, eventually, perhaps, forming a new tree.

In close forests deodars flower and seed rather sparsely; for good seed bearers we have to look to the old trees on dry ridges, where they can get a large amount of sunlight.   When the seeds are ripe the cones break up and the scales fall; the winged seeds are then carried by the wind for a short distance.   It may be interesting to record the result of the examination of an average cone by Mr. B. B. Osmaston in October 1900.   He found in the top part 25 scales, with 50 bad seeds;

---

[1] Webber, in *Forests of Upper India*, 331 (1902), says: "I have seen deodars 40 feet in girth and 250 feet high, the age of which must be 1000 years or more"; and Pakenham Edgworth informed Bunbury that he had measured deodars 46 feet in girth.   Cf. Lyell, *Life of Sir C. J. F. Bunbury*, ii. 238 (1906).

[2] *Western Himalaya and Tibet*, 64 (1852).

in the middle 100 scales, with 90 good and 110 bad seeds; in the lower part 94 scales, with 188 bad seeds—the whole cone, therefore, giving 219 scales, with 438 seeds, of which 90 were good.

CULTIVATION

The best account we have of the introduction of the deodar is given by Ravenscroft, who states that the Hon. Leslie Melville sent seeds[1] in 1831 which were sown at Melville in Fifeshire, at Dropmore, and elsewhere.

Lord H. Bentinck sent some to Welbeck in 1832, but it was not until 1841 that the Right Honourable T. F. Kennedy, then at the head of the Woods and Forests, took steps to procure seed in large quantities from the Himalayas. His proceedings are described at great length in the *Thirty-first Report of the Commissioners of Woods*, pp. 168-172, and pp. 440-454 (1853), and further in the *Thirty-fourth Report* (1856), pp. 87, 88, and pp. 120-122. From this it appears that 60,000 seedlings were distributed in the spring of 1856 amongst the New, Dean, and Delamere forests, and a further 40,000 were sent out in the following autumn.

I am indebted to Mr. E. Stafford Howard, C.B., for information as to the results of these experiments as given in letters from the Hon. Gerald Lascelles and the late Mr. P. Baylis. The former says:—" I have made search for any records of the planting of the deodars, but can find nothing worthy of quotation. It is a fact that it was very largely planted here, as we can see for ourselves,—more, however, as an avenue or ornamental tree than, strictly speaking, for timber. Large quantities were raised in the nursery at Rhinefield, which at that time was managed by one Nelson, who in a small book speaks of the very large experience he has had in raising and transplanting deodars. The tree is, however, a failure by reason of the way in which it suddenly dies off, unaccountably, when it is about forty or fifty years old. There are some notable successes, such as the grove at Boldrewood[2] and others, but I must have cut hundreds which had died off suddenly."

Mr. Baylis wrote on 8th May 1905: "I cannot give much definite information on the subject, though Crown Keeper Smith remembers some deodars being planted about 1857 along the sides of the rides in the High Meadow estate; but large numbers of these have perished, and there are no very fine trees among those that are left. A ride along the top of the Churchill enclosure was also planted about the same time with similar trees; but many of these also have died, and I cannot say that any of them have thriven well, though one tree has occasionally borne cones. I think that the climate here is too cold and damp for them to thrive, and that they cannot stand the damp cold of our winters in the Forest, though on the slopes of the Malvern hills they flourish fairly well."

This liability of the deodar to die after attaining considerable size has been often noticed, and, so far as I have observed, is most common on soils which are poor in lime.

---

[1] A tree raised from these seeds was planted near the Director's Office at Kew, and had attained a height of 32 feet in 1864. It became diseased and was removed in 1888. Cf. *Kew Hand List of Coniferæ*, xiv. (1903).

[2] The best deodar at Boldrewood is now 64 feet high.

The Earl of Ducie informs me that in 1854, and for several years afterwards, he planted many deodars at Tortworth, both on the old red sandstone and on the mountain limestone.   Many of these have perished after thirty to forty years' growth, without any apparent reason, except that in one case where only six out of about ninety remain, it is probable that they were infected with disease by the dead roots of beech trees which previously occupied the ground.   Very few deodars at this place seem likely to attain a great age, and contrast unfavourably with the Cedar of Lebanon.   But at Miserden Park, in the same county, on a dry oolite limestone, at an elevation of at least 600 feet, a line of deodars about sixty years old have remained healthy, though their growth here is much slower than at Tortworth.

At Poltimore, near Exeter, there is a fine avenue of deodars which were planted in 1851-52, and have grown to an average height of 70 to 80 feet in 1906, most of them being extremely vigorous, but there are several blanks in this avenue.

The cause of these deaths is explained by Mr. R. L. Anderson in a note published in the *Quarterly Journal of Forestry*, i. 216, who states that the fungus now known as *Armillaria mellea*, Vahl., was present on the roots of one of these deodars; and as the best means of checking its spread to other healthy trees, recommends trenching the ground round the affected tree, digging up and burning its roots, and scattering gas lime over the ground where they have been.

At Castle Menzies, in Perthshire, of a number of deodars, which were planted by the late Sir R. Menzies about 1852 to commemorate the birth of his son, on soil which was too wet to suit them, though *Tsuga albertiana* and *Picea sitchensis* have succeeded very well close by, several are dead, and all are more or less stunted, though one of these trees measuring $7\frac{1}{2}$ feet in girth was successfully transplanted in February 1907, and had not lost a leaf when I saw it in the following July.

I have not myself gathered any ripe seed of the deodar in England, but there is a tree growing in Kew Gardens between the main gate and the Director's office which measures 37 feet by 4 feet 8 inches, and was raised from seed produced in 1861 or 1862 by a tree at Killerton, and sent by the late Sir Thomas Acland to Kew in February 1868.   Mr. Smith, the then Curator of Kew, was so much impressed by the good quality of the soil from the top of Killerton Hill in which this tree was raised, that two truck loads of it were sent to Kew.

The earliest record[1] of the deodar producing fruit in England is of a tree at Bury Hill, near Dorking, which produced cones in 1852, when it was 28 feet high. Cones have also been borne on trees at Dropmore,[2] Sunninghill,[2] Bishopsteignton[2] near Teignmouth, Enys[2] in Cornwall, and Fota[2] in Ireland.   Seedlings have been raised from home-grown seed at Rozel Bay[2] in Jersey and at Bicton.[3]

A deodar in Kew Gardens produced cones in 1887, according to a note in *Gardeners' Chronicle*, ii. 248 (1887), where it is stated that the production of cones on this species in this country has hitherto been a rare occurrence.   At The Coppice, Henley, the seat of Sir Walter Phillimore, Bart., and at Shiplake House, the

---

[1] *Gard. Chron.* 1852, p. 582, and x. 279 (1891).
[2] *Ibid.* x. 423, 435, 436, 492, 679 (1891).   [3] *Ibid.* 1869, p. 1279.

residence of Miss Phillimore, there are deodars coning profusely at present, probably on account of the hot summer of 1906. At White Knights Park, Reading, there is a seedling now about 8 feet in height, and supposed to be 16 years old, which germinated on a vine border, the seed having come from a tree which measures 75 feet in height and 10 feet in girth.

In India the cones are often much damaged by the larvæ of a Pyralid moth which eats out the seeds, and the saplings are attacked by the well-known fungus *Trametes radiciperda*, which spreads underground through the roots from tree to tree. The leaves are also attacked by Uredinous fungi, especially by *Æcidium cedri*, Barclay, which forms small yellow spots and causes them to fall.

As regards the comparative hardiness to severe winter frosts of the three cedars we have valuable evidence [1] collected by Mr. Palmer in 1860-61. Reports were received from no less than 211 places in England, Scotland, and Ireland. "The winter of 1860-61 was the most severe that has happened since its introduction. It was a winter such as had scarcely any parallel for severity in the memory of man, and unless some general change of climate should take place, it may be looked upon as exceedingly improbable that any cold of greater intensity should again visit us. The effect of that winter upon the deodar may therefore be taken as a safe guide in judging of its suitableness for our climate; what the effect was we are, as already mentioned, enabled, through the kindness of Mr. Palmer, to state with accuracy.

Mr. Palmer's record of observations shows that the deodar is by no means so hardy a tree as the larch, and also that it is the least hardy of any of the cedars. There is no instance of any of the larches reported to him having been injured by the cold of 1860; while out of the deodars growing at 211 places in Great Britain and Ireland, plants were killed at 55, and were uninjured only at 80, having been more or less injured at the remaining 76, a percentage of frailty much greater than we should have anticipated. The Cedar of Lebanon and the *Cedrus atlantica* proved more hardy, and about equal between themselves. The following summary will show the actual results of Mr. Palmer's report on all three :—

	Total Places reported on.	Not injured.	Injured.	Much injured.	Killed.	Proportion of Killed and Much injured.
Cedrus Deodara .	211	80	50	26	55	1 in 2½
Cedrus Libani .	81	51	19	6	5	1 in 7½
Cedrus atlantica .	74	48	19	2	5	1 in 10½

It may be interesting to notice in what proportion the three different parts of the kingdom suffered. It was as follows :—

	Total Places reported on.	Not injured.	Injured.	Much injured.	Killed.	Proportion of Killed and Much injured.
Scotland . .	64	19	26	14	5	1 in 3½
England . .	142	61	24	13	50	1 in 2½
Ireland . .	4	3	1			"

[1] Published in Ravenscroft, *Pinet. Brit.* iii. 242 (1884).

REMARKABLE TREES

The two finest deodars, as regards size and symmetry, that I have seen in Great Britain are at Bicton, where cones were produced, according to *Pinet. Brit.*, as long ago as 1858.   One of these on the lawn measured in 1902 was 80 feet by 11 feet 8 inches (Plate 140).   The other is near the ornamental water in a more sheltered situation, and was then 90 feet by 9 feet 1 inch.

Another of about the same height at Beauport has an erect top, and looks as if it might become much taller.   The tallest reported at the Conifer Conference was at Studley Royal, and was then 70 feet by 7½ feet; but when I visited that place I saw no very large tree of the kind.

At Dropmore there is a handsome tree which in 1905 was 77 feet by 8 feet 10 inches, and had many of the woody knots embedded in the bark that are sometimes seen in the cedars.   It is said to have been planted[1] in 1834.

At Westonbirt, a tree, planted by the late Mr. Holford, about 85 feet by 8 feet 9 inches, is one of the largest and best shaped that I have seen.   A deodar of peculiar habit at Linton Park, Kent, reported to be 79 feet high, is figured in *Gardeners' Chronicle*, December 12, 1903, fig. 159.

At Barton there is a fine tree branched to the ground, which in 1904 was 76 feet by 9½ feet.   At Highclere there is a handsome tree about 75 feet by 8 feet 4 inches, which was planted by the then King of Spain in 1844.   At Williamstrip, on rather heavy soil, there is a healthy tree of 72 feet by 8 feet.

At Ombersley Court, near Worcester, there is a very fine tree 84 feet by 8 feet 4 inches, which has the erect habit of *atlantica*; but the drooping branchlets show it to be a deodar.

At the Frythe, near Welwyn, Herts, a large deodar was cut down some years ago; and from the side of the stump there is now (1906) a young tree springing up, quite vigorous and healthy, and about 25 feet high.   At Chart Park, Surrey, there is a tree 89 feet by 8 feet 11 inches; and adjoining this place, in the Tunnel Park, Deepdene, there is another fine tree 77 feet by 9 feet, both measured by Henry in 1905.   At Fulmodestone, Norfolk, a tree planted in 1861 was in 1905 66 feet by 7 feet 4 inches in girth.   At Shiplake House, near Henley, a tree, planted in 1852, was 73 feet by 7 feet 9 inches in 1905, and is bearing numerous cones in the present year.   A deodar, growing on Haddington Hill, near Wendover, at 800 feet elevation, is 63 feet by 5 feet 10 inches.

There are many trees of from 60 to 70 feet in other parts of England, but we have seen none which call for special notice.

In Scotland the deodar is only hardy in the warmer parts of the country, and does not seem to have attained anything like the same dimensions as in England or Ireland.   At Poltalloch, notwithstanding the wet and windy climate, it grows fairly well and has attained over 50 feet.   At Rossdhu, on Loch Lomond, it is even taller.

In Perthshire there are good specimens at Abercairney, Castle Menzies, and Dunkeld, which seem to have been planted after the great frost of 1860-61, which

---

[1] *Gard. Chron.* xxv. 138 (1899).

destroyed so many of this tree in the north. The tree at Abercairney is remarkably weeping in habit, and measured, in 1904, 51 feet high by 4 feet 8 inches in girth. The best that we know in this county is perhaps one at Murthly, which is older and bore cones in 1892. It grows well at Gordon Castle, where there is a tree about 50 feet high, and as far north as Dunrobin in Sutherlandshire. At Conan House, Ross-shire, there is a healthy tree 47 feet by 9 feet 9 inches. At Leny, near Callander, there is a very old-looking but rather stunted deodar, which may have been introduced by the distinguished Indian naturalist Buchanan Hamilton, grandfather of the present owner, but when I saw it in 1906 it was only about 45 feet by 7 feet.

At Smeaton-Hepburn, a tree[1] planted in 1841, when it was 2½ feet high, measured in 1902, 55 feet in height and 6 feet 7 inches in girth.

The finest deodar in Ireland is growing at Fota, Co. Cork, and measured, in 1903, 84 feet high by 7 feet 2 inches in girth. At Coollattin, Wicklow, there are two trees, one of which measured, in 1906, 53 feet by 6 feet 10 inches. At Hamwood, Co. Meath, a tree, supposed to have been planted in 1844, was 74 feet by 7½ feet in 1905. At Mount Shannon, Limerick, there is a tree 66 feet by 8 feet 5 in. in 1905. At Emo Park, Portarlington, a tree measured, in 1907, 61 feet by 7 feet 4 inches, and was thriving; but in the dry climate of Queen's County, the deodar as a rule is not a satisfactory tree.

TIMBER

The timber is the most important of any in North-Western India, and supplies a large quantity of railway sleepers, bridge, and building timber. Gamble says that it is rather brittle to work, and does not take paint or varnish well. It has also a very strong odour which, although pleasant in the open air, is not so in a room. It is extremely durable, probably with cypress (*Cupressus torulosa*) the most durable of Himalayan woods. Stewart mentions the pillars of the Shah Hamadin Mosque at Srinagar in Kashmir, which date from 1426 A.D., and were quite sound when he wrote. Its grain is so straight that the logs can be split into boards, which are afterwards trimmed with an adze; and shingles for roofing, according to Webber,[2] stand the changes of climate for centuries without any sign of decay.

The weight of well-seasoned dry wood of average growth is about 35 pounds per cubic foot, branch wood being very much heavier and more full of resin.

Oil is extracted from it by distillation, which is a dark brown, strong, and unpleasant smelling fluid, said to be a good antiseptic, and serves to coat the inflated skins known as "mussucks" used for crossing the Himalayan rivers. (H. J. E.)

[1] Sir Archibald Buchan-Hepburn in *Proc. Berwick Nat. Club*, xviii. 210 (1904).
[2] *Forests of Upper India*, 41 (1902).

LIBOCEDRUS

Libocedrus, Endlicher, *Syn. Conif.* 42 (1847); Bentham et Hooker, *Gen. Pl.* iii. 426 (1880); Masters, *Journ. Linn. Soc. (Bot.)*, xxx. 19 (1892), and *Gard. Chron.* xxx. 467 (1900).

Heyderia, Koch, *Dendrologie*, ii. 2, p. 179 (1873).

Calocedrus, Kurz, *Journ. Bot.* xi. 196 (1873).

Thuya, Baillon, *Hist. Pl.* xii. 34 (1892).

EVERGREEN trees with aromatic odour, belonging to the tribe Cupressineæ of the order Coniferæ, closely resembling Thuya in habit and other characters, the branches as in that genus ending in frondose "branch-systems," which are flattened in one plane and three- to four-pinnately divided, with their axes bearing scale-like leaves in four ranks. On the main axes the leaves are often remote by the lengthening of the nodes; on the lateral axes they are closely imbricated, and vary in the different species in size and form, as detailed in the three sections below. In seedling plants the leaves are always linear-lanceolate and spreading.

Flowers: monœcious with those of the two sexes on different branchlets, or rarely diœcious, solitary, terminal. Male flowers oblong, subsessile, with six to twenty stamens decussately opposite on a slender axis; filaments short, dilated into broadly ovate or orbicular scale-like peltate connectives, which bear usually four sub-globose anther-cells, two-valved and opening on the back. Female flowers oblong; subtended at the base by several pairs of leaf-like scales, which persist slightly enlarged under the fruit; composed of four or six decussately opposite acuminate bracts; lowest pair small, unfertile; next pair above fertile, bearing at the base two erect ovules on a minute accrescent ovular scale; uppermost pair when present unfertile.

Cones small, pendulous or erect, ripening and letting out the seed in the first year, persistent empty on the branchlets in the second year. Scales decussate, four or six; the lowermost pair short, thin, often reflexed; the next pair long, thickened, woody, widely spreading at maturity, marked externally close to the apex by the shortly acuminate or long-beaked tip of the bract; third pair, when present, connate into an erect median partition. Seeds, two or one by abortion on each of the two fertile scales, with two lateral wings, one broad, oblique, nearly as long as the scale; the other short, narrow, or rudimentary; cotyledons two.

Eight species of Libocedrus have been described, remarkable for their distribution over widely separated areas in the two hemispheres. Three sections may be distinguished :—

I. Ultimate branchlets on mature trees tetragonal, bearing leaves all alike and uniform in size.

    1. *Libocedrus tetragona*, Endlicher.   Chile, Patagonia.

    Leaves spreading.

    2. *Libocedrus Bidwilli*, Hooker.   New Zealand.

    Leaves closely appressed.

II. Ultimate branchlets flattened, with leaves of two kinds; lateral boat-shaped, median flat and appressed.

A. *Median and lateral leaves equal in length.*

    3. *Libocedrus decurrens*, Torrey.   Oregon, California, W. Nevada.

    Leaves green on both surfaces.

    4. *Libocedrus macrolepis*, Bentham et Hooker.   China, Formosa.

    Leaves glaucous on the lower surface, with white stomatic bands.

B. *Lateral leaves much longer than the median leaves.*

    5. *Libocedrus chilensis*, Endlicher.   Chile.

    Median leaves minute, rounded at the apex, with a conspicuous gland.

    6. *Libocedrus Doniana*, Endlicher.   New Zealand.

    Median leaves ovate, acute, mucronate, scarcely glandular.

The two following species, imperfectly known and not introduced, will only be mentioned here.   They belong to the last section :—

    7. *Libocedrus papuana*, F. v. Mueller.[1]   New Guinea.

    8. *Libocedrus austro-caledonica*, Brongniart et Gris.[2]   New Caledonia.

LIBOCEDRUS TETRAGONA

Libocedrus tetragona, Endlichler, *Syn. Conif.* 44 (1847); Lindley and Paxton, *Flower Garden*, i. 47,
    f. 32 (1850); Kent, Veitch's *Man. Coniferæ*, 256 (1900).
Libocedrus cupressoides, Sargent, *Silva N. Amer.* x. 134 (1896).
Thuya tetragona, Hooker, *London Journ. Bot.* iii. 148, t. 4 (1844).
Pinus cupressoides, Molina, *Saggio Sulla Storia Naturale del Chile*, 168 (1782).

A tree[3] attaining in South America, though rarely, a height of 160 feet. Branchlets tetragonal.   Leaves equal in size and uniform in shape in the four ranks; those on the ultimate branchlets about $\frac{1}{12}$ inch long, adnate only at the base, the remaining part free and spreading; ovate, acute, or rounded at the apex, keeled on the back, concave and glaucescent above; those on primary axes larger, adnate for the most of their length, the apices only being free and spreading.

Cones on long branchlets, less than $\frac{1}{2}$ inch long, brown.   Scales four, minutely

---

[1] *Trans. Roy. Soc. Victoria*, i. 32 (1889).   [2] *Bull. Soc. Bot. France*, xviii. 140 (1871).

[3] This tree has been confused by travellers with *Fitzroya patagonica*, which has very similar foliage when old. In the former, the leaves gradually taper to a rounded or acute apex; in the latter they are broadest in their upper third, close to the rounded apex. The cones are entirely different.

pubescent on the margin, each bearing above the middle on the back a lanceolate, subulate, erect, incurved spine; the two smaller scales lanceolate; the two larger scales oblong, each bearing a solitary seed; the larger wing oblique, obovate, obtuse, twice as long as the seed, the shorter wing narrow. (A. H.)

This tree inhabits the western slopes of the Andes of Chile from latitude 35° southwards, and was collected by me in February, 1902, on the west end of Lake Nahuel-Huapi at two to three thousand feet. It was growing both on swampy ground, where it attained a considerable size, and on the steep hill-sides above Puerto Blest. The natives of the district call it Alerce,[1] which is the usual name in South Chile for *Fitzroya patagonica*, and use it for making long straight thin shingles, which seem to be extremely durable. Owing to the inaccessible nature of the country and the scarcity of inhabitants, little or no timber has as yet been cut in the dense forests which clothe the shores of this large and picturesque lake. Judging from the climate, which is severe in winter, this beautiful tree should be hardy in the west and south-west of Great Britain and Ireland. According to Dusen and Macloskie,[2] it is common in Western Patagonia, extending through Fuegia to Cape Horn, rising up to the snow-line in the mountains, and met with of all sizes, from 2 to 160 feet high. As a rule it never forms forests, but grows either in small thin groves or sparingly mixed with *Nothofagus betuloides* and *Drimys Winteri*.

It was introduced by W. Lobb[3] in 1849, but is excessively rare in cultivation, the only specimen we have seen being a small tree 15 feet high, in 1906, at Kilmacurragh, Co. Wicklow. This tree is narrowly pyramidal in habit, with bark scaling off in long papery ribbons. (H. J. E.)

LIBOCEDRUS CHILENSIS

Libocedrus chilensis, Endlicher, *Syn. Conif.* 44 (1847); Lindley, *Journ. Hort. Soc.* v. 35 (1850); Lindley and Paxton, *Flower Garden*, i. 48, f. 33 (1850); Kent, Veitch's *Man. Coniferæ*, 252 (1900).
Thuya chilensis, Don, in Lambert, *Pinus*, ii. 19 (1824); Hooker, *London Journ. Bot.* ii. 199, t. 4 (1843).
Thuya andina, Poeppig et Endlicher, *Nov. Gen. et Spec.* iii. 17, t. 220 (1845).

A tree, attaining in Chile 50 feet in height, usually with a short trunk branching into a compact pyramidal head, or becoming at high altitudes a dense shrub. Branchlets compressed, slender; leaves scale-like in four imbricated ranks, those of the lateral ranks much longer than the others, boat-shaped, free at the apex, and spreading for one-third their length, keeled, acute, marked above and below with a white stomatic band; median leaves, minute, appressed, rounded at the apex, the dorsal with a prominent gland.

---

[1] Sir W. T. Thiselton-Dyer suggests that this is no doubt a Spanish corruption of the Arabic *El Arz*, a name which seems to include any coniferous tree, *e.g. Cedrus Libani* and *Pinus halepensis*. According to Pearce, the tree producing the valuable alerce timber is *Fitzroya patagonica*. Cf. *Hortus Veitchii*, 46 (1906).
[2] Scott, *Princetown Univ. Exped. Patagonia*, viii. 6, 18, 142 (1903).   [3] *Gard. Chron.* 1849, p. 563.

Cones[1] on short branchlets, $\frac{1}{2}$ inch long; scales four, each with a minute projecting point below the apex, bright brown, two larger fertile and two smaller unfertile. Seeds one or two on each of the larger scales, oblique, with a narrow short wing on one side below, and an oblique broad oval wing on the other side above, the two wings being upper and lower, rather than lateral in position.   (A. H.)

A tree, said by Bridges—who was the first to send home seeds to Low of Clapton in 1847—to attain occasionally 80 feet in height.   It grows on the lower slopes of the Andes of Southern Chile, from lat. 34° southward to Valdivia; and was collected by me in the valley of the Rio Limay below Lake Nahuel-Huapi at 3500 to 4500 feet.   Here it grows scattered on grassy hillsides or in open groves, and is a graceful tree of 50 to 60 feet in height.   A photograph of our camp in this valley, taken by Mr. Calvert, gives a good idea of its appearance (Plate 141).

Though from the climate of the region in which it grows, this tree ought to be hardy in the warmer parts of England, and though in Mr. Palmer's tables a small number of trees seem to have survived the frost of 1860-61, as at Bishopstowe, Nettlecombe, Southampton, and even at Keir in Perthshire, yet by far the greater number of the plants introduced in 1847 were killed; and it is now very rare in cultivation; but seems, though slow in growth, to thrive at several places.   By far the largest specimen I have seen is at Whiteway near Chudleigh, Devon, the property of Lord Morley, which in 1907, according to the measurements of the gardener, Mr. Nanscawen, was 46 feet 8 inches by $5\frac{1}{2}$ feet.   We have also seen specimens in England at Blackmoor, Hants, the seat of Lord Selborne; and in Ireland at Castlewellan, the largest tree there being 20 feet high in 1903; at Powerscourt, where in 1906 there was a tree 28 feet high by 3 feet 3 inches, with the bark scaling off in long, narrow, papery slips, the habit being much wider than that of *L. decurrens*, with ascending branches; and at Kilmacurragh, Wicklow, where there is a tree 25 feet in height.   (H. J. E.)

LIBOCEDRUS DONIANA

Libocedrus Doniana, Endlicher, *Syn. Conif.* 43 (1847); Kirk, *Forest Flora New Zealand*, 157, tt. 82, 82A (1889); Kent, Veitch's *Man. Coniferæ*, 254 (1900); Cheeseman, *New Zealand Flora*, 646 (1906).
Libocedrus plumosa, Sargent, *Silva N. Amer.* x. 134 (1896).
Dacrydium plumosum, D. Don, in Lambert, *Pinus*, ed. 2, Appendix 143 (1828).
Thuya Doniana, Hooker, *London Journ. Bot.* i. 571, t. 18 (1842).

A tree, attaining in New Zealand 100 feet in height and 15 feet in girth, with reddish, stringy bark scaling off in ribbons.   Branchlets flattened, with leaves similar in shape and arrangement to those of *L. chilensis*; lateral leaves adnate in the lower half, free and spreading in the upper half, acute, mucronate, green and shining above, glaucescent with a white band below; median leaves appressed, ovate, acute, mucronate, scarcely glandular.

Cones about $\frac{1}{2}$ inch long; scales four, each with a lanceolate acuminate, erect,

[1] Cones ripened on young trees at Les Barres in France in 1900.   Pardé, *Arb. Nat. des Barres*, 31 (1906).

incurved spine above the middle on the back; two lower scales half the size of the others, acute; two upper scales rounded at the apex, each bearing one seed, which has two lateral wings, one short and narrow, the other broad and entire or sub-dentate.

This tree occurs in the North Island of New Zealand, in forests from Mongonui southward to Hawke's Bay and Taranaki, at elevations from sea-level to 2000 feet, usually rare and local. Kawaka is the native name, and it is also known as the New Zealand *Arbor Vitæ*, the dark red wood, beautifully grained and durable, being used in cabinet-making.

It is occasionally seen in conservatories; the only tree growing in the open, that we know of, being one at Powerscourt, which was 20 feet high and 18 inches in girth in 1903. (A. H.)

LIBOCEDRUS BIDWILLI

Libocedrus Bidwilli, J. D. Hooker, *Flora New Zealand*, i. 257 (1867); Kirk, *Forest Flora New Zealand*, 159, tt. 82A, 83 (1889); Cheeseman, *New Zealand Flora*, 647 (1906).

A tree similar to *L. Doniana*, but smaller, attaining a maximum of 80 feet in height and 12 feet in girth; but often bushy at high altitudes and on peat-bogs.

Branchlets on young trees like those of *L. Doniana*, but more slender; on old trees tetragonal, $\frac{1}{20}$th to $\frac{1}{10}$th inch in diameter, clothed with densely imbricated, minute, scale-like leaves, uniform in size and shape in the four ranks, closely appressed, boat-shaped, ovate, acute, green in colour. Cones like those of *L. Doniana*, but smaller, $\frac{1}{4}$ to $\frac{1}{3}$ inch long.

This tree occurs both on the North and South Islands of New Zealand, from Te Aroha mountain and Mount Egmont southward to the Foveaux Strait, not uncommon in hilly and mountain forests at 800 to 4000 feet elevation. It is known as cedar or Pahautea, and has soft, red, and rather brittle wood. This species has not apparently been introduced, though judging from its occurrence higher in the mountains and more southerly in latitude than *L. Doniana*, it ought to be hardy in the milder parts of the British Isles. (A. H.)

LIBOCEDRUS MACROLEPIS

Libocedrus macrolepis, Bentham et Hooker, *Gen. Pl.* iii. 426 (1880); Kent, Veitch's *Man. Coniferæ*, 255 (1900); Masters, *Gard. Chron.* xxx. 467 (1901); Henry, *Garden*, lxii. 183, with figure of tree (1902).
Calocedrus macrolepis, Kurz, *Journ. Bot.* xi. 196, t. 133 (1873).

A tree, attaining in China 100 feet in height, broadly pyramidal in habit, with whitish, scaly bark. This species resembles *L. decurrens* in foliage—the frondose branch-systems being, however, more flattened, and the leaves thinner in texture and larger at the corresponding stages of growth than in that species—the best mark of distinction being the glaucous tint of the leaves beneath. Staminate flowers oblong, tetragonal; stamens **sixteen** to twenty. Cones on **very** slender branchlets (which are modified in being tetragonal, with minute appressed leaves uniform in the four ranks), about ¾ inch long, purplish or dark brown, roughened externally by longitudinal ridges; scales six, resembling those of *L. decurrens*, but smaller and with blunter minute processes. Seed, one on each of the two middle scales; two-winged, with the larger wing broader in the middle and more obtuse than in the Californian species.

This species occurs in the forests of Southern Yunnan in China, at 4000 to 5000 feet, but is rarely met with wild, and only in ravines near water-courses. It was discovered by Anderson near Hotha in 1888; and was subsequently seen by me wild, near Talang, and frequently planted in temples. It is known to the Chinese in Yunnan as Poh or Peh; and the wood is much esteemed, especially that of logs often found buried, the result of inundations in past times. Specimens of this species, so far as one can judge by the foliage alone, have been sent to Kew from North Formosa by Bourne.

The Chinese Libocedrus was introduced by Mr. E. H. Wilson, who collected seeds when he was paying me a visit at Szemao in the autumn of 1899. Young plants,[1] raised at the Coombe Wood Nursery, have beautiful, glaucous, large, flat foliage, the apices of the leaves being tipped with very fine, long, cartilaginous points. They may also be seen in the temperate house at Kew. The tree would probably be hardy in Cornwall and the south-west of Ireland, and being highly ornamental, is worth a trial in warm, sheltered spots. (A. H.)

LIBOCEDRUS DECURRENS, INCENSE CEDAR

Libocedrus decurrens, Torrey, *Smithsonian Contrib.* vi. 7, t. 3 (1854); Sargent, *Silva N. Amer.* x. 135, t. 534 (1896), and *Trees N. Amer.* 73 (1905); Kent, Veitch's *Man. Coniferæ*, 253 (1900); Mayr, *Fremdländ. Wald- u. Parkbäume*, 315 (1906).

Thuya Craigana, Murray, *Botan. Exped. Oregon*, 2, t. 5 (1853).

Thuya gigantea, Carrière, *Rev. Hort.* 1854, p. 224 (in part) (not Nuttall).

Heyderia decurrens, Koch, *Dendrologie*, ii. 2, p. 179 (1873).

A tree, attaining in America 180 feet in height and 21 feet in girth, with a straight stem tapering from a broad base. Bark nearly an inch thick, light cinnamon-red, irregularly fissuring into ridges covered with appressed flat scales.

Leaves shining green, each set of four equal in length, adnate for most of their length to the branchlets, but free at the tips, which end in fine cartilaginous points; about ⅛ inch long on the conspicuously flattened secondary and tertiary axes, increasing to ½ inch on the main axes, which are only slightly flattened: those of the lateral ranks boat-shaped, gradually narrowing to an acuminate apex, keeled and glandular on the back, covering in part the median leaves, which are obscurely glandular and flattened, with broadly triangular cuspidate apices.

---

[1] A seedling is figured in *Ann. of Bot.* xvi. 557, fig. 30 (1902), concerning which Sir W. Thiselton Dyer says :—"The primitive leaves are not very different from the cotyledons, with which they are serially continuous; but after a time there is a complete change in the form and disposition of the foliar organs."

Flowers appearing in January at the end of short lateral branchlets of the previous year; staminate, $\frac{1}{4}$ inch long; pistillate, with ovate, acute, greenish-yellow scales, subtended at the base by two to six pairs of slightly altered leaves, which persist yellowish, sharp-pointed and membranous at the base of the fruit.

Cones about an inch long, pendulous, reddish-brown, on short branchlets with ordinary leaves. Scales six; lower pair short with a reflexed process; middle pair long, lanceolate, gradually narrowing to a rounded apex, below which is a minute deltoid spreading or reflexed process, and concave on the inner surface at the base, with depressions for the seeds; upper pair connate into a thick, woody, median partition, slightly longer than the fertile scales, crowned by three minute spines. Seeds four, two collateral on each of the middle scales; body, $\frac{1}{3}$ to $\frac{1}{2}$ inch, lanceolate, pale brown, containing liquid resin, marked with a white hilum on each surface at the base; wings two lateral, one short and narrow, the other oblique, produced above the seed, nearly as long as the scale, rounded at the narrow apex, and about one-third as broad as long in the middle widest part.

Seedling.—Seedlings sown at Colesborne in spring were about 3 inches high in August, and had a slender tap-root, about 5 inches long. Caulicle, $1\frac{1}{4}$ inch long, terete, brownish, glabrous. Cotyledons, two, $1\frac{5}{8}$ inch long, linear, nearly uniform in width, rounded at the apex, green beneath, marked above with numerous inconspicuous stomatic lines. Primary leaves variable in number, first pair opposite, succeeded by three or four whorls in sets of four each, or only one or two whorls are produced; linear, $\frac{3}{4}$ inch long, tapering to an acuminate apex, glaucous on both surfaces with indistinct stomatic lines. Above the primary leaves the stem gives off branches, and produces scale-like small leaves, arranged in four ranks, and intermediate in character between the primary leaves and the adult foliage.   (A. H.)

DISTRIBUTION

Libocedrus decurrens was discovered by Fremont in 1846 on the upper waters of the Sacramento river. It was introduced in 1853 by Jeffrey, who collected for the Oregon Botanical Association of Edinburgh; and his specimens were named by Murray *Thuya Craigana* in honour of Sir W. Gibson Craig, one of the members of the association. Carrière confused the tree with *Thuya gigantea*; and for some time there was great confusion in the nomenclature of the two species. *Libocedrus decurrens* is the name now universally adopted.

According to Sargent, the distribution extends from the north fork of the Santiam river in Oregon, lat. 44° 50', southward along the western slopes of the Cascade Mountains, and through the Sierra Nevada in California, occasionally crossing the range into Western Nevada; also along the Californian coast ranges from Mendocino county to the San Bernardino, San Jacinto, and Cuyamaca Mountains, reaching its most southerly point on Mount San Pedro Martin, half-way down the peninsula of Lower California. Sargent states that it is rather rare in Oregon, ascending to 5000 feet, and in the Californian coast ranges, where it rises to 5000 to 7000 feet; being most abundant and of largest size in the sierras of Central California at 3000 to 5000 feet, thriving best on warm, dry hillsides, plateaux, and the floors of open valleys, usually growing singly or in small groves, often mixed with *Pinus ponderosa* and black oak.

Henry saw it in Oregon on the eastern spurs of the coast range near Kerby; and found it common on the road from there south-west across the Siskiyou range into Northern California, where it grew near Gasquet's Inn, about twenty miles inland from Crescent City on the coast. In these localities it occurred scattered on dry, sunny hills, in situations similar to that of *Pinus ponderosa*, at 2000 to 3000 feet altitude, and was not seen in shaded, moist ravines. The trees here are broadly pyramidal in habit, not assuming the columnar form of English cultivated trees, and of no great size, the largest measured being 123 feet by 11 feet 1 inch.

Plummer, in his report on the Cascade Forest Reserve, where a good illustration is given, on p. 102, of a grove of this tree, says:—" The incense cedar is almost always hollow-trunked or dry-rotted at the heart, even though the tree may have every outward appearance of perfect health. The wood has been very little used for any purpose but fuel or fencing, and is not cut when better is obtainable. It is said by Rothwell and Rix to ascend the mountains as high as 5750 feet."

Sudworth in his report on the Stanislaus and Lake Tahoe forest reserves[1] says that it is here an abundant tree at between 3500 and 5500 feet, but extends from 2000 to 7000 feet. Mature trees are from 80 to 100 feet high, and 4 to 7 feet in diameter, attaining these dimensions in from 100 to 200 years. Large trees, as shown by a photograph (plate cxiii. of Sudworth), are almost always rotten at heart. Reproduction by seed is good and abundant almost everywhere, especially in the drier situations.

The largest trees I have seen were on the lower slope of Mount Shasta at about 4000 feet, where I measured a tree 130 feet by 12 feet 7 inches which had been left standing when the surrounding forest was cut. Here it grew in company with Douglas fir, *Abies concolor*, and *A. magnifica*, on dry soil, and though the fruit on 1st September was fully formed the seeds were not ripe. The average size of the trees here was 90 to 100 by 8 or 9 feet. Prof. Sheldon says that it attains 100 to 150 feet high by 3 to 7 feet diameter, but such dimensions are not common.

CULTIVATION

When raised from seed it is somewhat slow in growth at first, but in good nursery soil soon makes a well-rooted plant, which is proof against the worst spring and winter frosts, and seems hardy on heavy soil and in damp situations, where *Thuya plicata* is sometimes injured when young. It produces very little seed in this country, and these do not always mature, and in consequence is usually propagated by cuttings.

Though it seems doubtful at present whether this tree can be looked on as a timber tree in England, yet on account of its rather stiff and formal habit it is well suited for the formation of small avenues, and when planted close together, as at Ashridge Park, forms a dense shelter without any clipping.

[1] Washington, 1900.

Remarkable Trees

The finest tree that I know in England is the one figured (Plate 142) which grows in the grounds at Frogmore. This was planted, as I am told by Mr. M'Kellar, by H.S.H. the Princess of Hohenlohe on 16th March 1857, and must be about 54 years old. It has been stated on a photograph taken for the late Hon. Charles Ellis to be 82 feet high, but when I measured it in 1904 I could not make it more than 65 feet, and being forked at about five feet from the ground its girth was about 9 feet.

Another very fine tree grows close to the house at Bicton, which in 1900 was 60 feet by 7 feet 7 inches; and at Killerton there is a tree 55 feet by 5 feet 5 inches. At Orton Hall, near Peterborough, the tree succeeds very well on rather heavy soil, which does not suit many conifers, and here a tree 60 feet by 6 feet 9 inches has borne fruit, from which Mr. Harding, gardener to the Marquess of Huntly, has raised seedlings, some of which are now 9 feet high; smaller ones which he sent me are growing at Colesborne. At Hardwicke, near Bury St. Edmunds, one of the healthiest young trees I have seen, which was only planted in 1873, is already 48 feet by 4 feet 5 inches. At Crowsley Park, Oxfordshire, a tree planted about 1850 was, in 1907, 53 feet high by 8 feet 1 inch in girth, dividing into two stems at 10 feet from the ground, but forming a very narrow column. At the Wilderness, White Knights, Reading, an extremely narrow tree is 60 feet high by 4½ feet in girth. At Nuneham Park, Oxford, there is a fine tree in the pinetum, which is 58 feet by 7 feet. At Bayfordbury, Herts, the best specimen is 52 feet by 5 feet 9 inches.

In Herefordshire the best specimen I know of is at Eastnor Castle, which forks at about 6 feet, and measured in 1906 53 feet by 7 feet 6 inches. There is a nice avenue of it in the grounds at Ashridge Park, and also a circle consisting of 32 trees at only 1 yard apart, which were planted by Earl Brownlow thirty-five years ago, and are now about 35 feet high.

Other remarkable trees which we have seen are at Fulmodestone, Norfolk, 58 feet by 5 feet 11 inches in 1905; at Highnam, Gloucester, 50 feet by 5 feet 3 inches in 1905; at Beauport, Sussex, 53 feet by 6 feet 2 inches at 2 feet up, dividing into two stems, a conical tree, with extremely dense foliage, in 1904; at Dropmore, a large tree not measured. At Coldrinick, Cornwall, there is a tree which Mr. Bartlett informs us, was, in 1905, 51 feet by 6½ feet. Mr. R. Woodward, jun., measured in 1906 a tree at Wexham Park, Stoke Pogis, 56 feet by 3 feet. At Salhouse, Norfolk, Sir Hugh Beevor measured, in 1904, a tree 57 feet by 6 feet 8 inches. A fine specimen at Tittenhurst, Sunninghill, is figured in *Gardeners' Chronicle*, xxxvi. 284, fig. 127 (1904).

In Scotland the tree is not so common, though specimens 40 to 50 feet high are growing in various places; the tallest reported at the Conifer Conference of 1891 was at Torloisk in Mull, and then measured 37 feet in height.

At Smeaton-Hepburn, East Lothian, a tree planted in 1843 was measured by Henry in 1905 as 53 feet by 5 feet 4 inches. A tree at Keir, Perthshire, seen by Henry, was 42 feet by 4 feet 10 inches in 1905. At Brahan Castle, Ross-shire, Col. Stewart Mackenzie of Seaforth informed us in 1904 that he had a tree 4 feet 10 inches in girth, height not stated.

In Ireland, *Libocedrus decurrens* is rare in cultivation. At Stradbally Hall, Queen's County, a fine tree measures 53 feet high by 5½ feet in girth. There is a tree at Fota 45 feet high, dividing into two stems at 2 feet from the ground. At Churchill, Armagh, a fine healthy specimen, growing in sand, was 45 feet by 4 feet 10 inches in 1905. At Adare a tree measured, in 1903, 47 feet high by 7 feet 9 inches in girth.

In North Italy this tree grows larger than in England and ripens seed freely, which it rarely does here. At Pallanza, in Rovelli's nursery, I measured a splendid tree over 90 feet high by 9 feet 3 inches in girth. Another on the Isola Madre was 90 feet by 9 feet 10 inches, from which I gathered seed in October 1906, which have produced a good crop of seedlings.

It also ripens seed and grows well in the climate of Paris, and also at Les Barres, and has produced self-sown seedlings at Thiollets (Allier).[1] The largest I have seen in France is at Verrières, near Paris, a handsome and well-shaped tree, which measured, in 1905, 50 feet by 5 feet 5 inches, and is figured on plate vii. of *Hortus Vilmorinianus* (1906). (H. J. E.)

---

[1] Pardé, *Arb. Nat. des Barres*, 32 (1906).

CUNNINGHAMIA

Cunninghamia, R. Brown, in Richard, *Conif.* 80, t. 18 (1826); Bentham et Hooker, *Gen. Pl.* iii.
    435 (1880); Masters, *Journ. Linn. Soc.* (*Bot.*) xxvii. 304, fig. 18 (1889), and xxx. 25 (1892).
Belis,[1] Salisbury, *Trans. Linn. Soc.* viii. 315 (1807).
Jacularia, Rafinesque, in Loudon, *Gard. Mag.* viii. 247 (1832).
Raxopitys, Nelson (Senilis), *Pinaceæ*, 97 (1866).

A GENUS, belonging to the Coniferæ, with only one known living species,[2] and
doubtfully represented in the fossil state.[3]

Cunninghamia is considered by Bentham and Hooker, and by Masters, to be a
member of the family Araucarieæ; but it is placed by Eichler[3] in Taxodineæ.
Seward and Ford, who have lately published an exhaustive monograph[4] of Araucaria
and its allied genus Agathis, agree with Eichler that it has no close relationship
with those genera. It appears, however, to be a connecting link between the
Araucarieæ and the Taxodineæ; and mainly differs from Araucaria, some species of
which it closely resembles, in foliage, in having three ovules on the bract, and not one
only, as in that genus.

The generic characters are given in the following detailed account of the
species :—

CUNNINGHAMIA SINENSIS

Cunninghamia sinensis, R. Brown, *loc. cit.* (1826); Lambert, *Genus Pinus*, ed. 2, t. 53 (1832);
    Loudon, *Arb. et Frut. Brit.* iv. 2445 (1838); Murray, *Pines and Firs of Japan*, 116, figs. 216-
    224 (1863); Kent, Veitch's *Man. Coniferæ*, 292 (1900); Shirasawa, *Icon. Ess. Forest. Japon*,
    text 23, t. 9, ff. 1-24 (1900).
Cunninghamia lanceolata, W. J. Hooker, *Bot. Mag.* t. 2743 (1827).
Pinus lanceolata, Lambert, *Genus Pinus*, ed. 1, t. 34 (1803).
Belis jaculifolia, Salisbury, *loc. cit.* 316 (1807).
Belis lanceolata, Sweet, *Hort. Brit.* 475 (1830).

---

[1] This name, though the earliest, is not adopted on account of its close resemblance to the genus *Bellis*, used for the daisies.

[2] While the above was passing through the press, there has been received at Kew a specimen of a new species of Cunninghamia, lately discovered in the mountains of Formosa at 7000 feet altitude. This species, which will shortly be published by Mr. Hayata, differs from *C. sinensis* in having shorter leaves, acute and not acuminate at the apex. Mr. Hemsley is inclined to think that a specimen, preserved in the Herbarium, which was collected on Mt. Omei, in Western China, by Faber, is possibly a third distinct species.

[3] Engler u. Prantl, *Natur. Pflanzenfamil.* ii. 85 (1889). *Cunninghamites*, an allied fossil genus, has been found in the Keuper and Chalk deposits in Saxony, Bohemia, Westphalia, Southern France, and Greenland. Cf. Schimper u. Schenck, *Palæontologie*, 283 (1890).

[4] *The Araucarieæ: Phil. Trans. Roy. Soc.*, vol. cviii. p. 308 (1906).

An evergreen tree, attaining in China 150 feet in height and 18 feet in girth of stem, with brownish bark scaling off in irregular longitudinal plates, and exposing a reddish cortex beneath. Branches at first in pseudo-whorls, afterwards given off irregularly. Young branchlets sub-opposite or in pseudo-whorls, covered with green epidermis; older shoots brownish except for the green leaf-bases. Leaves persistent alive five to seven years, afterwards remaining dry and dead for many years on the branches and even upon the stem; densely and spirally arranged on the branchlets, but twisted on their bases so as to be thrown into two lateral spreading ranks; narrowed at the base and decurrent on the shoot to the insertion of the next leaf; rigid, more or less curved, narrowly lanceolate, acuminate, 1 to 2 inches long; upper surface dark green, concave with slightly raised margins; lower surface convex, with a green midrib and two white stomatic bands, the stomata in several regular lines; sharply and finely serrate; with one resin-canal beneath the single unbranched fibro-vascular bundle.

Staminate flowers, five to ten in an umbel at the apex of a branchlet; the umbel surrounded at its base by numerous triangular imbricated serrulate bracts; each flower a spike-like cylindrical column of spirally crowded stamens; each stamen consisting of a slender stalk with an ovate serrulate connective, from which hang three longitudinally-dehiscing anther-cells. Female flowers, single or three or four together at the apex of a branchlet; erect ovoid cones, composed of numerous spirally imbricated lanceolate mucronate bracts in a continuous series with the leaves; lower bracts sterile, resembling leaves but with thickened bases; ovular scale on the upper fertile bracts visible only as a slight projection; ovules three on the base of each bract, reversed.

Fruit, an ovoid-globose brownish cone, about 1½ inch long, composed of thin woody scales, which are the bracts of the flowers increased in size and hardened, but otherwise little altered; loosely imbricated, serrate in margin, broadly ovate or reniform, with a cusped apex often reflected outwards. Seed-scale visible only as a transverse narrow membranous fimbriated projection on the inner surface of the woody bract, below its centre and above the seeds. Seeds three on each bract, about ¼ inch long, brown, oblong compressed, surrounded by a membranous narrow wing. Cotyledons two. The cones persist for a year or more on the branchlets after the escape of the seed; and are occasionally proliferous, the elongated shoot above the cone producing leaves and growing to be several inches in length.[1]

Seedling.—Seedlings sown at Colesborne in spring were about 3 inches high in August, and had a short flexuose tap-root, provided with a few lateral fibres. Caulicle brownish, terete, glabrous, 1 inch long. Cotyledons two, about ½ inch long, coriaceous, entire, linear, with a median groove beneath. Young stem glabrous, ridged by the decurrent bases of the leaves. Leaves numerous, spirally arranged on the stem, ½ to 1½ inch long, soft in texture, linear, curved, broad at the base, whence they taper gradually to a fine bristle-pointed apex, serrulate in margin, green above, marked beneath with two narrow white stomatic bands.

In Cunninghamia, as in Araucaria, root-suckers are often produced, which grow

---

[1] Cf. *Woods and Forests*, 1884, p. 593, and *Garden*, xxix. 173 (1886).

into young trees, close to the parent stem.   Coppice shoots are also produced freely from the stools in China, when the trees are felled.

Cunninghamia [1] was discovered in 1701 by J. Cunningham in the island of Chusan; and his specimens, preserved in the British Museum, were early described by Plukenet.[2]   The first accurately scientific description, however, is due to Lambert, and was based on specimens brought home by Sir G. Staunton, who accompanied Lord Macartney's embassy to China in 1793.

The tree has been known to the Chinese from the most ancient times, being mentioned in their earliest classical writings.   It is called *sha*, a name, however, which is often applied also to Cryptomeria and other conifers yielding valuable timber.   It was introduced [3] by William Kerr from Canton into Kew Gardens in 1804; but no trees of that date now exist there.   Probably most of the existing trees in England were raised from seed collected by Fortune about 1844.

Cunninghamia is widely spread throughout the central, western, and southern provinces of China, extending southwards from Szechwan, Hupeh, Kiangsi, and Kiangsu to Yunnan and Kwangtung.   It is usually a tree of mountain valleys, requiring a hot summer and considerable humidity to thrive; and ranges in altitude from sea-level to 5000 feet, occurring in Central China below the zone, which, in the high mountains, is covered by silver fir and spruce.   There appear to be large forests of it in the interior of Hunan and Fokien, judging from the vast quantities of its timber which are exported from there.   In Fortune's time it was abundant on the islands of Chusan and Pootoo, but was rare in Hongkong, where the only wild trees of this species grew as isolated specimens in the Happy Valley.   Fortune,[4] in 1849, passed through fine forests of Cunninghamia in the mountains of Northern Fokien, many of the trees being 80 feet in height, and perfectly straight; and he noticed variations in the tint of the foliage.   He met with dense woods in the Snowy Valley and other parts of Chekiang, but the trees were usually young, and not remarkable for size.

Mr. E. H. Wilson informs me that there are magnificent forests of Cunninghamia in Western Szechwan.   One which he specially noted in the Upper Ya Valley extended for fifty miles between 2000 and 5000 feet altitude, the best trees ranging from 100 to 150 feet in height, and from 15 to 18 feet in girth; and when growing in close stands, with straight stems clean to 40 feet or more, the branches above being short, slender, and horizontal.   In the open the trees have much longer pendulous branches.   The foliage is occasionally glaucous.   Where trees had been cut down, new growth was being everywhere produced by shoots from the stools. Mr. Wilson mentions the common use of the timber in China for house-building purposes generally, and for the masts and planking of native craft.   The bark is also used in the mountains for roofing houses.   In the Chien Chang Valley in

---

[1] In a note in King's *Survey of the Coasts of Australia*, ii. 564 (1826), R. Brown states that he requested Richard to change the name *Belis*, given by Salisbury, into *Cunninghamia*, in honour of both J. Cunningham, the discoverer of the tree, and of the collector Allan Cunningham.

[2] *Amaltheum Botanicum*, i. t. 351, f. 2. (1705).   [3] Aiton, *Hortus Kewensis*, v. 320 (1813).

[4] *Wanderings in China*, 379 (1847); *Tea Countries*, ii. 178, 212 (1853); *Residence among the Chinese*, 189, 277 (1857).

Szechwan, owing, according to tradition, to earthquakes some two centuries ago, landslips occurred which have buried whole forests in certain places beneath the soil. The dead timber is now being dug out, and is in an excellent state of preservation, being redder and more fragrant than the ordinary timber. It is known to the Chinese as fragrant Cunninghamia, *hsiang-sha*, and sells for extraordinary prices, selected thick planks for coffins often being worth £12 to £60 a piece.

The wood, according to Mayr,[1] is extraordinarily light, with a broad sap-wood and a dark yellow heart-wood. It is used extensively in the coast ports of China for making tea-chests.

Cunninghamia appears to be confined as a wild tree to China; but it is occasionally planted[2] in Japan, the Loochoo Islands, and Formosa. M. Hickel has lately received seeds from Tongking, but these may have been gathered from cultivated trees. (A. H.)

REMARKABLE TREES

The growth of this tree in England depends mainly on the amount of heat in summer, which in most places is evidently insufficient; and though it endures severe winter frosts without much injury on well-drained soil, it suffers much from wind and frost in spring. It rarely ripens seed in this country, the only case I know of being a tree at Penrhyn Castle which is now dead, but from whose seed some young trees were raised. The best of these, when I saw them in 1906, was about 10 feet high.

The tallest trees of this species that we know are at Killerton, where, in 1904, there were two which measured 62 and 60 feet in height by 4 feet in girth. One of these has since been cut down, its branches having become ragged, and a section sent to the Kew Museum shows the age to be at least 63 years. Another, at Bicton, was, in 1906, 56 feet by 4 feet 10 inches, also rather ragged in its branches. There is a tree at Highnam, in Gloucestershire, about 25 feet high.

At Heanton Satchville, the seat of Lord Clinton, in North Devon, there is a slender but healthy-looking tree 50 feet by 3 feet, and another one which has thrown up a shoot from the stool. At Escot in South Devon, the seat of Sir John Kennaway, Miss Woolward measured one in 1905, 45 feet high. At Pencarrow in Cornwall Mr. Bartlett showed me a tree, planted by Sir W. Molesworth in 1850, which was in 1905 40 feet by 4 feet 8 inches, and one of the healthiest that I have seen; and there is a smaller tree, 30 feet by 4 feet, at Coldrinick, in the same county.

Coming farther east there is a splendid tree at Bagshot Park, the seat of H.R.H. the Duke of Connaught, which, when I saw it in May 1907, was no less than 47 feet high by 7 feet in girth, and 48 yards in the circumference of its branches. Being on very well drained soil, and well sheltered by other trees, it has suffered

---

[1] *Fremdländ. Wald- u. Parkbäume*, 285 (1906).   [2] Cf. Hayata, in *Tokyo Bot. Mag.* xix. 50 (1905).

little from frost and wind, and is the handsomest and best-shaped tree[1] I have seen (Plate 143).

At Beechlands, near Lewes, the seat of Captain Rose, I am told by Mr. Chisholm that there is a Cunninghamia 50 feet high by 5 feet 1 inch in girth, forked near the top. It bears many cones, which, however, do not produce fertile seeds.

At Grayswood, Haslemere, a tree planted in 1882 is 30 feet high by 2 feet 7 inches in girth, but has not a very thriving appearance. Another at Redleaf, near Penshurst, Kent, the seat of Mrs. E. Hills, though forked near the ground, has one good trunk 47 feet by 5 feet 4 inches, and healthy foliage. At Langley Park, near Norwich, the seat of Sir Reginald Beauchamp, there is a tree 35 feet by 3 feet, which though healthy looking has grown but little for many years. At Tittenhurst,[2] near Sunninghill, there is a fine healthy tree over 20 feet in height. At Bayfordbury, Herts, Cunninghamia, though planted several times, has never succeeded, being much injured by spring frosts, and only one specimen, a few feet high, survives.

The most northern point at which I have seen the tree growing in England is in the sheltered Duchess' garden at Belvoir Castle. This, I was told by Mr. Divers, was planted in 1844, and in 1907 measured 39 feet by 3 feet 2 inches; but Mr. Fenner informs me that there is one 32 feet high at Holker Hall, Lancashire.

In Scotland, as might be expected, there are no trees of any great size. At Brodick Castle, in the Isle of Arran, a tree, which was planted about the year 1858, had only attained, according to the Rev. Dr. Landsborough,[3] 10 feet high in 1895, and never throve. There was formerly a tree at Smeaton-Hepburn, East Lothian, which died about five or six years ago after a drought.

In Ireland, Cunninghamia is a very rare tree. There is one in Mr. Walpole's garden at Mount Usher in Co. Wicklow, which was in 1903 31 feet high by 2 feet 2 inches in girth. It was supposed to be then about 28 years old. In Mr. Acton's arboretum at Kilmacurragh, in the same county, there is a thriving specimen, which Henry measured in 1903 as 25 feet high by 1½ feet in girth.

Around Paris[4] the tree always looks suffering, the leaves turning yellowish and assuming a burnt aspect; but it grows well at Les Barres,[5] and fructifies annually. In North Italy the climate evidently suits the tree much better, as I saw, in the grounds of the Villa Ceriana near Intra, a tree 76 feet by 7 feet 4 inches, producing cones freely in 1906, from which I have raised healthy seedlings. At Locarno,[6] on the northern end of Lake Maggiore, a tree planted fifteen years is 23 feet in height. (H. J. E.)

---

[1] John Smith, in *Records of Kew Gardens*, 290 (1880), states that a Cunninghamia, possibly the same tree as the one mentioned above, bore cones at Bagshot in 1838.

[2] *Gard. Chron.* xxxvi. 284 (1904).　　　　　　　　　　[3] *Trans. Bot. Soc. Edin.* 1896, xx. 527.

[4] Mouillefert, *Traité des Arbres*, ii. 1336 (1898).　　　[5] Pardé, *Arb. Nat. des Barres*, 57 (1906).

[6] Christ, *Flore de la Suisse*, 77 (1907).

LIQUIDAMBAR

Liquidambar, Linnæus, *Gen. Pl.* 463 (1742); Bentham et Hooker, *Gen. Pl.* i. 669 (1865); Engler u. Prantl, *Pflanzenfam.* iii. pt. 2, 123 (1891).

DECIDUOUS trees belonging to the order Hamamelideæ. Leaves alternate on long shoots, crowded and almost fascicled on short shoots, long-stalked, simple, palmately lobed, glandular-serrate. Stipules two, attached to the petiole near its base, lanceolate or subulate, caducous or persisting throughout the summer.

Flowers monœcious, or in rare cases polygamous, in heads subtended at the base by caducous bracts. Staminal heads, globose or elongated, several in a raceme on an erect axis, which is subterminal; each head composed of numerous stamens, interspersed with minute scales, without corolla or calyx; filaments slender; anthers basi-fixed, oblong-obcordate, dehiscing longitudinally. Pistillate heads solitary, on long pendulous stalks, arising in the axils of the uppermost leaves, composed of numerous confluent flowers, the ovaries embedded in the axis of the inflorescence; calyces minute, united together and with the ovaries, and bearing on their summits each four or more stamens, with usually aborted anthers; corolla absent; ovary two-celled, each cell with numerous ovules; styles two, recurved, stigmatic above on their inner surface.

Fruit: a woody spherical head, composed of numerous capsules, consolidated together. Capsule with two valves, opening above to let out the seeds, each valve terminating in a beak (the hardened woody persistent style); calyx persistent, either minutely tuberculate or produced above into long spines. Perfect seeds, angled, winged above, one or two in a capsule, the remaining ovules having aborted. Most of the capsules, however, contain only numerous minute unfertile seeds without wings.

The leaves of Liquidambar resemble strongly those of certain maples; but in the latter they are always opposite, and not alternate or fascicled as in the former. Moreover, stipules or their scars are present on the petiole near its base in Liquidambar, and are absent entirely in Acer.

Three species of Liquidambar are well known, and occur in cultivation. Besides these there are apparently two species,[1] wild in China, which are imperfectly known and not introduced.

[1] These are :—

1. *Liquidambar Rosthornii*, Diels, *Flora von Central China*, 380 (1901), a small tree occurring in Szechwan; flowers and fruit unknown. It resembles in foliage *L. orientalis*.

2. *Liquidambar* sp., Hemsley, *Journ. Linn. Soc.* (*Bot.*) xxiii. 292 (1887). Specimens, consisting of detached leaves and fruits, were sent to Kew from Hankow by Consul Alabaster. Judging from the imperfect material, this is a distinct species. Mr. E. H. Wilson has recently observed a species of Liquidambar, growing on the plain near Kiukiang, in Kiangsi, which is probably the same. Cf. *Gard. Chron.* xlii. 344 (1907).

The species in cultivation are :—

1. *Liquidambar styraciflua*, Linnæus.   North America.

Shoots glabrous.   Leaves large, usually five-lobed, only occasionally lobulate in margin ; under surface glabrous, except for dense tufts of pubescence in the axils of the main nerves at the base, and occasional minute tufts at the junctions of the lateral and main nerves.

2. *Liquidambar orientalis*, Miller.   Asia Minor.

Shoots glabrous.   Leaves small, five-lobed, margin with large lobules ; under surface quite glabrous.

3. *Liquidambar formosana*, Hance.   China, Formosa, Tonking.

Shoots pilose.   Leaves large, usually three-lobed ; under surface pilose, without conspicuous axil-tufts.

LIQUIDAMBAR STYRACIFLUA, SWEET GUM

Liquidambar styraciflua, Linnæus, *Sp. Pl.* 999 (1753); Loudon, *Arb. et Frut. Brit.* iv. 2049 (1838); Oliver, in Hooker, *Icon. Plant.* xi. 13 (1867); Sargent, *Silva N. Amer.* v. 10. t. 199 (1893), and *Trees N. Amer.* 340 (1905).
Liquidambar macrophylla, Oersted, *Am. Cent.* xvi. t. 10 (1863).

A tree, attaining in America 160 feet in height and 17 feet in girth.   Bark deeply and longitudinally fissured, with broad ridges covered by thick corky scales.

Young shoots green, glabrous.   Leaves (Plate 199, Fig. 7) large, averaging 6 inches broad and 5 inches long, variable in form, cordate or almost truncate at the base, five-nerved, palmately and deeply cut into five oblong-triangular acuminate lobes, the terminal lobe largest, the basal lobes smallest, rarely lobulate ; serrations shallow, non-ciliate ; upper surface dark green, shining, glabrous ; lower surface light green, shining, glabrous except for dense tufts of pubescence in the axils of the nerves at the base and occasional minute tufts at the junctions of the lateral and main nerves.   Petiole glabrous, slightly grooved on its upper side, dilated at the base, near which are two scars indicating where the lanceolate stipules have fallen off in early summer.

Fruiting heads, about $1\frac{1}{2}$ inch in diameter, hanging on the tree during winter after the fall of the seeds in autumn, calyx margins with irregular small tubercules ; capsules with two stout style appendages, forming woody spines, one terminating each valve.   Perfect seeds few, with short terminal wing ; imperfect seeds numerous, minute, angled, without wings.

The branchlets[1] of many trees of this species are remarkable for their corky wings, which begin to develop in the second season and increase in width and thickness for many years.   These wings occur on lateral branches, on the upper side only, in three or four parallel ranks ; but on vertical branches they are borne irregularly on all sides.   Trelease[2] observed in the case of Liquidambar trees

---

[1] See Miss Gregory in *Botanical Gazette*, xiii. 282 (1888).   [2] *Garden and Forest*, 1890, p. 195.

growing in Tower Grove Park, St. Louis, that about half the trees either showed no sign of the corky wings or in some cases only a slight trace of them. In Kew Gardens the same difference is noticeable in trees of the same age growing close together, some being without corky-winged branchlets, while others have them much developed.

The leaves usually turn a most brilliant colour in autumn, the tint being red purple, or yellow.

IDENTIFICATION

In summer the maple-like but alternately-placed leaves are unmistakable. In winter (Plate 200, Fig. 2) the following characters are available : Twigs moderately stout, slightly angled, greenish, glabrous; lenticels scattered, prominent. Leaf-scars alternate, obliquely set on projecting pulvini, arcuate or semicircular, marked by three bundle-dots. Terminal bud about ⅜ inch long; lateral buds smaller, varying in size, and directed outwards from the twig at an angle of about 45°; all ovoid, acute at the apex, and composed of six to seven imbricated scales, which are green with brown margins, vaulted on the back, shining, glabrous, ciliate, and often minutely cuspidate at the apex.

Short shoots are numerous in this species, and, unlike the long shoots, are pubescent. All the shoots show at the base ring-like marks, indicating where the accrescent scales of the terminal bud of the preceding year have fallen off in spring.

VARIETIES

Though Oersted considered the Mexican and Guatemalan trees to constitute distinct forms, no varieties have been clearly made out. The species occurs over a wide extent of territory and in diverse climates; and certain differences are observable in the shape, size, and pubescence of leaves in wild specimens; but these scarcely warrant the division of the species into geographical forms. In dry regions in Mexico the under surface of the leaf is covered with dense pubescence. Leaves with only three lobes occur on adult trees in Mexico and Guatemala; but as three-lobed leaves are frequently borne on young shoots of the common form, this peculiarity scarcely merits the rank of a variety.   (A. H.)

DISTRIBUTION

The Liquidambar or Sweet Gum,[1] as it is usually called in the United States, has a very wide range of distribution. Its most northerly station is, according to Sargent,[2] near Newhaven, Connecticut, where it only grows near the coast as a small tree, 40 to 60 feet high. Farther south it extends westwards as far as S.E. Missouri and Arkansas, and in the south to Florida and Texas, reappearing on the mountains of Mexico and Guatemala. In the maritime region of the South Atlantic States and in the Lower Mississippi basin it is one of the most abundant

---

[1] Also known as Red Gum.   [2] *Garden and Forest*, ii. p. 232.

forest trees, but only attains its full size and perfection in deep rich swamps and river bottoms. I have seen it of immense size in the Lower Wabash Valley in Southern Illinois, where Ridgway measured a tree no less than 164 feet high by 17 feet in girth with a clear stem 80 feet long, and another 137 feet high by 11½ feet in girth, which was 94 feet to the first branch. Plate 144 A, taken from a photograph for which I am indebted to the U.S. Bureau of Forestry, represents the tree (Example M) mentioned in *Proc. U.S. Nat. Mus.* v. 67, by Ridgway, which was 12½ feet in girth at the base, 78 feet to the first limb, and contained 7888 feet board measure. It grew two miles from Mount Carmel on land now cleared. Such trees, however, are now hardly to be found except in very inaccessible places. On the coast region of North Carolina, Ashe and Pinchot give its dimensions as 100 feet high and 5 or 6 feet in diameter.

The largest that I saw in the Eastern States was a tree in the Clifton Park, near Baltimore, which was 71 feet by 5 feet 9 inches. In New England, near Boston, Sargent says that it suffers from frost in severe winters, and I saw none in cultivation so large as those in England. I found it in a very different and more beautiful form in the mountains near Jalapa, Mexico, at about 4000 feet elevation, where in the month of March in open forests its leaves were conspicuous by their scarlet colour, but the trees were not of extraordinary dimensions. In America it grows mixed with Nyssa, Liriodendron, maples, and oaks. Ashe says that it fruits annually or every other year, but that much of the seed is abortive, and that it springs up commonly on damp hillsides and bottom lands, and also shoots from the stool after the trees have been felled.

HISTORY AND CULTIVATION

According to Loudon, this tree was first mentioned by Francis Hernandez, a Spanish naturalist, who published a work on the natural history of Mexico in 1651 at Rome. In 1681 it was sent home by Banister to Bishop Compton, who planted it in the Palace Gardens at Fulham. It had become common in cultivation in Michaux's time, but he says that even in France it had never produced seed. In Northern Italy it grows well, and I found a good-sized tree on the Isola Madre in Lake Maggiore, which bore seed, from which I have raised plants.

Though this tree will grow to considerable size in the warmest parts of England, and on account of its beautiful autumnal tints is highly ornamental, yet it requires a much greater degree of heat and moisture than our climate affords to bring it to perfection, and has been somewhat neglected by nurserymen on account of its tenderness when young. I have raised it from imported seeds, which do not keep well when extracted from the fruits, but the seedlings grow so slowly that the more common way of raising it is from layers. It does not transplant well, and requires a good deal of moisture in the soil and a warm, sheltered situation. Its branches are easily broken by the wind, and though it does not come early into leaf, is often injured by late frosts.

REMARKABLE TREES

The largest trees mentioned by Loudon were at Strathfieldsaye (64 feet) and at Syon (59 feet), the latter tree being reported in 1849 to measure 84 feet by 4 feet. We cannot identify either of them now; but at Syon there is a tree, leaning considerably to one side, which was about 75 feet by 6 feet in 1904. The tallest which I have seen is at Godinton, the property of G. Ashley Dodd, Esq., near Ashford, Kent, which in 1907 was 82 feet by 6 feet, a piece estimated at 12 feet long having been broken off the top; and the next to it is one at Petworth, which Sir Hugh Beevor measured in 1894, 84 feet by 5 feet 7 inches; another tree at the same place, 7 feet 6 inches in girth, has been damaged at the top by wind.

Miss Woolward tells me of a fine tree at Escot, Devonshire, the seat of Sir John Kennaway, which was referred to by Bunbury as the largest known to him, and in 1905 measured 75 feet by 7 feet 8 inches. At Cobham Hall, Kent, there is one which I measured as 80 feet by 5 feet 9 inches; and at Broom House, Fulham, there are two trees on the lawn of about the same height and over 6 feet in girth.

At Barton,[1] Suffolk, there are four trees, which were planted in 1825-26, the two largest measuring, in 1904, 71 feet by 5 feet 6 inches and 52 feet by 3 feet 2 inches. At Arno's Grove, Middlesex, a tree drawn up in a plantation, measured by Henry in 1904, was 83 feet by 3 feet 10 inches. A large tree which we have not seen was reported[2] to be growing on the lake side at Chevening Park, near Sevenoaks, Kent. At Arley Castle there is a tree 65 feet by 4 feet 3 inches.

In Scotland we have no records worth mentioning, though the species exists in the south-west.

In Ireland there is a good tree at Fota, which in 1903 measured 57 feet high by 8 feet in girth.

TIMBER

Though neglected until recent years this tree is now very largely cut for timber in the Mississippi valley, and has been introduced to Europe under the name of satin walnut. Owing to its low price it has been tried, under the name of red gum, for street paving with very bad results, though, according to Stone,[3] it is very resilient, and if creosoted may be a useful wood for this purpose.

A careful investigation of the mechanical properties of this wood was made by A. K. Chittenden of the U.S.A. Bureau of Forestry in 1905,[4] from which I take the following:—"Red Gum is perhaps the commonest timber tree in the hardwood bottoms and drier swamps of the Southern States, growing best on alluvial soil of great fertility, which is liable to heavy floods in winter and spring, and often covered with water from January till May. In the best situations it reaches a height of 150 feet and a diameter of 5 feet. It reproduces well only where there is sufficient light, as the seedlings will not bear shade. It also sprouts readily from the stump up to about fifty years of age, but such shoots rarely form large trees. The demand for

---

[1] Bunbury, *Arboretum Notes*, 28.   [2] *Garden*, xxxviii. 208 (1890).   [3] *Timbers of Commerce*, 113 (1904).
[4] *U.S. Dept. of Agriculture, Bureau of Forestry, Bulletin*, No. 58 (1905).

the timber has increased rapidly during the last few years, owing to the increasing scarcity of better timber, and about 75 per cent. of the best grades, 'Nos. 1 and 2 clear heart, are exported to Europe for furniture and inside fittings. It is said to make very good flooring, and is now largely used for railway waggon box boards, the price in the U.S.A. being about 27 dollars per 1000 feet for firsts and seconds, as compared with 41 dollars for cypress. From 1900 to 1902 much of the wood was cut into 3-inch by 9-inch planks, to be used for cutting paving blocks in London, but in 1902 the market for this gave way, and the mills are now trying to introduce this wood as a paving-block material in the United States, where several large cities were in 1905 considering the use of this wood. The qualities necessary for a good paving block are durability, close grain, and the power of resisting abrasion. These qualities are found in red gum."

A very unfortunate experiment was made in Whitehall in the autumn of 1901, when the Corporation of Westminster accepted the tender of an American contractor to pave this street with "red gum." The surveyor seems to have supposed that red gum in America was the same as red gum in Australia, where the name is applied to several species of eucalyptus, which have a good reputation for street paving. Be this as it may, the paving wore out so soon that a large proportion was taken up again in July 1902, and a long and costly lawsuit followed. The contractor alleged (1) that the defects arose from the bad foundations of the road; (2) from excessive watering; (3) from stones having been forced into the pavement; and the case was not finally settled till October 1905.

Mr. Weale tells me that an inferior quality of this wood containing much sapwood is also known in the trade as "hazel pine." "Satin walnut" is worth wholesale from 2s. to 2s. 3d. per cube foot, and "hazel pine" only 1s. 3d. to 1s. 6d. In colour the former is a light fawn, often marked with a rich dark stripe; but is so deficient in strength and durability, and even when well seasoned is so liable to warp and twist, that it is only used for the cheapest classes of furniture.

Michaux says that though much inferior to black walnut and cherry, it was used a good deal in his time in America for picture-frames, bedsteads, coffins, and furniture. Red gum is now much used for veneer in the United States. It furnishes 17 per cent of all the veneer produced, the quantity in 1905 being over 187 million square feet.[1] I brought from St. Louis a slab of this timber cut from a tree of 30 inches diameter, of which the sapwood was about 6 inches thick and much paler in colour. Though cut 4 inches thick this plank cracked badly in drying; and it will evidently be a very difficult wood to dry without warping. It has a very close, fine grain, and takes a good polish.   (H. J. E.)

[1] *U.S. Dept. Agric. Forest Service Circular*, No. 51 (1906).

LIQUIDAMBAR ORIENTALIS

Liquidambar orientalis, Miller, *Gard. Dict.* No. 2 (1768); Oliver, in Hooker, *Icon. Plant.* xi. 13, t. 1019 (1867); Hanbury, *Science Papers*, 139, with figure (1876); Bentley and Trimen, *Medicinal Plants*, ii. No. 107, t. 107 (1880).

Liquidambar imberbe, Aiton, *Hort. Kew.* iii. 365 (1789); Loudon, *Arb. et Frut. Brit.* iv. 2053 (1838).

A tree attaining in Asia Minor 40 to 60 feet in height. Bark longitudinally fissured, with corky irregularly quadrangular scales on the ridges, the orange-coloured inner bark visible in the fissures. Young shoots glabrous. Leaves (Plate 199, Fig. 6) small, averaging 3 inches wide by 2½ inches long, palmately cut about half-way into five oblong triangular acute lobes, the upper three lobes usually with one to four lobules; base truncate or widely cordate; margin with shallow glandular serrations; upper and lower surfaces quite glabrous in cultivated trees, but with axil tufts of pubescence at the base of the under surface in wild specimens. Petiole glabrous, swollen at the base, and bearing near its insertion two minute triangular stipules.

Flowers and fruit similar to those of *Liquidambar styraciflua*, but smaller. Fruiting head about 1 inch in diameter; capsules with more slender beaks than in the preceding species; calyx slightly tuberculate and not spiny.

In winter the twigs resemble those of the American species, but are more slender, with smaller leaf-scars and buds, which are reddish and have six glabrous ciliate scales; short shoots glabrous.

This species does not apparently develop corky ridges on the branches.

DISTRIBUTION

Liquidambar orientalis is known to occur wild only in the south-western part of Asia Minor lying opposite to the island of Rhodes, and in Cilicia, near Alexandretta. It forms woods of considerable extent in the district of Sighala, near Melasso, and in the vicinity of Budrum, Mughla, Djova, Ughla, Marmoriza, and Isgengak. According to Maltass, who obtained specimens for Hanbury, there is a fine forest of this species between the village of Caponisi and the town of Mughla, many trees attaining 40 feet in height, while in other forests, according to native report, they were as high as 60 feet.[1]

Liquid storax, a balsamic resin, obtained from the inner bark of the tree by boiling it in water, is exported in considerable quantity from Smyrna and other Levantine ports, the bulk of this product going to China and India, where it is known in commerce as rose maloes.[2] Liquid storax is used to a small extent by druggists in this country, and is one of the ingredients of " Friar's Balsam."

[1] Elwes passed through this district in 1874 on the way from Makri to Ephesus, but saw no trees of any size. This is a very hot country in summer, myrtle, oleander, and arbutus being the common shrubs.

[2] Rose maloes is a corruption of rassamala, the Javanese and Malay name for *Altingia excelsa*, Noronha, a tree allied to *Liquidambar*, which yields by incisions in the bark a sweet-scented resin. Cf. Bretschneider, *Bot. Sinicum*, iii. 464 (1895).

CULTIVATION

The Oriental Liquidambar was introduced into France about the middle of the eighteenth century by the French Consul at Smyrna, and speedily passed into England, where it was cultivated in 1759 by Miller.

It grows very slowly in this country, where it is very rarely seen in cultivation. There is a tree in Kew Gardens, about 15 feet high, the age of which is unknown. According to Nicholson it was 10 feet high in 1884. It has a twisted, crooked trunk, dividing about 6 feet up into two main stems. The branches are numerous and drooping, the habit of this tree being in marked contrast to that of a tall *Liquidambar styraciflua* close beside it, and probably results from the young branchlets being continually killed by the frost.

A larger and very old tree at White Knights, near Reading, in the grounds of Mr. J. Heelas, was in 1904 about 25 feet high by 3 feet 4 inches in girth, and was decayed at the top, with many dead branches and a hole in the butt close to the ground.

This tree is commonly cultivated in the Mediterranean region; and Mr. Hickel, Inspector in the French Forest Service, informs us that there is a very large specimen, rivalling in size the American species, in the square near the railway station at Montpellier. In the park at Baleine[1] (Allier) there is a tree 75 feet high by 7 feet in girth.

Elwes measured a tree in the Jardin des Plantes, Paris, which was 40 feet high; but was told that it did not ripen seed; and in the Botanic Garden at Padua he saw a tree about 50 feet high by 4 feet in girth, which in May had abundant fruit of the preceding year upon it, but could find no seeds in them.   (A. H.)

LIQUIDAMBAR FORMOSANA

Liquidambar formosana, Hance, *Ann. Sc. Nat.* 5me série, v. 215 (1866), and *Journ. Bot.* 1870, p. 274; Oliver, in Hooker, *Icon. Plant.* xi. 14, t. 1020 (1867); Hemsley, *Journ. Linn. Soc.* (*Bot.*) xxiii. 291.
Liquidambar acerifolia, Maximowicz, *Mél. Biol.* vi. 21 (1866) and viii. 419 (1871).
Liquidambar Maximowiczii, Miquel, *Ann. Mus. Bot. Lugd. Bat.* iii. 200 (1867).

A tree[2] attaining, in China, 80 feet in length and 15 feet in girth. Young shoots with scattered long hairs. Leaves (Plate 199, Fig. 8) widely cordate at the base, usually with three broad oblong-triangular acute or acuminate lobes, the outer lobes occasionally giving off two short additional lobes; margin, occasionally lobulate, sharply serrate, ciliate; palmately three-nerved with two strong lateral nerves; upper surface dull with scattered long hairs; lower surface light green

---

[1] Pardé, *Arbor. Nat. des Barres*, 205, note 1 (1906).
[2] The peculiarities of the buds, leaves, and stipules have been fully described by Lubbock, in *Journ. Linn. Soc.* (*Bot.*) xxx. 495 (1894).

with dense long pubescence. Petiole pilose, with two subulate, persistent, pubescent, glandular stipules.

Fruiting heads spiny, 1½ inch in diameter, each capsule surrounded by several long spines arising from the calyx, and resembling the two indurated styles which terminate the valves. Perfect seeds few, or absent in many capsules, with narrow short wings.

This species is widely distributed over the central and southern provinces of China, and occurs also in Tonking, Hainan,[1] and Formosa. In Hupeh, where it has not been seen over 1000 feet altitude, the tree is valuable, as its timber is used for making the Hankow tea-chests. The Chinese call it Fêng tree.[2]

It is doubtful if it will prove hardy, and is extremely rare in cultivation in Europe, the only plant known to us being one in Kew gardens, which is trained against a wall, and is interesting for its beautiful foliage, which lasts till late in November. It was introduced by seeds sent by Consul Alabaster from Hankow in 1884. (A. H.)

[1] Swinhoe, *Journ. Bot.* i. 257, says it is the commonest tree in the mountain forests of Hainan. Hance, *loc. cit.*, says that at Canton old stumps buried beneath the soil sucker freely.

[2] It yields a resin, *Fêng-hsiang*; and a caterpillar, which feeds on its leaves, produces a coarse kind of silk, used for fishing-lines.

NYSSA

Nyssa, Linnæus, *Gen. Pl.* 308 (1737); Bentham et Hooker, *Gen. Pl.* i. 952 (1867); Harms in Engler u. Prantl, *Pflanzenfam.* iii. 8, 257 (1898).
Tupelo, Adanson, *Fam. Pl.* ii. 80 (1763).
Ceratostachys, Blume, *Bijdr. Fl. Ned. Ind.* 644 (1825).
Agathisanthes, Blume, *loc. cit.* 645.
Daphniphyllopsis, Kurz, *Journ. Asiat. Soc.* 1875, ii. 201.

DECIDUOUS trees or shrubs belonging to the order Cornaceæ. Leaves alternate simple, stalked, with margin entire or remotely one- to four-toothed, without stipules. Branchlets with discoid pith.

Flowers small, diœcious or polygamous, borne at the summit of axillary peduncles, the staminate flowers numerous in heads, umbels, or short racemes, the pistillate and perfect flowers solitary or aggregated in two- to eight-flowered heads, umbels, or short racemes. Staminate flowers: calyx short, flat or cup-shaped, five- to seven-toothed or entire; petals five to seven or ten to fourteen; stamens five to ten, inserted on the margin of an entire or lobed disc; filaments slender, anthers oblong. Pistillate flowers: calyx campanulate or urceolate, five-toothed or entire; petals four to five, seldom three or six to eight; stamens absent or equal in number to the petals and alternating with them, bearing fruitful or barren anthers; ovary coalesced with the receptacle, crowned above by a disc, one- rarely two-celled, one ovule in each cell; style one, recurved, stigmatic along one side near the apex. Fruit a drupe, oblong or ovoid, urceolate at the apex; flesh thin, oily; stone bony, thick-walled, terete or compressed, ridged or winged, one- or rarely two-celled, containing one seed, which has a membranous testa and copious albumen. Cotyledons flat and leafy.

The alternate stalked simple leaves, entire and ciliate in margin; and the branchlets with true terminal buds, without stipules or their scars, showing on section the peculiar discoid pith, are characteristic of Nyssa.

Seven species of Nyssa have been described:—*Nyssa sessiliflora*, Hooker, a tree attaining 60 feet in the Himalayas and Java; has not been introduced and would probably not be hardy in England. *Nyssa sinensis*, Oliver, has recently been introduced from Central China. The remaining five species are natives of Eastern North America. *Nyssa acuminata*, Small, a species imperfectly known, is a small shrub growing in pineland swamps in Georgia. *Nyssa Ogeche*, Marshall, a tree of moderate size, occurring in river swamps in South Carolina, Georgia, and Florida, is unknown in cultivation outside of its native home, and would probably not grow in England. *Nyssa biflora*, Walter, a small tree, growing in ponds, from North Carolina to

Louisiana, is probably only a variety of *Nyssa sylvatica*, Marshall; and no trees referable without doubt to it are known to us in England. *Nyssa sylvatica*, Marshall, and *Nyssa aquatica*, Marshall, occur rarely in cultivation in England.

NYSSA SYLVATICA, TUPELO, PEPPERIDGE, BLACK GUM

Nyssa sylvatica, Marshall, *Arbust. Am.* 97 (1785); Sargent, *Silva N. Amer.* v. 75, t. 217 (1893), and *Trees N. Amer.* 707 (1905).

Nyssa multiflora, Wangenheim, *Nordam. Holz.* 46, t. 16, f. 39 (1787).

Nyssa villosa, Michaux, *Fl. Bor Am.* ii. 258 (1803); Loudon, *Arb. et Frut. Brit.* iii. 1317 (1838).

A tree, occasionally attaining in America 100 feet in height and 15 feet in girth. Bark thick and deeply fissured longitudinally. Young shoots glabrous or with short, erect pubescence. Leaves (Plate 199, Fig. 2, leaf from a tree in Arnold Arboretum, U.S.; and Fig. 9, leaf from a tree at Kew) extremely variable in shape and size, obovate, oval or elliptical; base tapering or rounded, apex acuminate or acute, margin entire or repand and ciliate; upper surface glabrous, dark green, usually shining; lower surface glabrous or with slight pubescence on the midrib and principal veins. Petiole channelled or winged, glabrous or pubescent, $\frac{1}{4}$ to 1 inch long. Flowers on pubescent peduncles, appearing after the leaves; staminate flowers numerous, stalked and in crowded clusters; pistillate flowers sessile, two to fourteen in a head. Fruit ovoid, bluish-black, $\frac{1}{3}$ to $\frac{2}{3}$ inch long; stone terete or more or less flattened, with ten to twelve indistinct ribs.

Seedling.—The caulicle, glabrous, terete, and about 2 inches long, ends in a long flexuose whitish tap-root, which gives off numerous lateral fibres. The cotyledons are ovate-lanceolate, rounded at both base and apex, about $1\frac{1}{2}$ inch long by $\frac{5}{8}$ inch broad, on petioles $\frac{1}{8}$ inch long, slightly coriaceous, entire in margin, pale beneath, glabrous, pinnately veined. The stem, reddish and pubescent, gives off alternately the true leaves, which are oval, with a cuneate base and acuminate apex, entire or one- to two-toothed and ciliate in margin, pale and glabrous on the under surface with the exception of some pubescence at the base of the midrib, and with a pubescent petiole. The preceding description was drawn up in the summer of 1905, from a seedling at Colesborne, raised from seed gathered by Elwes at Boston at the end of the preceding September.

IDENTIFICATION

Nyssa sylvatica, with leaves quite glabrous or pubescent only on the midrib and principal veins beneath, is readily distinguishable from *Nyssa aquatica*, with leaves grey and pubescent all over the under surface, and with one or two teeth often on the margin. *Nyssa sinensis*, which resembles in foliage *Nyssa sylvatica*, is distinguished by the appressed pubescence of the shoots.

In winter *Nyssa sylvatica* (Plate 200, Fig. 5) shows the following characters :—

Twigs slender, glabrous, or with slight pubescence near the tip only; stipule-scars absent.   Leaf-scars small, crescentic, set somewhat obliquely on slightly prominent pulvini, surrounded by a narrow raised rim, marked with three bundle dots.   Buds conical, pubescent, and acute ; scales five or six, imbricated, pubescent, ciliate, reddish or greenish ; terminal bud larger than the lateral buds which arise at an angle of about 45°.   Pith solid, but interrupted by transverse woody partitions, showing on longitudinal section a ladder-like appearance.

The inner scales of the bud are accrescent ; and the base of the shoot is marked by ring-like scars, indicating where these scales have fallen off in the preceding spring.

VARIETIES

This species is extremely variable in leaf, both in wild specimens and cultivated trees.   This is well shown in the Strathfieldsaye tree, the leaves of which vary from a long elliptical acuminate to a short broad obovate obtuse outline ; some are quite glabrous, whilst others are pubescent on the midrib and principal veins beneath. Usually the leaves are very shining above and coriaceous ; but in a tree growing at Kew in a wood, they are dull above and thin in texture.   In some specimens there are numerous glands on the under surface of the leaf; whilst in others, as in a specimen growing in the Arnold Arboretum collected by Elwes, no glands are visible. The fruit is also variable, being either terete or flattened.   The tree occurs in America in very diverse stations, both on wet soils and on dry mountain slopes ; and this may explain the remarkable extent of its variation.

Var. *biflora*, Sargent, *Silva N. Amer.* v. 76, t. 218 (1893).

Nyssa biflora, Walter, *Fl. Carol.* 253 (1788); Loudon, *Arb. et Frut.* iii. 1317 (1838); Sargent, *Trees N. Amer.* 709 (1905).

Leaves smaller than in the type, very narrow, glabrous and glandular beneath, quite entire in margin.   Fruit with an oval, flattened stone, narrowed at both ends and prominently ribbed.   This variety is a small tree, rarely more than 30 feet high, growing in ponds on the pine barrens near the coast from N. Carolina to Louisiana. It usually has a trunk with a swollen base, and appears to be a form of the species which has adapted itself to life in water.

The cultivated trees mentioned by Loudon as being *Nyssa biflora* were all probably *Nyssa sylvatica* of the typical form.   (A. H.)

DISTRIBUTION

Nyssa sylvatica is found in North America from Southern Ontario, where it grows to a good size near Niagara, and in New England, where I saw it in the neighbourhood of Boston 60 or 70 feet high, westwards to Central Michigan and South-Eastern Missouri, and southwards to Florida and Texas.   It attains its largest size, according to Sargent, in the southern Appalachian Mountains, growing as high as 100 feet with a maximum girth of about 15 feet.[1]   It is found generally in wet soil on

[1] But Ridgway measured a black gum in Wabash Valley, 125 feet high by 13 feet in girth, and 64 feet to the first limb.

the borders of swamps ; but in the south grows also on high wooded mountain slopes. It is very variable in form, sometimes branching close to the ground ; but oftener has a stout straight trunk, covered with light brown deeply furrowed bark, which is often curiously divided into hexagonal scales. Plate 144 B shows the trunk of a tree in America. The upper branches are twiggy and usually crooked. The glossy green leaves are rarely disfigured by fungi or insects, and turn to deep red in autumn. An excellent illustration of a group of trees growing near a pond in Massachusetts is given in *Garden and Forest*, iii. 491, which resemble in habit the Siberian or Japanese larch ; and this is the form which the trees often assume in low swampy ground in New England. Another figure in the same journal, vii. 275, fig. 46, shows the habit of a tree growing in drier ground in Pennsylvania.

CULTIVATION

Nyssa sylvatica was in cultivation at Whitton, near Hounslow, in 1750. It is, when well grown, a very distinct and beautiful tree, the brilliant scarlet assumed by its leaves in autumn rendering it a very desirable ornament for the park or pleasure ground.

Sargent says that one reason why this tree is not more generally planted is that its long roots with few rootlets make it difficult to transplant, and that it must be either planted out when quite young or frequently transplanted in the nursery. Those which I have raised from seed grew slowly the first year, but seemed to ripen their young wood better than many American trees. When pricked out singly into pots in the following spring they all died.

We have seen very few specimens in this country, the only one of great size being the tree[1] at Strathfieldsaye, which measured in 1897 74 feet high by 5 feet 5 inches in girth. It grows on rather heavy soil. This tree was reported by Loudon to be about 30 feet high in 1838, and is probably over 100 years old (Plate 145). It produced seed in 1906 which appeared to be mature.

There is a tree at Munden, near Watford, the seat of the Hon. A. Holland Hibbert, which has a short bole of 4½ feet long with a girth of 3 feet 3 inches, dividing into two stems, the branches of which are very spreading, forming a crown of foliage 38 feet in diameter ; the total height is only 20 feet. Mr. Daniel Hill of Watford, who kindly sent these measurements, says that the fork has been leaded over ; and it is possible that the tree lost its leader early from some accident, and in consequence subsequently assumed its present peculiar habit.

At White Knights, near Reading, there was a large tree of this species which was cut down some years ago ; and there are now many suckers arising from the roots.[2] There is another tree at Bicton about 35 feet high by 3½ feet, which in August 1906 had full-sized fruit upon it which seemed likely to ripen.

[1] The girth of this tree given in *Gard. Chron.* xxvi. 162 (1899), is evidently erroneous, 14 feet 10½ inches being a misprint for 4 feet 10½ inches.

[2] Schenck, in *Biltmore Lectures on Sylviculture*, 56 (1905), says that in the forest old trees are often surrounded by an abundance of seedlings ; but on abandoned fields it seems to come up from sprouts and not from seeds.

There are three small trees in Kew Gardens, the largest about 20 feet high, growing in a densely wooded part close to the Arboretum Nursery.

A tree growing in the garden at Harpton in Radnorshire, at an elevation of 700 feet above sea-level, was in 1905 27½ feet high by 2 feet 8 inches in girth. The owner, Sir Herbert E. F. Lewis, Bart., who kindly sent us particulars, has not noticed during the last forty years any considerable increase in the size of this tree. Its leaves turn bright yellow in autumn.

Timber

The wood seems to be unknown in commerce, and is not mentioned by any of the English writers, but Sargent says it is very durable under water and used for keels of boats, and being extremely difficult to split, is also used for yokes, rollers, wheel-hubs, and pumps. Sections of it in Hough's *American Woods*, Pt. I. No. 9, show a pale or reddish-brown wood of very close texture, somewhat resembling sycamore in appearance.   (H. J. E.)

NYSSA AQUATICA, Cotton Gum, Tupelo Gum

Nyssa aquatica, Marshall, *Arbust. Am.* 96 (1785); Linnæus, *Sp. Pl.* 1058 (*ex parte*) (1753); Sargent,
    Silva N. Amer. v. 83, t. 210 (1893), and *Trees N. Amer.* 711 (1905).
Nyssa uniflora, Wangenheim, *Nordam. Holz.* 83, t. 27, f. 57 (1787).
Nyssa denticulata, Aiton, *Hort. Kew.* iii. 446 (1789).
Nyssa tomentosa, Michaux, *Fl. Bor. Am.* ii. 259 (1803).
Nyssa angulisans, Michaux, *loc. cit.*
Nyssa grandidentata, Michaux f., *Hist. Arb. Am.* ii. 252, t. 19 (1812); Loudon, *Arb. et Frut. Brit.*
    iii. 1319 (1838).

A tree, attaining in America 100 feet in height, with a trunk 12 feet in girth above the greatly enlarged base. Bark thick, longitudinally fissured, and roughened on the surface by small scales. Young shoots pubescent towards the tip, becoming glabrous below in summer. Leaves (Plate 199, Fig. 10) elliptical or ovate-oblong, base rounded or tapering, apex long-acuminate; margin entire or repand, ciliate, often with one to three or more triangular teeth, usually ending in a bristle; upper surface dark green, glabrous; lower surface greyish in colour and with a scattered, fine pubescence; petioles more or less pubescent, 1 to 1½ inch or more in length.

Flowers on long, slender, pubescent peduncles: staminate flowers pedicellate in dense clusters, with a cup-shaped, obscurely five-toothed calyx and oblong short petals rounded at the apex; pistillate flowers solitary, with long, tubular calyx, ovate minute spreading petals, and included stamens with small mostly fertile anthers. Fruit solitary, on long, drooping stalks, oblong, dark purple, about an inch long; stone obovate, rounded at the apex, pointed at the base, flattened, with about ten wing-like ridges.

Identification. (See under *Nyssa sylvatica*)

In winter, specimens from the tree at White Knights showed the following characters:—Twigs stout, pubescent near the tip, glabrescent elsewhere. Leaf-scars slightly oblique on prominent pulvini, almost orbicular or obcordate, notched in the upper margin, surrounded by a slightly raised rim, and marked by three conspicuous bundle-dots. Lateral buds minute, globose, two-scaled, reddish, shining, glabrous, arising in the notch of the leaf-scar. Terminal buds nearly globose, short and broad, with four to five thick, pubescent, reddish scales, keeled on the back and apiculate at the apex; in December the three outermost scales had dropped the apiculus and showed a truncate apex with a terminal scar. The base of the shoot is marked by ring-like scars as in *Nyssa sylvatica.*

Distribution

Nyssa aquatica is found growing in swamps throughout the coast region of the United States, from Southern Virginia to Texas, and in the Mississippi valley, in Arkansas, Southern and South-Eastern Missouri, Western Kentucky, and Tennessee, and in the valley of the lower Wabash River in Illinois.

An interesting account of the peculiar habit of this tree, as observed in the swamps of Arkansas, is given by Coulter.[1] Occurring in company with *Taxodium distichum*, wherever the ground is inundated with water, the trunk develops an enlarged, dome-like base, often of immense size. A tree only 45 feet high, of which a figure is given, had a swollen base 55 feet in girth at the point where the roots entered the ground. When the water-supply is scanty the base is only slightly enlarged; and trees growing in dry soil show no swelling of the trunk. Coulter saw numerous seedlings of Nyssa, and concludes that it is gradually ousting from the swamps the Deciduous Cypress, which rarely seeds itself. Wilson[2] states that around the swollen base of these trees in the swamps there are masses of roots extending 6 to 8 inches above high-water line, each root going vertically up out of the water, and after a sharp bend going down into the water again. He compares these roots, rising above the water for purposes of aeration, with the knees of Taxodium.

Cultivation

Nyssa aquatica was cultivated[3] by Collinson near London in 1735. It is now scarcely known in cultivation in England, the only tree which we have found being one at White Knights Park, Reading, the residence of T. Friedlander, Esq. It is a slender tree, about 36 feet by 2 feet 2 inches, which looks of considerable age and is not vigorous in growth. Loudon[4] states that most of the trees which he saw at White Knights in 1833 were planted between 1790 and 1810; and one was a fine specimen[5] of *Nyssa aquatica*, perhaps identical with the tree now living.

[1] *Report Missouri Bot. Garden*, 1904, xv. 56, plates 18, 19.   [2] *Proc. Philadelphia Acad. Nat. Sc.* 1889, p. 69.
[3] Aiton, *Hort. Kew*, iii. 446 (1789).   [4] *Gardeners' Magazine*, ix. 664 (1833).
[5] This tree is not referred to by Loudon in his large work, published in 1838.

Michaux states that it endures the climate of Paris, and does not exact in Europe as moist a soil as it constantly requires in the United States.   (A. H.)

TIMBER

According to Holroyd,[1] it has only recently been possible to market the timber of this tree, and under a fictitious name, so great has been the prejudice against this and others known as gums.   Formerly when lands bearing tupelo and cypress were logged, the cypress alone was taken, and tupelo trees from 2 to 3 feet in diameter were left, because the lumbermen considered them to be worthless.   At present, however, tupelo timber is extensively cut in Alabama, near Mobile, as well as in Southern and Central Louisiana.   The best grades closely resemble the Yellow Poplar (*Liriodendron*).   The wood has a fine uniform texture, is moderately hard and strong, not elastic, but very tough and hard to split, and easy to work with tools.   It is not durable in contact with the ground, and requires much care in seasoning.   It is now extensively used for house flooring and indoor finish.   Mr. Weale informs me that it has a tendency to warp and split which cannot be prevented by any known process of seasoning; and only a small quantity has as yet been imported to England, in the form of boards, which are worth from 1s. 9d. to 2s. per cubic foot, and are used by the makers of cheap furniture.   But he thinks that if it was sent in boards as well planed as those of the so-called Hazel Pine, it would be more attractive, and its consumption would increase.

(H. J. E.)

NYSSA SINENSIS, CHINESE TUPELO

Nyssa sinensis, Oliver, in Hooker, *Icon. Plant.* t. 1964 (1891).

A tree, attaining in China 40 feet in height.   Young shoots covered with a dense appressed white short pubescence, retained in the second year.   Leaves (Plate 199, Fig. 1) elliptical, base tapering, apex acuminate, margin entire and ciliate; upper surface dull, dark green, and glabrous except for some slight pubescence on the midrib towards the base; lower surface light green, shining, pilose on the midrib and chief veins and occasionally on the veinlets; petiole, $\frac{1}{4}$ to $\frac{3}{8}$ inch long, pilose.

Flowers, on long slender axillary peduncles, pedicellate, crowded in racemose clusters.   Staminate flowers with a minute calyx, narrow oblong petals, and five to ten stamens on a fleshy disc.   Pistillate flowers imperfectly known, but with bifid style and glabrous ovary.   Fruit in clusters of about three, on short pedicels at the ends of long (two to three inches) erect or ascending pubescent peduncles; oblong, bluish, $\frac{3}{8}$ inch long; flesh scanty; stone with ten inconspicuous longitudinal ribs.

This is a rare tree, occurring in mountain woods in Central China, in the western part of Hupeh, and on the Lushan Mountains, near Kiukiang, in Kiangsi.[2] It was discovered by me in 1888, and was subsequently found by Mr. E. H. Wilson, who sent home seed to Messrs. Veitch in 1902, from which a single plant has been raised at Coombe Wood, where it is perfectly hardy so far.   (A. H.)

[1] *U.S. Dept. Agric., Forest Service Circular*, No. 40 (1906).   [2] E. H. Wilson in *Gard. Chron.* xlii. 344 (1907).

SASSAFRAS

Sassafras, Nees ab Esenbeck u. Ebermaier, *Handb. Med. Pharm. Bot.* i. 418 (1830); Bentham et Hooker, *Gen. Pl.* iii. 160 (1880).

DECIDUOUS trees belonging to the order Lauraceæ, with alternate pinnately-veined simple leaves without stipules. Flowers diœcious or rarely perfect, in few-flowered racemes in the axils of bud-scales at the ends of the previous year's shoots. Calyx six-lobed, the lobes in two series, imbricated in bud; petals absent. Staminate flowers; stamens nine in three series, the three inner ones each with two stalked glands at the base; anthers opening with four valves. Pistillate flowers with flattened ovate pointed or slightly two-lobed staminodes, or occasionally with fertile stamens like those of the male flowers; ovary ovoid, glabrous, superior, one-celled; ovule solitary, suspended; one style elongated with a capitate stigma. Fruit an oblong-ovoid, one-seeded dark-blue berry, surrounded at the base by the enlarged and thickened calyx-limb, and supported on pedicels much thickened above the middle.

The genus comprises only two species, one occurring in North America and the other in China.

SASSAFRAS TZUMU, CHINESE SASSAFRAS

Sassafras Tzumu, Hemsley, in *Kew Bull.* 1907, p. 55, and in Hooker, *Icon. Plant.* t. 2833 (1907).
Litsea laxiflora, Hemsley, *Journ. Linn. Soc. (Bot.)* xxvi. 383, t. 8 (1891).
Lindera Tzumu, Hemsley, *op. cit.* 392 (1891).

This species grows sparingly in China in mountain woods at 3000 to 5000 feet elevation, south-west of Ichang, in the province of Hupeh; near Kiukiang in Kiangsi; and inland from Ningpo in Chekiang. It attains a height of 50 feet and yields a timber esteemed by the mountaineers, who call it the *tzu-mu* or *huang ch'iu* tree. Resembling very closely the American species in the characters of the foliage and inflorescence, it was considered by Prof. Sargent[1] and Mr. E. H. Wilson to be indistinguishable. Mr. Hemsley, however, points out certain differences in the floral organs, which entitle it to rank as a distinct species. The flowers are slightly smaller than those of the American tree, and are pubescent within and not glabrous as in that species. The male flowers have three staminodes alternating with the glandular row of stamens and a prominent pistillode, which are wanting in *Sassafras*

---

[1] *Trees N. Amer.* 336 (1905).

officinale. The female flowers have twelve staminodes in three rows of six, three, and three; only six staminodes in two rows of three each occurring in the American species.

There is a tree of this species, 10 feet high, growing in the Coombe Wood nursery, which was raised from seed sent by Wilson in 1900. It has made wonderful growth during the past summer, and is very handsome. It differs from the American species in having glabrous non-ciliate leaves, which are very lustrous on the upper surface; and the young branchlets are also devoid of pubescence.

SASSAFRAS OFFICINALE, Sassafras

Sassafras officinale, Nees ab Esenbeck u. Ebermaier, *loc. cit.*; Bentley and Trimen, *Medicinal Plants*, iii. 220 (1880).

Sassafras Sassafras, Karsten, *Pharm. Med. Bot.* 505 (1882); Sargent, *Silva N. Amer.* vii. 17, tt. 304, 305 (1895), and *Trees N. Amer.* 337 (1905).

Sassafras variifolium, O. Kuntze, *Rev. Gen. Pl.* ii. 574 (1891); Sargent, in *Bot. Gazette*, xliv. 226 (1907).

Laurus Sassafras, Linnæus, *Sp. Pl.* 371 (1753); Loudon, *Arb. et Frut. Brit.* iii. 1301 (1838).

A tree, attaining in America 90 feet in height and 18 feet in girth. Bark,[1] according to Sargent, dark red-brown, deeply and irregularly divided into broad scaly ridges. Young shoots green or reddish, pubescent when young, becoming glabrous, remaining green in the second year. Leaves (Plate 199, Fig. 5) deciduous, entire, or two- to three-lobed; the entire leaves oval with an obtuse apex and cuneate base; the others obovate, with a large triangular or oblong lobe on one or both sides, directed forwards and outwards; margin entire or repand, ciliate; upper surface dark green with a scattered short pubescence; lower surface pale with a long pubescence, often falling by the end of summer; petiole, 1 to 2 inches long, pubescent. The nerves are pinnate, the two lowest arising near the base of the leaf, running nearly parallel with the margin, and ending in the lobes when these are present.

Berry[2] gives an account with illustrations of the extraordinary variation which occurs in the leaves of wild trees growing in America. He has found leaves with four, five, and even six lobes.

Seedling

Out of some seed gathered by Elwes at the Arnold Arboretum late in September and sown at Colesborne in October 1904, only one germinated in the following June, and the seedling showed the following characters in August:—The cotyledons remain in the seed-case, the young stem emerging between them after the splitting of the seed into two halves. The terete glabrous and reddish stem first gives off alternately two minute scales, which are succeeded by true leaves; the first, $\frac{1}{2}$ inch long, arising $1\frac{1}{2}$ inch above the ground, is half-oval in shape, one side of the leaf

---

[1] In cultivated trees in England the bark is grey and fissured into longitudinal narrow ridges.

[2] *Bot. Gazette*, xxxiv. 426 (1902).

being scarcely developed, entire in margin, and on a short stalk about ⅓ inch long. The second leaf, ¾ inch long, is obovate-spathulate, entire in margin, very unequal-sided, rounded at the apex, and tapering at the base. Succeeding leaves (six in all being produced by August) are oval, 1½ to 2½ inches long, stalked, unequal-sided, pinnately-veined, slightly undulate in margin; pale green and glabrous, with a raised midrib beneath.

IDENTIFICATION

In summer Sassafras is readily distinguishable by the aromatic leaves of different shapes, entire and two- to three-lobed, and by the branchlets, without stipules or their scars, remaining green for two or three years. In winter (Plate 200, Fig. 6) the following characters are available :—Twigs glabrous, green, shining, brittle, and strongly aromatic in odour when broken; lenticels few and inconspicuous; pith wide and mucilaginous. Leaf-scars alternate, oblique on prominent pulvini, very small, semicircular with a raised rim, and showing a transverse band of minute coalesced bundle-dots. Terminal buds ovoid, with a long sharp beak; external scales, four to five, imbricated, slightly pubescent, ciliate, green, often ridged or veined. Lateral buds minute, arising from the twigs at about an angle of 45°. Base of the shoot marked by ring-like scars, indicating where the scales of the previous season's terminal bud have fallen off. (A. H.)

DISTRIBUTION

Sassafras occurs usually in rich, sandy, well-drained soil; and is widely spread in the eastern half of the United States, crossing into Canada in Southern Ontario. The northern limit passes through the southern parts of Maine, Vermont, and Ontario to Central Michigan, whence the western limit is continued through Eastern Kansas and the Indian Territory, to the valley of the Brazos river in Texas. On the eastern side it extends from Maine to Central Florida. In the South Atlantic and Gulf States it often takes possession of abandoned fields.

In America the tree is very handsome at all seasons of the year, the light green foliage of summer turning delicate shades of yellow, orange, and red in autumn. The fruit, which is abundantly produced in some years, is showy, the berries dark blue in colour contrasting with the scarlet cups in which they sit. The tree produces root-suckers very freely.

In New England the Sassafras does not often become a tree of considerable size. Emerson[1] states that it rarely reaches 30 feet in height by a foot in diameter, and Michaux says that near Portsmouth, N.H., it is only a tall shrub rarely exceeding 15 to 20 feet high. But near Boston it sometimes grows much larger, and Emerson mentions one which grew at West Cambridge in 1842, and measured nearly 60 by 8 to 9 feet, with a clean straight stem 30 feet long. This tree was felled in order, as he says, "to allow a wall to run in a straight line." But such vandalism as this, which a generation ago was common in New England, is now disappearing; and great care is taken of the few surviving old trees of the original forest. Tree

[1] *Trees and Shrubs of Massachusetts*, p. 359.

wardens are appointed in most parishes, who are often ladies ; and I am indebted to one of the most enthusiastic and active of them, Miss Emma G. Cummings of Brookline, Mass., for showing me some of the large Sassafras trees which still survive in the suburbs of Boston. These form a group on a slope on the south side of Covey Hill, the smallest being 6 feet in girth, and the largest 9 feet 7 inches and over 50 feet high. But these are far inferior to the trees in the forests of the south and west, where Ridgway measured, in the Wabash valley, a Sassafras 95 feet high by 7½ in girth, and where, he says, it sometimes attains 12 feet in circumference.

CULTIVATION

The Sassafras was one of the earliest American trees introduced into England, having been cultivated in 1633 in a garden near London.[1] The tree is propagated by seeds, which should be sown as soon as ripe, and by suckers and root-cuttings. When large it is difficult to transplant, as the thick fleshy roots are scantily provided with rootlets.

Cobbett,[2] who gave an interesting account of the Sassafras, and was very enthusiastic in its praise, found that the seeds rarely if ever come up in the first year, and apparently often lie over for two years. Fresh seeds gathered by me in the Arnold Arboretum and sown in autumn, only produced one seedling in the first year, and no more have since germinated. This seedling though kept in a green-house grows very slowly, and at three years old is only 10 inches high. But though the tree is now rare in England there is no reason why it should not be grown on rich sandy soil in those districts where the summers are warm and dry, if young trees can be procured and established.

REMARKABLE TREES

The only really fine specimen of this species that we have seen in England is in the garden at Claremont, the seat of H.R.H. the Duchess of Albany. This is a handsome, healthy tree which in 1907 measured 48 feet by 6 feet 8 inches at 1 foot from the ground. It forks low down, and the main stem is 4 feet 10 inches at 5 feet. This tree flowers freely in the month of May, but Mr. Burrell has observed no seeds on it (Plate 146). A tree formerly grew at Beeston Hall, near Norwich, which Grigor states to have been 38 feet high in 1840, but this, as I am informed by Mr. Wall, the gardener there, died and was taken down about 1898.

There are four small trees in Mr. Friedlander's garden at White Knights Park, Reading, which appear to be suckers from the roots of an older one now dead ; and in the adjoining properties, White Knights and the Wilderness, there are also trees of which the tallest is about 35 feet by 2 feet 10 inches. There is a younger tree in Mrs. Robb's grounds at Goldenfield, Liphook, and a small one in Kew Gardens planted by Sir W. T. Thiselton-Dyer. There is also a healthy young tree at Tortworth.

---

[1] Gerard, *Herball* (ed. Johnson), 1524 (1633).   [2] *Woodlands*, Nos. 489 *seq.* (1825).

The trees reported by Loudon to be growing in his time at Syon and at Croome cannot now be found.

MEDICINAL PROPERTIES

An interesting article on this tree by Prof. Sargent, with a figure of the trunk of an old one on Long Island showing the peculiar bark, is given in *Garden and Forest*, vii. 215; and from this I take the following :—

The Sassafras is one of the most interesting trees of eastern North America. The last survivor of a race which at an earlier period of the earth's history was common to the two hemispheres, it is the only tree in a large family which has been able to maintain itself in a region of severe winter cold. Towards the middle of the sixteenth century the French in Florida heard from the Indians wonderful accounts of the curative properties of a tree which they called *Pavame*, and which for no obvious reasons the Europeans called Sassafras. The tree and its virtues were first described by the Spanish physician, Nicholas Monardes, in his *Natural History of the New World*, published in Seville in 1569.

The reputation of the roots and wood as a sovereign cure for most human maladies soon spread through Europe, and extraordinary efforts were made to procure them. To collect Sassafras was one of the objects of the English expedition which landed in Massachusetts in 1602, and eight years later Sassafras is mentioned among the articles to be sent home, in the instructions of the English Government to the officers of the young colony in Virginia.

For nearly two centuries the reputation of Sassafras was maintained, and many medical treatises have extolled its virtues, though now it is generally recognised as simply a mild aromatic stimulant. Recently the thick pith of the young branches has been found to yield a mucilage useful to oculists, as it can be combined with alcohol and subacetate of lead without causing their precipitation. The oil of Sassafras, obtained from the wood and roots by distillation, is used to perfume soap and other articles; and perhaps after all the most useful product of the Sassafras tree is the yellow powder prepared from the leaves by the Choctaw Indians of Louisiana, used to give peculiar flavour and consistency to "Gumbo filé," one of the best products of the Creole kitchen.

TIMBER

The wood has little or no economic value and is unknown in Europe. Michaux says that it was never seen in the lumber yard, and was only occasionally used for joists, rafters, and bedsteads; and that it is not attacked by beetles on account of the odour, which it preserves as long as it is kept dry. Ashe says it is light, soft, weak, brittle, and coarse-grained, very durable in contact with the soil, and apt to crack in drying. But the unusual orange-brown colour of the heartwood seems to me to give it a value for ornamental carpentry, if it can be procured of sufficient size.

(H. J. E.)

CORYLUS

Corylus, Linnæus, *Sp. Pl.* 998 (1753); Bentham et Hooker, *Gen. Pl.* iii. 406 (1880); Winkler, in Engler, *Pflanzenreich*, iv. 61, *Betulaceæ*, 44 (1904).

DECIDUOUS trees or shrubs, belonging to the order Betulaceæ. Leaves alternate, distichous on the branchlets, stalked, simple, penninerved, doubly serrate; stipules two, caducous. Buds composed of numerous imbricated scales, corresponding to stipules.

Flowers monœcious, arising from buds on the branchlets of the previous year. Male flowers in cylindrical catkins, appearing in autumn; fascicled, or two to five on a common peduncle; composed of numerous imbricated bracts, each bearing on its inner side two partly adnate bracteoles and four stamens, without a perianth; filaments bifid, each branch bearing a single anther cell, tufted with hairs at its apex. Female flowers in buds resembling those which contain leaves only, but distinguishable in spring by the projecting styles. The lower scales of the buds bear leaves in their axils, the flowers, few in number, arising only in the axils of the uppermost scales, each scale bearing two flowers. Each flower, surrounded at the base by two minute bracteoles, more or less deeply cut and forming an involucre, consists of a two-celled ovary, surmounted by a short, denticulate perianth and two long styles; each cell containing one ovule.

Fruit, in clusters at the end of the short leafy branch into which the bud has developed; a one-celled, one-seeded nut, the remains of the other cell and ovule, which have aborted, being visible in its upper part. The nut is contained in a leafy involucre, open at the summit, and variously lobed or dentate. Seed without albumen; cotyledons thick, fleshy, containing oil, remaining on germination underground.

Eight or nine species of Corylus are known, all natives of northern temperate regions, and mostly shrubs or small trees. Only one species, *Corylus Colurna*, attains the dimensions of a timber tree, and comes within the scope of our work.

CORYLUS COLURNA, CONSTANTINOPLE OR TURKISH HAZEL

Corylus Colurna, Linnæus, *Sp. Pl.* 999 (1753); Loudon, *Arb. et Frut. Brit.* iii. 2029 (1838); Willkomm, *Forstliche Flora*, 377 (1887); Winkler, *op. cit.* 50.

A tree of moderate size, attaining 60 to 80 feet in height and 7 to 8 feet in girth of stem. Bark of trunk grey, thick, and scaling off in small irregular plates.

Twigs brittle, the young shoots glandular pubescent, those of a year old glabrous and brown in colour, the bark of older shoots becoming corky. Leaves 3 to 5 inches long by 2 to 4 inches wide, broadly oval, ovate, or obovate, deeply cordate at the base, acuminate at the apex, doubly serrate or with large serrate teeth, dark green above, lower surface lighter green and sparingly pubescent, with glandular hairs on the principal nerves and midrib; nerves usually eight pairs; petiole ½ to 1 inch long, glandular pubescent or glabrescent. Catkins[1] 1½ to 3 inches long. Fruits crowded, three to ten in number, long, compressed, pubescent towards the apex. Involucres tomentose with intermixed glandular hairs, deeply and irregularly divided into linear, acute, stiff, long-pointed segments, which are either entire or toothed, exceeding in length two to three times the nut.

SEEDLING

The germination resembles that of the oak, the cotyledons, which are short-stalked, plano-convex and obovate, remaining in the seed and not being carried above ground. Caulicle stout, terete, tapering, ending in a long tap root with numerous branching fibres. Stem stout, terete, covered with numerous scattered glandular hairs, giving off an inch above the cotyledons a pair of opposite leaves, which are about 2 inches long, broadly ovate, acute at the apex, cordate at the base, with three to five pairs of lateral lobes, unequal in size, toothed and ciliate in margin; petiole ¾ inch, glandular-pubescent. Succeeding leaves are alternate and larger in size.

VARIETIES

In addition to the typical form described above, several geographical varieties occur, as the species is distributed over a wide area.

1. Var. *glandulifera*, A. de Candolle, *Prod.* xvi. 2, p. 132 (1864).—Occurs with the type in Europe and western Asia. In this variety the pubescence on the petioles, peduncles, and fruit-involucres is intermixed with glandular bristles; and the segments of the involucres are less acute and often dentate.

2. Var. *lacera*, A. de Candolle, *op. cit.* 131 (*Corylus lacera*, Wallich, *List*, 2798). —Leaves obovate, larger, up to 7 inches long, with ten to twelve pairs of nerves. Involucre-segments linear-lanceolate with glandular hairs. This variety occurs in the western Himalayas, from Kashmir to Nepal, at elevations of 6000 to 10,000 feet, and in many places is gregarious. Sir George Watt informs me that it is a handsome tree, usually growing in the mixed forests, and often attaining 80 feet in height.

3. Var. *chinensis*, Burkill, *Journ. Linn. Soc.* (*Bot.*) xxvi. 503 (1899) (*Corylus chinensis*, Franchet, *Journ. de Bot.* 1899, xiii. 197).—Leaves large, up to 7 inches long, with ten to twelve pairs of nerves, broadly ovate, unequal, acuminate; petioles bristly. Involucres striate and constricted above the fruit, lobes forked, lobules

---

[1] Abnormal male flowers with enlarged bracteoles are figured in *Gard. Chron.* xxvi. 691, fig. 135 (1886).

acute and falcate.   This variety occurs in China, and grows to about 40 feet high in mixed forests in Yunnan, Szechwan, and Hupeh.

Apparently no varieties have originated in cultivation, but a hybrid has been obtained between this species and the common hazel, viz. :—

Corylus intermedia, Loddiges, *Catalogue* (1836) (*Corylus avellana* × *Corylus Colurna*, Rehder, *Mitth. Deuts. Dendrol. Gesell.* 1894, p. 43).—This is a tall shrub or small tree with the bark of the common hazel, *i.e.* darker and less scaly and fissured than that of *C. Colurna*.   The fruit resembles that of the last species, but is shorter and scarcely glandular.   Specimens of this are growing in the Botanic Gardens of Jena and Göttingen and in the Forestry Garden at Münden, but we know of none in England.

Identification

In summer the Turkish hazel is readily distinguishable by the scaly bark and the obovate leaves deeply cordate at the base and distichously placed on the branchlets.   In winter (Plate 126, Fig. 6) the following characters are available :—Twigs : brittle, shining, brownish-yellow, with few and inconspicuous lenticels and scattered glandular pubescence, usually, however, dense near the base of the shoot, which is ringed with the scars of the previous season's bud-scales, one or two of the lowermost scales often persisting dry and darkened in colour ; second year's shoot with corky bark, which fissures and exfoliates slightly.   True terminal bud absent, a small oval scar at the apex of the twig, on the side opposite to the highest leaf-scar, indicating where the tip of the shoot fell off in summer.   Leaf-scars semicircular with three to six bundle-dots,[1] somewhat obliquely set on prominent pulvini.   Stipule-scars small, transverse, lunate, one on each side of the leaf-scar. Buds pretty uniform in size, alternate and distichous on the twig, from which they arise at a wide angle, ovoid, rounded at the apex ; scales about ten, imbricated, pubescent, ciliate in margin.   Pith small, circular.   Male catkins present in winter on flower-bearing trees.

Distribution

The Turkish hazel has a wide distribution, extending from south-eastern Europe, through Asia Minor and the Caucasus, to the Himalayas and Western China.   In Europe it is found growing wild in Banat, Slavonia, Herzegovina, Bosnia, Servia, Roumania, and Greece.[2]   In Banat, according to Willkomm, it sometimes forms pure woods in the mountains ; and in Northern Albania it ascends as a bush to 3000 feet altitude.[3]   It occurs in Asia Minor in Bithynia, Paphlagonia, and Anatolia.   According to Radde,[4] it grows in small groups on the south side of the main chain of the Caucasus and in many localities in Georgia, at 3500 to 5000 feet elevation, where it is a stately tree 50 to 70 feet in height, and with a

---

[1] The cicatrices left by the leaf-bundles on the leaf-scar are very irregular in number and shape, being circular dots or curved lines.

[2] In Thessaly and Acarnania, according to Halacsy, *Consp. Fl. Græcæ*, iii. 135 (1904).

[3] Beck, *Veg. Illyrischen Länder*, 300 (1901).   [4] *Pflanzenverb. Kaukasusländ.* 187 (1899).

stem diameter of 18 inches. The nuts of the wild tree are small, with a thick and hard shell. It also grows in the mountains of Karabagh, but does not occur in the Talysch district. (A. H.)

CULTIVATION

The Turkish hazel was first cultivated in western Europe by Clusius, who obtained it from Constantinople in 1582. Linnæus states that in 1736 the finest specimen known was a tree in the Botanic Garden at Leyden, which had been planted by Clusius. It was apparently first cultivated in England about the year 1665 by John Rea,[1] who states that he had then " many goodly trees of the filbeard of Constantinople." He grafted these upon ordinary hazel stocks.

The Turkish hazel is now a rare tree in England, seldom to be got from a nursery, though perfectly hardy and easy to grow from seed, which it ripens in most seasons in the southern half of England. I have raised many from a tree at Tortworth Court, and the Earl of Ducie has done the same. The seed usually germinates in the following spring if sown when ripe, but if kept till spring, sometimes not until the next year. The seedlings, on my soil at least, have more inclination to become bushes than to make a single stem, but, if cut down two or three years after planting, will throw up strong suckers which may be trained into a tree, and should be planted in half-shady places or in an opening in a wood, as they are liable when young to be injured by spring frosts.

REMARKABLE TREES

No other place can show so many fine trees as Syon, where there are in the grounds at least five, all apparently of about the same age. The largest of these stands near the east bridge over the lake, and is about 75 feet high, with a bole about 30 feet long and 6 feet 9 inches in girth. Near the gardener's house is another fine tree more spreading in habit, about 70 feet by 7 feet 6 inches, which is probably not the same as one figured by Loudon, which was then 61 feet high. This has been figured by the Hon. S. Tollemache as the Hazel.[2]

At Bute House, Petersham, Henry measured a well-shaped tree which, in 1904, was 56 feet by 6 feet 7 inches.

At Corsham Court there is a remarkable tree about 50 feet high, which divides near the base into two stems, one of which is quite decayed, and the other, which has the appearance of having originated as a sucker from it, is quite sound and 6 feet 8 inches in girth. Lord Methuen tells me that he can remember this tree as formerly producing fruits which were sent up to table, but now it no longer bears any nuts.

At White Knights I saw a grafted tree from which seedlings had sprung up in the shrubbery, and one of these, growing at the base of a stump, is 10 feet high at about ten years old.

At Arley Castle there is a good tree which, in 1904, was by Mr. Woodward's measurement 60 feet by 5 feet 7 inches.

---

[1] *Flora*, 225 (1665).   [2] *British Trees with Illustrations*, 9 (1901).

At Wollaton Hall, near Nottingham, the property of Lord Middleton, there is a tree 43 feet high which at 5 feet girths 7 feet 10 inches, and at 10 feet, where it forks, 8 feet. It has a spread of not less than 78 feet, which for this tree is very unusual (Plate 147). It is perhaps the most symmetrical of its kind that I have seen anywhere. In the Botanic Gardens at Oxford and Kew there are fair-sized specimens.

In Scotland and Ireland we know of no trees of great size, and none were recorded by Loudon; but at Glasnevin there is one about 35 feet in height, which divides into three stems close to the ground, and has very pendent wide-spreading branches.

TIMBER

Little or nothing is known of the timber in England, but a wood has been imported to France under the name of " Noisetier," which I believe to belong to this species, and which, as exhibited by M. Hollande of Paris, is very handsome. I purchased some very handsome veneer from Mr. Witt of London, which he told me had come to him direct from Constantinople, and which I believe was cut from the root of *C. Colurna*. Two good-sized logs of this tree were in the collection of Servian timbers shown at the Balkan States Exhibition in London in 1907; one of them is now in the Kew Museum. Gamble[1] says that in the Himalaya it is a well-grained timber, which does not warp, of a pinkish-white colour, and often shows a fine shining grain resembling that of bird's-eye maple.   (H. J. E.)

[1] *Man. Indian Timbers*, 684 (1902).

CARPINUS

Carpinus, Linnæus, *Gen. Pl.* 292 (*ex parte*) (1737); Bentham et Hooker, *Gen. Pl.* iii. 405 (1880); Winkler, in Engler, *Pflanzenreich*, iv. 61, *Betulaceæ*, 24 (1904).
Distegocarpus, Siebold et Zuccarini, *Flor. Jap. Fam. Nat.* ii. 103, t. 3 (1846).

DECIDUOUS trees belonging to the order Betulaceæ. Leaves, alternate, distichous on the branchlets, stalked, ovate, doubly-serrate, penninerved, the nerves ending in the points of the teeth; stipules scarious, caducous or persistent. Flowers appearing in early spring with the unfolding of the leaves, unisexual, monœcious, without petals. Staminate flowers in pendulous, cylindrical catkins, arising from buds produced near the ends of lateral branches of the previous year; stamens, three to twenty, crowded on a pilose receptacle adnate to the base of a concave scale; filaments short, two-branched, each branch bearing a one-celled anther, tipped with a cluster of long hairs. Pistillate flowers, in loose, semi-erect catkins, which are terminal on the branchlets of the year; in pairs at the base of an ovate, acute, deciduous scale; each flower subtended by a small bract and two minute bracteoles, and consisting of a two-celled ovary, surmounted by a minute epigynous calyx and two elongated styles; each cell containing one ovule.

Fruit, in pendent, stalked strobiles, composed of imbricated, foliaceous or membranous involucres, resulting from the developed bract and bracteoles of the flower, each with a nutlet at its base. Nutlet, ovoid, compressed, longitudinally ribbed, crowned by the calyx and remains of the style, one-seeded, and falling from the involucre in autumn. Seed, filling the cavity of the nutlet, without albumen; cotyledons fleshy, carried above ground in germination.

The genus consists of about eighteen species inhabiting the temperate regions of North America, Europe, and Asia. Two sections are distinguished:—

I. DISTEGOCARPUS, Sargent, *Silva N. Amer.* ix. 40 (1896).

Scales of the staminate catkins lanceolate, stalked. Fruit-involucres, membranous, infolded below, completely covering the nutlet, closely imbricated in the strobile. Trees with scaly bark. Two species, *C. japonica*, Blume, and *C. cordata*, Blume.

II. EU-CARPINUS, Sargent, *loc. cit.*

Scales of the staminate catkins ovate, sub-sessile. Fruit-involucres, foliaceous, open or only slightly infolded over the nutlets, loosely imbricated in the strobile. Trees usually with smooth bark. This section includes the remaining species.

Carpinus and Ostrya are very similar in foliage; and the following key, based upon the characters of the leaves and branches (Plate 201), distinguishes all the species of both genera which are in cultivation in England.   *Carpinus laxiflora*, though not yet introduced, has been included, as it has been much confused with the other Japanese hornbeams.

KEY TO CARPINUS AND OSTRYA

I. *Leaves not exceeding 2 inches in length.*

  1. *Carpinus orientalis*, Miller.   South-eastern Europe, western Asia.

    Leaves 1½ inch long, acute, deeply plicate.

  2. *Carpinus polyneura*, Franchet.   Central China.

    Leaves 2 inches long, acute, smooth and only slightly plicate.

II. *Leaves exceeding 2 inches in length.*

A. *Leaves lanceolate.*

3. *Carpinus japonica*, Blume.   Japan.

    Leaves about 4 inches long, much longer in proportion to their width than in the other species, with numerous (eighteen to twenty-four pairs) nerves.

B. *Leaves ovate, acute at the apex.*

4. *Carpinus yedoensis*, Maximowicz.   Central China.   Cultivated in Japan.

    Leaves 2½ inches long, rounded at the base, with conspicuous bands of appressed pubescence on the upper surface.   Branchlets pilose.

C. *Leaves ovate, acuminate-at the apex.*
* *Leaves deeply cordate at the base.*

5. *Carpinus cordata*, Blume.   China, Korea, Manchuria, and Japan.

    Leaves 4 to 5 inches long, broad in proportion to their length, with fifteen to twenty pairs of nerves.

** *Leaves rounded or only slightly cordate at the base.*
† *Under surface glabrous between the nerves.*

6. *Carpinus laxiflora*, Blume.   China, Japan.

    Leaves 2½ inches long, rounded at the base, abruptly contracted into a very long acuminate apex.   Branchlets with scattered long hairs.   Buds minute, $\frac{1}{16}$ inch long.

7. *Carpinus Betulus*, Linnæus.   Europe, western Asia.

    Leaves 3 inches long, slightly cordate at the base, turning yellow in autumn. Branchlets with scattered long hairs.   Buds fusiform, $\frac{1}{4}$ to $\frac{1}{3}$ inch long.

8. *Carpinus caroliniana*, Walter.   North America.

    Leaves as in *C. Betulus*, but turning red in autumn.   Branchlets with scattered long hairs.   Buds ovoid, $\frac{1}{8}$ inch long.

†† *Under surface pubescent between the nerves.*

9. *Ostrya carpinifolia*, Scopoli.   Southern Europe, Asia Minor, Syria.

    Leaves 3 inches long, not velvety to the touch above, rounded at the base. Branchlets with dense appressed pubescence.

10. *Ostrya japonica*, Sargent.   China, Japan.

Leaves 3 to 4 inches long, velvety to the touch above, slightly cordate at the base.   Branchlets with dense, scarcely appressed, pubescence.

11. *Ostrya virginica*, Willdenow.   North America.

Leaves 3 to 4 inches long, not velvety to the touch above, slightly cordate at the base.   Branchlets glandular-pubescent.   (A. H.)

CARPINUS ORIENTALIS

Carpinus orientalis, Miller, *Gard. Dict.* ed. 7, No. 3 (1759); Loudon, *Arb. et Frut. Brit.* iii. 2014 (1838); Winkler, *Betulaceæ*, 37 (1904).

Carpinus duinensis, Scopoli, *Fl. Carniol.* ii. 243, t. 60 (1772); Boissier, *Fl. Orient.* iv. 1177 (1879); Willkomm, *Forstliche Flora*, 366 (1887).

Carpinus nigra, Moench, *Verz. Ausländ. Bäume u. Staud.* 19 (1785).

A small tree or large shrub, rarely attaining 50 feet in height; bark smooth and greyish.   Young branchlets covered with a very minute dense pubescence, with which are intermixed scattered long hairs.   Leaves (Plate 201, Fig. 7) small,[1] strongly plicate, the nerves being deeply impressed above, about 1½ inch long by ¾ inch wide, ovate or ovate-elliptical, acute at the apex, unequal and slightly cordate at the base; margin sharply bi-serrate, ciliate; upper surface dark green, shining, with scattered long hairs; lower surface light green, pilose on the midrib and nerves, glabrous between the nerves, with minute axil-tufts; nerves nine to thirteen pairs; petioles, ¼ to ⅜ inch, pilose; stipules linear-lanceolate, pubescent at the apex, ¼ inch long, often persistent during summer.   Fruit: strobiles, up to 2 inches long; bracts densely imbricated, ¾ inch long, obliquely ovate, not lobed, sharply and irregularly serrate.

This species is a native of south-eastern Europe and western Asia.   It occurs in Italy and Sicily, reaching its northern limit in Istria, Croatia, Slavonia, Banat, and Transylvania, and extending southwards through the Balkan States to Macedonia and Greece.   It is also met with in the Crimea, Asia Minor, and the Caucasus.

It was introduced into cultivation in England in 1739 by Miller.   It appears to be exceedingly rare, the only specimens we have seen being at Kew, where there are several small trees, one of which, planted in 1878, is now about 20 feet high.   (A. H.)

CARPINUS POLYNEURA

Carpinus polyneura, Franchet, *Journ. de Bot.* xiii. 202 (1899); Burkill, *Journ. Linn. Soc. (Bot.)*, xxvi. 501 (1899).

Carpinus Turczaninowii, Hance, var. *polyneura*, Winkler, *Betulaceæ*, 38, f. 12 (1904).

A small tree, attaining 30 feet in height; bark greyish, slightly fissuring and scaly.   Young branchlets with scattered long hairs.   Leaves (Plate 201, Fig. 5)

[1] In wild specimens the leaves are often larger, 2 to 2½ inches in length.

small, weakly plicate, the nerves being only slightly impressed above, about 2 inches long by $\frac{7}{8}$ inch broad, ovate, acute at the apex, unequal and slightly cordate at the base; margin bi-serrate, ciliate; upper surface dark green, shining, with scattered, appressed hairs; lower surface as in *C. orientalis*; nerves nine to twelve pairs; petiole, $\frac{1}{4}$ to $\frac{3}{8}$ inch, pilose; stipules linear, pubescent along the margins, $\frac{1}{4}$ inch long, persistent during summer.   Fruit: strobiles 2 inches long; bracts loosely imbricated, obliquely ovate, $\frac{1}{2}$ inch long, outer margin slightly serrate, inner margin sub-entire, not lobed, without a basal auricle.

This species is a rare tree in the mountains of Eastern Szechwan and Western Hupeh in China; and is closely allied to, if not a mere variety of, *C. Turczaninowii*, Hance, which is common in Northern China.   *C. polyneura* differs little in technical characters from *C. orientalis*, but is very distinct in appearance owing to the leaves being smooth and flat and not deeply plicate, as in the other species of hornbeam.

It is only represented in cultivation by a single tree, about 15 feet high, in Kew Gardens, which was raised from seed sent by me in 1889.   (A. H.)

CARPINUS JAPONICA

Carpinus japonica, Blume, *Mus. Bat. Lugd. Bot.* i. 308 (1850); Shirasawa, *Icon. Ess. Forest. Japon.*, text 47, t. 24, ff. 1-17 (1900); Winkler, *Betulaceæ*, 25 (1904).

Carpinus Carpinus, Sargent, *Garden and Forest*, vi. 364, f. 56 (1893); *Forest Flora Japan*, 64, t. 21 (1894).

Distegocarpus Carpinus, Siebold et Zuccarini, *Fl. Jap. Fam. Nat.* ii. 103 (1846).

A tree attaining in Japan 50 feet in height and 5 feet in girth; bark furrowed and scaly.   Young branchlets with scattered long hairs, which fall off in autumn. Leaves (Plate 201, Fig. 1) ovate-oblong, up to 4 inches long by $1\frac{1}{2}$ inch broad, acuminate at the apex, oblique at the base, which is rounded or slightly cordate; margin finely bi-serrate, non-ciliate; upper surface dark green, pubescent on the midrib and nerves; lower surface pale green, with scattered long hairs on the midrib and nerves and slight axil tufts; nerves, eighteen to twenty-four pairs, impressed above; petiole $\frac{1}{2}$ inch long, pubescent; stipules $\frac{1}{2}$ inch long, linear-lanceolate, pubescent, persistent during summer.   Fruit: strobiles $2\frac{1}{2}$ inches long; bracts densely imbricated, $\frac{3}{4}$ to $\frac{7}{8}$ inch long, ovate, sharply serrate; nutlet covered by a minute orbicular lobe, attached merely by its base to the bract, the outer margin of the latter being slightly infolded below.

This species is a native of central and southern Japan, and, according to Sargent, is common on the Hakone and Nikko Mountains between 2000 and 3000 feet elevation.   It was collected near Nikko by Elwes, and at Nagasaki by Oldham.

It was introduced by Maries in 1879; but no trees of this date are now to be found, there being only small plants about 3 feet high in the Coombe Wood Nursery. It is perfectly hardy in New England, where it produced fruit for the first time in 1891 in the Arnold Arboretum, where it had been introduced a few years previously.

Young plants were sent by Prof. Sargent in 1895 and 1897 to Kew, which have now attained about 10 feet in height. At Tortworth a young tree has produced fruit.

The foliage of this species is remarkably distinct and handsome.   (A. H.)

CARPINUS YEDOENSIS

Carpinus yedoensis, Maximowicz, *Mél. Biol.* xi. 314 (1881); Burkill, *Journ. Linn. Soc. (Bot.)*, xxvi. 502 (1899); Franchet, *Journ. de Bot.* xiii. 203 (1899); Winkler, *Betulaceæ*, 35 (1904).

A small tree. Young branchlets densely covered with long hairs. Leaves (Plate 201, Fig. 3), 2½ inches long by 1½ inch wide, ovate, acute at the apex, rounded at the base; margin biserrate and ciliate; upper surface with conspicuous bands of long appressed pubescence in the intervals between the lateral nerves; lower surface pilose on the midrib and nerves, glabrous or with scattered long hairs in the intervals between the nerves; nerves ten to twelve pairs; petiole, ⅜ to ½ inch long, pilose; stipules, linear-lanceolate, caducous. Fruit: strobiles, 2½ inches long; bracts loosely imbricated, ¾ inch long, semi-ovate, coarsely serrate on the outer side, subentire on the inner side, which is slightly infolded at the base, forming a small auricle partly covering the nutlet.

This species is only cultivated in Japan, where it was first seen by Maximowicz. It was discovered growing wild in the mountains of North-Eastern Szechwan in China by Père Farges, and may have been brought to Japan by Buddhist monks in early days, like many other Chinese plants. Young plants were raised from Japanese seed in 1901 by Purpus, in the Botanic Garden at Darmstadt. In the nursery at Kew there are two or three plants, growing vigorously, and about 3 feet in height, which were obtained from Simon Louis in 1904.   (A. H.)

CARPINUS CORDATA

Carpinus cordata, Blume, *Mus. Bot. Lugd. Bat.* i. 309 (1850); Sargent, *Garden and Forest*, vi. 364 (1893), viii. 294, f. 41 (1895), and *Forest Flora Japan*, 65 (1894); Burkill, *Journ. Linn. Soc. (Bot.)*, xxvi. 501 (1899); Shirasawa, *Icon. Ess. Forest. Japon*, text 46, t. 24, ff. 18-32 (1900); Winkler, *Betulaceæ*, 26 (1904); J. H. Veitch, *Hortus Veitchii*, 359 (1906).
Distegocarpus (?) *cordata*, De Candolle, *Prod.* xvi. 2, p. 128 (1864).

A tree, attaining in Japan and China a height of 50 feet and a girth of 6 feet; bark, dark grey, deeply furrowed and scaly. Young branchlets covered with a very minute pubescence, intermixed with scattered long hairs. Leaves (Plate 201, Fig. 2), ovate, up to 5 inches long and 2¾ inches wide, acuminate at the apex, unequally and deeply cordate at the base; margin finely bi-serrate, non-ciliate; upper surface dark green, with scattered long hairs; lower surface light green, pubescent between the nerves, pilose on the midrib and nerves, without axil tufts; nerves fifteen to twenty pairs, impressed above; petiole, $\frac{3}{4}$ inch long, with scattered long hairs; stipules caducous. Fruit: strobiles, 3 to 6 inches long, long-stalked; bracts densely imbricated, membranous, 1 to $1\frac{1}{8}$ inch long, irregularly serrate; the inner margin furnished below with an orbicular lobe, infolding and concealing the nutlet; the outer margin slightly inflected at the base. The basal lobe is much larger than in *C. japonica*, and is united to the bract, not only by its base, but also along one side.

Var. *chinensis*, Franchet, *Journ. de Bot.* 1899, p. 202.—Leaves, ovate-oblong, 3 inches long by $1\frac{3}{4}$ inch broad, with eighteen to twenty pairs of nerves, slightly cordate and unequal at the base, shortly acuminate at the apex. This variety strongly resembles in the shape of the leaf certain forms of *C. japonica*, but has the fruit of *C. cordata*. It seems to be intermediate between the two species, and is found in the mountains of Eastern Szechwan in China. It was introduced into cultivation by Mr. E. H. Wilson in 1901, and young plants are growing in the Coombe Wood Nursery.

According to Sargent, *Carpinus cordata* is one of the largest and perhaps the most beautiful of the hornbeams. It grows on the main island of Japan only at high altitudes, its true home being in the deciduous-leaved forest of central and northern Yezo. It is also a native of Korea and Manchuria; and occurs in China, in the typical form, in the province of Shensi,[1] the variety *chinensis* growing more to the south.

This species was introduced from Japan by Maries in 1879, and produced fruit in 1886 in the Coombe Wood Nursery, where the largest specimen now living is only 15 feet in height. A tree at Tortworth is about 20 feet, and has borne fruit, from which, however, Elwes did not succeed in raising seedlings. There is also a small tree at Grayswood, Haslemere. It seems to be very rare in cultivation, and there are no specimens growing in the Hornbeam Collection at Kew.   (A. H.)

CARPINUS LAXIFLORA

Carpinus laxiflora, Blume, *Mus. Bot. Lugd. Bat.* i. 309 (1850); Oliver, in Hooker, *Icon. Plant.* t. 1989 (1891); Sargent, *Garden and Forest*, vi. 364 (1893), and *Forest Flora Japan*, 64 (1894); Burkill, *Journ. Linn. Soc.* (*Bot.*), xxvi. 501 (1899); Shirasawa, *Icon. Ess. Forest. Japon*, text 48, t. 25, ff. 15-30 (1900); Winkler, *Betulaceæ*, 33 (1904).
Carpinus Fargesii, Franchet, *Journ. de Bot.* 1899, p. 202.

A tree, attaining in Japan 50 feet in height and 5 feet in girth; bark smooth, grey, sometimes almost white in colour. Young branchlets with scattered long hairs. Leaves (Plate 201, Fig. 8), $2\frac{1}{2}$ inches long by $1\frac{1}{2}$ inch broad, ovate or ovate-elliptical, contracted above into a long acuminate apex, rounded or slightly cuneate at the base; margin, bi-serrate, non-ciliate; upper surface with scattered long appressed hairs; lower surface with long appressed hairs on the midrib and nerves, glabrous between the nerves; nerves thirteen to fifteen pairs; petiole, $\frac{1}{2}$ inch long, pilose;

[1] Burkill, *loc. cit.*

stipules caducous. Fruit: strobiles, up to 3 inches long; bracts very loosely imbricated, about ⅛ inch long, semi-ovate, outer side irregularly serrate, inner side sub-entire, with a lobe near the base, which is infolded, but does not conceal the nutlet.

This species is a native of China and Japan. According to Sargent, it is very like the European hornbeam in habit, fluted stem, and smooth bark. It is common in all the mountain forests of Hondo, where it is most abundant at elevations between 2000 and 3000 feet. Near Agematsu, in Shinshu, at 2000 feet altitude, it was collected by Elwes, who saw no tree of any great size or beauty, though the leaves turn red and yellow in autumn. In Yezo, it descends to sea-level on the southern shores of Volcano Bay, where, near the town of Mori, it is common in oak forests, and grows to its largest size. In China, this species grows in the mountains of Hupeh, Eastern Szechwan, and Kiangsi; but is rare, displaying considerable variation in the character of the leaves and fruit.[1]

It has not yet, apparently, been introduced into cultivation. Plants at Kew, sent under the name of *C. laxiflora*, from the Arnold Arboretum in 1895, are *C. japonica*.   (A. H.)

CARPINUS CAROLINIANA, AMERICAN HORNBEAM

Carpinus caroliniana, Walter, *Fl. Carol.* 236 (1788); Sargent, *Silva N. Amer.* ix. 42, t. 447 (1896), and *Trees N. Amer.* 190 (1905); Winkler, *Betulaceæ*, 31 (1904).
Carpinus americana, Michaux, *Fl. Bor. Amer.* ii. 201 (1803) Loudon, *Arb. et Frut. Brit.* iii. 2013 (1838).
Carpinus Betulus, Koehne, *Deutsche Dendrologie*, 116 (1893).

A bushy tree, attaining, in America, rarely 40 feet in height and 6 feet in girth, with stem and bark like the common hornbeam. Young branchlets with a few scattered long hairs, the minute glandular pubescence often seen in *C. Betulus* never being present. Leaves (Plate 201, Fig. 6) as in the common hornbeam, but usually with fewer nerves, nine to ten pairs; and unequal, rounded, or slightly cuneate at the base. Stipules lanceolate, ⅛ inch long, caducous. Fruits: strobiles, 2 to 3 inches long; bracts loosely imbricated, triangular-ovate, ¾ to 1 inch long, with two short unequal lateral lobes, and a much longer middle lobe, which is usually serrate on only one margin; pedicels of each pair of bracts united only at the base.

In the absence of fruit, this species is difficult to distinguish from *C. Betulus* from which Koehne could not distinguish it even as a variety. In autumn, the beautiful red tint of the foliage of the American species is diagnostic. The best mark of distinction lies, however, in the buds, which are small, ovoid, acute, ⅛ inch long, with glabrous ciliate scales; those of *C. Betulus* being large, fusiform, ¼ to ⅓ inch long, with pubescent ciliate scales.

This species, which is known in America as the blue beech or water beech, is found along the borders of streams and swamps, from Southern and Western Quebec

---

[1] Three varieties are distinguished by Burkill, *loc. cit.*

to Florida, extending westward to Northern Minnesota, Eastern Nebraska, Kansas Indian Territory, and Eastern Texas. It is also met with in a slightly modified form[1] in the mountainous regions of Southern Mexico and Guatemala. It is most abundant and of its largest size in the southern Alleghany mountains and in Southern Arkansas and Texas.

It was introduced into England by Pursh in 1812; but is very rare in cultivation, the best specimen we have seen being at Arley Castle. It has no claim to be considered as a forest tree, its only merit being the scarlet colour of the foliage in autumn. Elwes gathered seeds of this species near Ottawa in 1904, which did not germinate.
(A. H.)

CARPINUS BETULUS, COMMON HORNBEAM

Carpinus Betulus, Linnæus, *Sp. Pl.* 998 (1753); Loudon, *Arb. et Frut. Brit.* iii. 2004 (1838); Willkomm, *Forstliche Flora*, 358 (1887); Mathieu, *Flore Forestière*, 396 (1897).
Carpinus vulgaris, Miller, *Gard. Dict.* ed. 8, No. 1 (1768).
Carpinus sepium, Lamarck, *Fl. Franç.* ii. 212 (1778).
Carpinus compressa, Gilbert, *Exerc.* ii. 399 (1792).
Carpinus ulmoides, Gray, *Nat. Arrang. Brit. Pl.* ii. 245 (1821).
Carpinus carpinizza, Host, *Fl. Austr.* ii. 626 (1831).
Carpinus intermedia, Wierzbicki, in Reichenbach, *Icon. Fl. Germ.* xii. f. 1297 (1850).
Carpinus nervata, Dulac, *Fl. Haut. Pyrén.* 141 (1867).

A tree, usually attaining only a moderate size, 60 or 70 feet in height and 8 feet in girth; but in England occasionally as large as 90 feet by 12 feet. Stem never perfectly circular in section, being more or less longitudinally fluted or ridged, with shallow rounded depressions between the ridges; bark smooth, thin, grey. Young branchlets with scattered long hairs, a very minute dense glandular pubescence being also often present. Leaves (Plate 201, Fig. 4) about 3 inches long by $1\frac{3}{4}$ inch broad, oval or ovate, acuminate at the apex; broad, unequal, and rounded or slightly cordate at the base; margin bi-serrate, non-ciliate; upper surface dark green, glabrous, or rarely pilose on the midrib and nerves; under surface light green, with appressed long hairs on the midrib and nerves and minute axil tufts; lateral nerves, ten to fifteen pairs, impressed on the upper surface, prominent beneath; petiole $\frac{1}{4}$ to $\frac{1}{2}$ inch long, pubescent; stipules narrow, lanceolate, $\frac{1}{2}$ inch long, caducous.

Male catkins, about $1\frac{1}{2}$ inch long; scales ovate, acute, entire, veined longitudinally; stamens, 4 to 12, with long yellow anthers. Female catkins, nearly 1 inch long; scales ovate, acuminate, ciliate. Fruit: strobiles up to 3 inches long; involucres loosely imbricated, in pairs, with their pedicels connate for the greater part of their length, three-lobed, the lateral lobes small and usually entire, the middle lobe, about $1\frac{1}{2}$ inch long, entire or minutely serrulate; nutlet, $\frac{1}{3}$ inch long, seven- to eleven-nerved, glabrous, with the apex umbonate and surrounded by a six-lobed calycine ring, within which are the remains of the style.

[1] Var. *tropicalis*, Donnell Smith, *Bot. Gaz.* xv. 28 (1890).

In winter, the **twigs** are smooth, shining, glabrous, with five-angled pith, and are marked at the **base** of the year's growth by ringlike scars, due to the fall of the accrescent scales of the bud of the previous season. Terminal bud not formed, the tip of the branchlet falling off in summer and leaving a small circular scar close to the uppermost axillary bud, the latter prolonging the shoot in the following season. **Leaf-scars** small, crescentic, three-dotted, with a short stipular scar on each side. Buds, distichous on the branchlets, unequal in size, on prominent leaf-cushions, appressed against the stem, fusiform, $\frac{1}{4}$ to $\frac{1}{6}$ inch long; scales, ciliate and pubescent towards the tips, brownish.

Seedling :[1] Primary root tapering, wiry, flexuose; caulicle terete, pubescent, $\frac{1}{2}$ inch long; cotyledons fleshy, rounded-obovate, $\frac{1}{5}$ inch long, auricled at the base, shortly stalked, glabrous, green above, whitish beneath; stem zigzag, pubescent, giving off alternate stalked bi-serrate leaves, which resemble those of the adult plant, but are smaller and occasionally lobulate in margin.

VARIETIES

The common hornbeam shows little variation in the wild state, the only form worth noticing being var. *carpinizza*, which is found in Transylvania. In this variety the leaves are often distinctly cordate at the base with only seven to nine pairs of nerves; and the fruit-involucre has very short lateral lobes.

Under cultivation, pyramidal,[2] fastigiate, pendulous, and variegated forms have originated. In var. *purpurea*, the young leaves have a reddish tint. Var. *incisa*, Aiton,[3] has leaves with large sharp serrated teeth. A wide-branching tree of this variety at Beauport, Sussex, is 6 feet 3 inches in girth; and there is also a fine specimen at Smeaton-Hepburn, East Lothian. In var. *quercifolia*, Desfontaines,[4] the leaves are smaller than in the type and are irregularly and deeply cut or lobed. In this variety, leaves of the ordinary form are often present on the same branch with those of the pinnatifid kind. Two remarkable trees of this variety are reported[5] to be growing on the bowling green of the Woodrow Inn, in Cawston Parish, near Aylsham, Norfolk.

DISTRIBUTION

The common hornbeam is indigenous in the south of England; but its true native limits cannot now be exactly determined. It is recorded[6] from Somerset, Wilts, Dorset, Hants, Berks, Oxford, Bucks, Herts, Surrey, Sussex, Kent, Essex, Cambridge, Suffolk, and Norfolk; but in many cases, especially in the south-western counties, the records are probably of planted and not really wild trees. In Dorset,[7] it is a very rare tree; and Townsend[8] considers it to be a doubtful native of Hampshire. Druce[9] considers it to be indigenous in Oxfordshire on the chalk, but always

---

[1] Cf. Lubbock, *Seedlings*, ii. 532, f. 667 (1892).

[2] A solitary wild specimen of the pyramidal hornbeam formerly grew in the forest of Gremsey, near Vic in France. Godron, *Les Hêtres Tortillards* (1869).   [3] Aiton, *Hort. Kew*, iii. 362 (1789).

[4] Desfontaines, *Tab. Écol. Bot. Mus. Hist. Nat.* 212 (1824).   [5] Rev. J. F. Noott in letter to Kew, March 1894.

[6] Watson, *Comp. Cybele Brit.* 311 (1870) and *Topog. Bot.* 355 (1873).

[7] Mansell-Pleydell, *Flora of Dorsetshire*, 246 (1895).

[8] *Flora of Hampshire*, 313 (1883).   [9] *Flora of Oxfordshire*, 268 (1886).

planted on other formations. There is no doubt, however, that it formed a considerable part of the ancient forests, which existed to the north and east of London; and in the Lea division of Hertfordshire [1] it still forms the chief portion of the underwood; whilst it is common in Essex and Kent, where it is usually treated as coppice.

The hornbeam has been found in the fossil state in Suffolk, in the interglacial strata at Hoxne, and in the preglacial strata at Pakefield. [2]

Carpinus Betulus is widely distributed on the continent of Europe, and occurs also in Asia Minor, the Caucasus, and Persia. In Europe, its northern limit, beginning in Norfolk in England, crosses over to Denmark and South Sweden, where it ascends on the west coast to lat. 56° 30', and on the east coast to 57° 13', reaching its extreme northerly point on the island of Gothland in lat. 57˝ 20'. In Norway, Schubeler says, it is not wild; but he has seen a tree at Christiania, planted in 1818, which in 1885 measured 36 feet by 4 feet. In Russia, the hornbeam occurs as far north as lat. 56° 10' on the coast of Courland, and is confined to the provinces which lie west of an irregular line drawn from near Riga to the Sea of Azov, its most easterly localities being in the governments of Vitebsk, Mohilef, Chernigof, and Poltava, and in the Crimea. South and west of the above limits, the hornbeam is spread through France, Belgium, Holland, Germany, Austrian Empire, Balkan Peninsula, Greece, Switzerland, and continental Italy; but is not found wild in Spain, Portugal, Corsica, Sardinia, and Sicily.

In France, the hornbeam is most common in the north and east, where it forms a large part of the coppice forests, and also occurs as undergrowth in the high forests of beech and oak. Its southerly limit in France is a curved line extending from Grenoble through Toulouse to near Bordeaux. Towards the west and south, it becomes a rare tree, and is totally absent from Brittany. It is rather a tree of the plains and low hills than of the mountains; but ascends in the Vosges to 2000 feet, in the Jura to 2300 feet, and in the French Alps to 2800 feet. Treated as coppice, its growth is very rapid in France, where it takes the first rank as firewood.

In Germany the hornbeam is widely spread in the plains and low hills, where it grows usually, as in France, in company with the beech and other deciduous trees, either as scattered individuals or in small groups. In east Prussia, where the beech does not occur, the hornbeam replaces it and grows to great perfection, often forming part of the spruce and pine forests. Pure woods are rare, though some of considerable extent occur, according to Willkomm, in Alsace, Baden, and South Bavaria. In Austria, Hungary, the Balkan States, and Greece, the hornbeam is no longer a tree of the plain, but grows in the mountains in the beech forests. It ascends in the Harz mountains to 1250 feet, in the Bavarian Alps to 2900 feet, and in the Swiss Alps to 3000 feet.

According to Radde, [3] it is met with through the whole region of the Caucasus, at elevations ranging from sea-level to 5600 feet. It is also recorded from the northern provinces of Asia Minor, and from Ghilan in Northern Persia.

(A. H.)

---

[1] Pryor, *Flora of Hertfordshire*, 373 (1887).   [2] C. Reid, *Origin Brit. Flora*, 144 (1899).
[3] *Pflanzenverb. Kaukasusländ.* 183 (1899).

CULTIVATION

The seeds of the hornbeam ripen in October, but with few exceptions do not germinate until after a second winter, and must be treated in the same manner as those of the ash. The seedlings though very hardy as regards spring frosts, grow slowly at first, and require about four years in the nursery before they are strong enough to plant out. Though on sandy soil the tree produces fruit freely and the seedlings bear shade as well as those of the beech, yet the hornbeam does not in England, as in some parts of France, tend to overpower the oak; and its economic value was formerly much greater than it is now, on account of its being one of the very best trees for firewood. It may, however, be used for underplanting, and as a nurse for other trees on soils too wet for beech, and is admirably suited for making clipped hedges. When the shoots are interlaced they form an impassable barrier, and bear clipping as well as any tree. It also bears pollarding and coppicing extremely well, some of the old pollards which are seen in the eastern counties being of very great age; but when not so treated it does not appear to be a very long-lived tree, and rarely exceeds 200 years. In France, Mouillefert says, it lives 100 to 120 years, and rarely over 150 years, but I think it must considerably exceed this age in some parts of England.

The hornbeam is more critical as to soil and climate than most of our native trees; and though Loudon says it is always found on stiff clay and on moist soils where scarcely any other timber tree will grow, this is hardly correct. I have never seen a really fine tree on any but fertile soils, and though it is the most abundant tree of Epping Forest, from which Loudon probably derived his idea; there is not, so far as I know, a really fine specimen in that district, though this may be partly due to their being nearly all pollards. I searched in vain for self-sown seedlings, with roots fit to transplant, and of fifty sent me by Mr. M'Kenzie, superintendent of Epping Forest, only one survived. He tells me that though large numbers of seedlings may be seen after a good seed year, yet most of them very soon disappear, as the deer and cattle bite them off when not protected by bushes. As a wild tree it is principally found in the south-eastern and eastern counties where the lowest rainfall occurs, but it grows well in the west and in Ireland, and even as far north as Morayshire. Mouillefert says that in France fresh and permeable sandy soils suit it best; and that sandy, gravelly, and flinty clays also suit it well, even when calcareous, but that it languishes or perishes on those which are too stiff, marshy, peaty, or very dry; and I think this is correct as regards England also. On account of its weak development of roots when young it requires shelter at first, and though it will stand shade fairly, it succeeds best as an isolated tree when adult.

As a forest tree it can only be considered of secondary importance, and Forbes does not include it in his *Estate Forestry*. As as ornamental tree, it has great value, both on account of the graceful pendent branches, which when in flower and fruit are very beautiful, and for the brilliant yellow colour of the leaves in autumn.

REMARKABLE TREES

Large hornbeams are not at all common, and exist so far as I have seen in comparatively few places, mostly old parks. The largest and finest that I know of, though by no means the tallest, is near the reservoir at Cornbury Park, Oxford, where there is a tree whose height I could not measure exactly, though it probably exceeds 75 feet, with a bole 11 feet 10 inches in girth and 12 to 14 feet long, which spreads out at that height into an immense number of branches covering a circle of 95 paces (Plate 148). There are two other trees of nearly similar size and habit on the north side of the beech avenue, one of them leaning very much on one side with drooping branches. Sir Hugh Beevor has recently measured a tree, 100 feet in height and 9 feet 8 inches in girth, on Sir Robert Dashwood's property near West Wycombe.

But there is no place where I have seen hornbeams so tall or so numerous as at Cobham Park, Kent, where there must be hundreds of trees 70 to 80 feet high, and many with clean boles 20 to 40 feet long. Among so many it is hard to say which are the largest, but one which I measured near the old heronry, and not far from the ash grove, was over 90 feet high, dividing at about 7 feet into four stems, each of which ran up straight and clean for about 40 feet. Another, a pollard, hollow on one side, measured 13 feet 6 inches in girth. These grow on a soil which suits the ash perfectly. Four shoots from a stool in a wood here measured 76 feet high and 2 to 3 feet in girth.

At Mersham-le-Hatch, Ashford, Kent, the seat of Sir Wyndham Knatchbull, Bart., I saw, in 1907, a remarkable wood called Bockhanger, composed of very old pollard hornbeams many of which are hollow and much decayed. They grow on a sandy loam, covered in spring with bluebells, and have for generations served to supply the mansion with firewood, of which the steward told me twelve to fourteen cords were annually consumed. The largest of these trees was about 16 feet to the crown, and had a very large kidney-shaped wen on one side, over which it measured 16 feet 4 inches in girth. Another tree here showed the remarkable power of the hornbeam in repairing wounds in its trunk. A large double-stemmed tree, widely split and hollow at the base, had higher up completely covered the open cleft with young healthy wood and bark in the same way that old yews often do.

A most remarkable hornbeam, on account of its very wide-spreading branches, grows in Fredville Park, Kent, and though not over 35 to 40 feet high, covers an area of no less than 103 paces round. It has about fifteen main branches which show the characteristic irregularities that old hornbeams always have. The branches are so thick that foxes often choose the crown of this tree as a lair, and when covered with fruit, as it was when I saw it in June 1907, it is a most striking and beautiful tree. It grows in a deep fertile loam overlying chalk, but rather wet in winter.

The hornbeam is, in Essex, especially in Epping Forest, most commonly seen as a pollard, the practice of lopping the branches for firewood having been very general in old times. A photograph showing the appearance of the tree when so treated was taken for me by Mr. Elsden of Hertford, at Waterhall, a farm on Mr. H. Clinton Baker's property near Bayfordbury, Herts, in January 1907 (Plate 149). At Essendon, Herts, Mr. Baker, in 1906 measured a tree, 81 feet by 11 feet 2 inches ; a pollarded tree at the same place being 56 feet high by 18 feet in girth. Sir Hugh Beevor measured in 1891 a hornbeam in Hatfield Park, Herts, which was 17½ feet in girth at about 4 feet from the ground.

The finest and largest examples of pollard hornbeams that I have seen are in Easton Park, Essex, the seat of the Earl of Warwick. A group of these trees, growing near the park-keeper's house, which was shown me by Mr. Rogers, agent for the Easton property, contains several trees of great beauty, which were in flower on 7th April. The largest of these measures no less than 28 feet round the head at about 8 feet from the ground, and 12 feet 2 inches at 2 feet (Plate 150). Another near it, dividing into two stems which are united at the crown, was 25 feet in girth at 7 feet and 17½ feet at 2 feet. A third, growing at some distance, has perhaps the finest head of all, and measures 26 feet round the head with a bole about 11 feet high. Mr. Shenstone tells me that the largest he has seen in Epping Forest is 27 feet in girth round the head, and he showed me another very old one in Braxted Park which was over 20 feet round.

Mrs. Delves Broughton has sent me a photograph (Plate 151) of a very fine group of hornbeams in Weald Park, Essex, the seat of C. J. H. Tower, Esq., in which, according to the measurements sent me by Mr. T. W. Bacon, the two largest trees are 75 feet by 16 feet 9 inches, and 88 feet by 15 feet 4 inches.

At Elveden, Suffolk, there is a very well-shaped and handsome tree in front of the house, which, as I was told by the late Prof. A. Newton, is probably not more than 140 years old, and measured, when I saw it in 1907, 75 feet by 10 feet.

At Nibley, Gloucestershire, there is a tree, of which Col. Noel has been good enough to send me a photograph, which measures about 80 feet by 11 feet 6 inches with a bole of 8 feet and a spread of 80 feet diameter. In Bitton churchyard, Gloucestershire, there is a tree planted since 1817 by Canon Ellacombe's father which is 65 feet by 8 feet 2 inches. At St. Pierre Park, near Chepstow, Major Stacey showed me a very fine hornbeam which, though not very tall, and with a bole only 10 feet high by 11 feet 7 inches in girth, spreads over an area 112 paces round.

In the wooded part of Kew Gardens, there are several fine trees, the best of which is 70 feet high and 10 feet in girth, dividing into three stems at 7 feet from the ground. One tree, 5½ feet in girth, has bark on the lower part of the trunk, divided into raised longitudinal ridges, which are covered with small scales. At Heron Court, Hants, there is a beautiful tree near the front entrance, 70 feet by 10 feet 5 inches with a spread of 25 yards.

At Brocklesby, Lincolnshire, Lord Kesteven measured, in 1906, a tree 77 feet high by 9 feet 4 inches in girth. At Castle Howard the hornbeam grows well and there are several large trees, the tallest being about 80 feet high, the thickest 9 feet 3 inches in girth. At Studley Park, Yorkshire, in the valley below Fountains Abbey, there are several very fine hornbeams, probably the same as those figured by Loudon (ff. 1933, 1934, 1935), which were in 1838 50 to 60 and one 73 feet high.

I measured three from 70 to 80 feet with a girth of 6½ to 8½ feet, one being covered with dense tufts of twigs, a kind of witches' broom, caused by *Exoascus Carpini.*

In Scotland the hornbeam is less common than in the south, but grows to a large size in the warmer districts; though, as it is not mentioned either by Hunter, or in the *Remarkable Trees of Scotland*, it is evidently looked on as a rare tree in the north.   Walker[1] speaks of one formerly growing at Bargally, which was 70 feet high, with a clear trunk of 20 feet.

The finest I have seen is a tree at Gordon Castle, perhaps the one mentioned by Loudon as being then 54 feet high; in 1904, it was 68 feet by 8 feet (Plate 152). At Murthly, in the lower park near the Tay, there is an old tree measuring, in 1906, 65 feet by 9 feet 8 inches; and Henry measured one at Scone of the same dimensions.

Mr. J. Renwick sends me particulars of a very remarkable hornbeam at Douglas Support, in Lanarkshire, which, in 1900, measured 78 feet by 8 feet 1 inch, with a bole of 17 feet long, and a spread of 60 feet, the branches having long pendulous twigs, which form a screen all round the tree and hang nearly to the ground.[2]

Another remarkable tree is at Eglinton Castle, Ayrshire, which separates into three stems near the ground, and measures at the narrowest point below the fork 14 feet in girth; its three stems girth 5 feet 9 inches, 5 feet 6 inches, and 4 feet 11 inches respectively.   Mr. Renwick sends me particulars of other fine hornbeams as follows:—at Househill, Renfrewshire, 10 feet girth, 72 feet spread; at Tulliechewan Castle, Dumbartonshire, 60 feet by 8 feet 3 inches; at Gargunnock House, Stirlingshire, 8 feet 11 inches girth, 83 feet spread.

The hornbeam is rarely planted in Ireland.   The largest tree, which Henry has seen, is growing beside the Killarney Lake, at Mahony's Point.   It measured, in 1904, 15 feet 8 inches in girth, at 18 inches above the ground, giving off six great stems, the three largest of which were—8 feet 4 inches, 7 feet 7 inches, and 6 feet 3 inches in girth.   This tree is about 70 feet in height, and the diameter of its spread is 80 feet.   It is in perfect health and bears fruit regularly.

At Adare, Co. Limerick, in 1903, Henry saw a fine tree, which measured 53 feet by 8 feet 8 inches, the spread of branches being 65 feet.   At Glenstal, in the same county, there is a tree of exactly the same dimensions, as regards height and girth.   At Kilrudderry, Co. Wicklow, a tree, which had been blown down, measured 8 feet 9 inches in girth; and here there is a very fine hornbeam hedge, about 15 feet in height.

TIMBER

The wood of the hornbeam is the hardest, heaviest, and toughest of our native woods, but though extremely strong, is not flexible; and as it is seldom found large enough and clean enough to cut into planks, it is little used in England except for fuel, for which it is one of the best woods known, burning slowly with a

---

[1] *Essays*, p. 95, *fide* Loudon.

[2] I am informed by Mr. Douglas that the peculiarity of this tree consists in the long drooping twigs, which are 20 to 30 feet in length, and hang like small cords to the ground on all sides, concealing the trunk, whilst the upper branches do not droop at all.   He thinks that this is due to its being a grafted tree.   A photograph, which he is good enough to promise me, will be given in a later volume.

bright flame, and making the best of charcoal. As it decays quickly when exposed to wet, it is of no use for outside work, and will not take creosote. The trunk of the tree is often very deeply furrowed, and the wood is said to be cross-grained and difficult to work. It is or was considered the best wood for cogs, mallets, and wooden screws for carpenters' benches, also for pulleys and butchers' blocks. Its value is uncertain, and depends largely on the locality, and on the size and age of the tree.

With regard to the use of this wood by pianoforte manufacturers, Mr. J. Rose, of Messrs. Broadwood and Sons, to whom I am indebted for much information, writes me as follows :—

" Hornbeam is still used for piano action work in England, though American maple has replaced it to a considerable extent. French hornbeam, and, I believe, Dutch also, are used for the purpose, because of larger size and more freely grown than the British product, and also because, when all charges are included, it is probably cheaper. There is a marked difference in the English hornbeam and that grown in France and elsewhere on the Continent. This is perhaps hardly perceptible in a small sample, but the English wood is smaller and more irregular, but of a distinctly firmer texture, so hard and close as sometimes to resemble ivory. It works beautifully with fine saws and small drills; but the waste is serious. The foreign timber is larger and more freely grown, producing much larger boards, but the grain is coarser, and the texture of the wood less firm, and more liable to split when in small pieces, such as are used in action work."   (H. J. E.)

OSTRYA

Ostrya, Scopoli, *Fl. Carniol.* 414 (1760); Bentham et Hooker, *Gen. Pl.* iii. 406 (1880); Winkler,
in Engler, *Pflanzenreich*, iv. 61, *Betulaceæ*, 24 (1904).
Carpinus, Linnæus, *Gen. Pl.* 292 (*ex parte*) (1737).

SMALL deciduous trees, belonging to the order Betulaceæ, agreeing with the genus
Carpinus in the characters of the branchlets, buds, foliage, and staminate flowers.
Pistillate flowers, in dense erect spikes, inserted in pairs on the base of ovate acute
leafy scales, each flower enclosed in a sac-like involucre, formed by the union of a
bract and two bracteoles, which is open at the apex at the time of flowering, afterwards becoming closed. Calyx dentate, adnate to the two-celled inferior ovary;
style short, divided into two linear subulate stigmatic branches; ovules solitary in
each cell. Fruits: disposed in stalked ovoid strobiles, composed of densely imbricated involucres, which are vesicular, closed, flattened, membranous, longitudinally
nerved, reticulate, pubescent at the apex, and hirsute at the base with sharp, rigid,
stinging hairs. Nutlet, sessile in the involucre, ovoid, compressed, longitudinally
ribbed, crowned by the remains of the calyx; seed solitary, pendulous.

Four species of Ostrya have been distinguished :—*Ostrya Knowltoni*, Coville,
a rare tree in Arizona, not yet introduced, and three species, occurring in North
America, Eastern Asia, and Europe and Asia Minor, which are so closely allied
that they have been considered by most botanists to be only geographical races of
one species. These three species are all in cultivation, and as they can be
distinguished (see Key to Carpinus and Ostrya, p. 526), will be treated by us
separately.

OSTRYA CARPINIFOLIA, HOP HORNBEAM

Ostrya carpinifolia, Scopoli, *Fl. Carniol.* ii. 244 (1772); Willkomm, *Forstliche Flora*, 368 (1887);
Mathieu, *Flore Forestière*, 403 (1897).
Ostrya vulgaris, Willdenow, *Sp. Pl.* iv. 469 (1805); Loudon, *Arb. et Frut. Brit.* iii. 2015 (1838).
Ostrya italica, Spach. *Ann. Sc. Nat.* sér. 2, xvi. 246 (1841).
Ostrya italica, sub-species *carpinifolia*, Winkler, *Betulaceæ*, 22 (1904).
Ostrya Ostrya, Sargent, *Silva N. Amer.* ix. 32 (1896).
Carpinus Ostrya, Linnæus, *Sp. Pl.* 998 (1753).

A tree attaining 60 feet in height and 10 feet in girth; stem cylindrical, bark
greyish, finely fissured, and scaly. Young branchlets with dense appressed

pubescence. **Leaves** (Plate 201, Fig. 11) about 3 inches long by 1¾ inch wide, ovate, shortly acuminate at the apex, unequal and rounded at the base; margin sharply bi-serrate and ciliate; covered above and below with appressed pubescence, spreading more or less over the whole surface, and not confined to the midrib and nerves, as in *Carpinus Betulus*, and with minute axil tufts on the lower surface; nerves twelve to fifteen pairs; petiole ¼ to ⅜ inch long, appressed pubescent; stipules persistent during summer. Nutlet ovoid, ⅙ inch long, crowned by a tuft of hairs; calyx-limb obsolete.

In winter the twigs are slender, zigzag, more or less pubescent. No true terminal bud is formed, the apex of the branchlet falling off in summer and leaving a minute circular scar at the side of the uppermost axillary bud. Buds small, $\frac{3}{16}$ inch long, ovoid, viscid, set obliquely on prominent leaf-cushions; scales 6 to 9, imbricated, greenish with a dark brown margin, more or less pubescent. Leaf-scar semicircular, with two bundle-dots above and one group of three smaller dots below.

Ostrya carpinifolia reaches its most westerly point in the extreme south-eastern corner of France, where it occupies a few isolated stations in the Basses-Alpes and Alpes-Maritimes Departments. In the forest of Miolans,[1] in the Basses-Alpes, which is mainly composed of *Pinus sylvestris*, it is found on a northern slope, over an area of about 400 acres, occurring chiefly as undergrowth and ascending to about 2700 feet altitude. In the Alpes-Maritimes it descends in some places to nearly sea-level. It extends eastward through Southern Switzerland, the Tyrol (where,[2] near Botzen, it ascends to 3500 feet altitude), Carinthia, and Lower Styria to Southern Hungary, and spreads southwards through Carniola, Croatia, and the Balkan States to Greece, growing usually in rocky situations, more commonly on limestone than on other formations. It is common throughout Italy and Sicily in the oak and chestnut regions, ascending to 3800 feet elevation; and forms woods of considerable extent around Lake Como, especially above Lecco, on the shores of Lake Lugano, and at Gaudria and Salvatore.[2] It occurs as a rare tree in Corsica and Sardinia. It is also met with in Asia Minor and in the Lebanon. It attains about a hundred years of age; and according to Pardé[3] produces coppice shoots like the hornbeam.

(A. H.)

CULTIVATION

It was introduced into cultivation in England some time before 1724, as it is mentioned in Furber's Nursery Catalogue published in that year. Though an ornamental tree which attains a good size and is perfectly hardy, it has always been very rare in this country. According to Mouillefert[4] its growth is about equal to that of the Hornbeam. I have raised plants from French seed which grow faster on my soil than those of the hornbeam, and seem at least as hardy, as they were

[1] Fliche, *Bull. Soc. Bot. France*, xlvi. 8 (1899). Cf. also *ibid.* xxxv. 160 (1888).

[2] Christ, *Flore de la Suisse*, 238 (1907). In the same work, p. 507, it is stated that this species has been found in the fossil state in miocene beds at Ardeche; and another species, probably a mere variety, has been found in the same strata at Var.

[3] *Arb. Nat. des Barres*, 281 (1906).

[4] *Principales Essences Forestières*, 148, note (1903). At Grignon in France, planted together in the arboretum, on calcareous soil with a chalky subsoil, at thirty years old the Hornbeam is 11 metres high by 70 centimetres in girth at 1 metre above the ground; and the Ostrya 11½ metres by 73 centimetres in girth. It bore here without injury the severe winter of 1879.

uninjured by the severe spring frost of May 21-23, 1905, and ripened their young wood well in October. They may be distinguished by the larger leaves with a pair of persistent linear stipules at the base.

REMARKABLE TREES

From the dimensions given by foreign authors I doubt whether in its native country the Hop Hornbeam ever attains a much larger size than the one which I figure (Plate 153). This remarkable tree is at Langley Park, near Norwich, the seat of Sir Reginald Beauchamp, and cannot be of great age, as it is not mentioned in an account of this place in Grigor's *Eastern Arboretum*, published in 1841. It is grafted on a stock of the hornbeam which measures 8 feet in girth below the graft, while the trunk above it is no less than 15 feet 8 inches. Its height is difficult to estimate, but may be about 50 feet.

A large tree formerly grew at Kew, on which Mr. J. G. Jack, in *Garden and Forest*, v. 602, remarks as follows :—"An unusually fine specimen of a hop hornbeam, 50 feet high, branching near the ground and spreading about 70 feet, with a trunk over 3 feet in diameter, was grafted on a stock of hornbeam at $2\frac{1}{2}$ feet from the ground, and is a good deal larger than its stock, with a swelling at the point of juncture. No one can help remarking the striking contrast between the rough bark of the Ostrya and the comparatively smooth bark of the Carpinus."

This tree was perhaps the one figured by Loudon[1] in 1838, which was then said to be 60 feet high, with a trunk 3 feet in diameter, and the finest specimen in England at that time. In 1890 it was figured in the *Gardeners' Chronicle*[2] as a handsome wide-spreading tree, but soon after began to decay, and was cut down in 1897,[3] when it measured 59 feet high by 9 feet 4 inches in girth at 3 feet. Fruit was abundantly produced; but no perfect seeds were ever developed. A part of its trunk is preserved in the Museum at Kew, and I am indebted to the Director for a sample of the timber, which somewhat resembles that of the pear. According to Mouillefert it has all the qualities of hornbeam wood in a superior degree.

There is a fine specimen in the Botanic Garden at Oxford, which measures about 40 feet by 4 feet. This tree, though quite healthy, is much infested by mistletoe. At Tortworth there is a tree about 40 feet high by 2 feet 7 inches in girth. At Munden, Watford, a tree, 32 feet by 2 feet 11 inches, is said to have been planted about 1830.

In Scotland we know of no tree of this species of large size now existing, though a large one formerly grew at Bargally,[4] a place between Gatehouse and Newton-Stewart, once the property of Andrew Heron, a celebrated planter, who died in 1729. Loudon went there in 1831, and gives the dimensions[5] of the *Ostrya*

---

[1] *Op. cit.* viii. 244 a.

[2] *Gard. Chron.* viii. 275, Fig. 47 (1890). Also figured in *Woods and Forests*, 1884, p. 318. The shapes of the trees figured in Loudon and in the *Gardeners' Chronicle* are very different.

[3] *Kew Bull.* 1897, p. 404.

[4] Walker, *Essays on Natural History and Rural Economy* (1812).

[5] Bargally is fully described by Loudon, *op. cit.* i. 95-99 (1838).

from a letter of the then owner, J. Mackie, as 60 feet high and 4 feet 1 inch in girth at 4 feet in 1835. Henry could find no trace of this tree when he visited Bargally in 1904. At Glasnevin, Dublin, there are two trees, 30 and 25 feet high, narrowly pyramidal in habit. These are 34 years old and are growing on the bank of a stream.

(H. J. E.)

OSTRYA VIRGINICA, Ironwood, American Hop Hornbeam

Ostrya virginica, Willdenow, *Sp. Pl.* iv. 469 (1805); Loudon, *Arb. et Frut. Brit.* iii. 2015 (1838).
Ostrya virginiana, Koch, *Dendrologie*, ii. pt. ii. 6 (1873); Sargent, *Silva N. Amer.* ix. 34, t. 445 (1896), and *Trees N. Amer.* 192 (1905).
Ostrya Ostrya, Macmillan, *Metaspermæ Minnesota Valley*, 187 (1892).
Ostrya italica, sub-species *virginiana*, Winkler, *Betulaceæ*, 22 (1904).
Carpinus Ostrya, Linnæus, *Sp. Pl.* 998 (1753) (in part).
Carpinus virginiana, Miller, *Dict.* ed. 8, No. 4 (1768).

A tree attaining 60 feet in height and 6 feet in girth, but usually smaller. This species, as seen in cultivation, is mainly distinguished from *Ostrya carpinifolia* by the presence on the young branchlets, petioles, and midrib of the leaf beneath, of short, erect, gland-tipped hairs. The leaves (Plate 201, Fig. 9) are usually larger, $3\frac{1}{2}$ inches long, slightly cordate at the base, with fewer nerves, about twelve pairs. The nutlet in this species is larger, $\frac{1}{4}$ to $\frac{1}{3}$ inch long, fusiform, flattened, without a tuft of hairs at the apex, surmounted by a plainly visible calyx-limb.

Two forms of this species occur in the wild state, which have been distinguished by Spach,[1] as follows :—

Var. *glandulosa*.—Young branchlets, petioles, and peduncles covered with gland-tipped short bristles. Specimens in the Kew herbarium from Ontario, Niagara Falls, and the Alleghany Mountains belong to this variety, which is the one known in cultivation in England.

Var. *eglandulosa*.—Glandular bristles not present on any part of the plant. Young shoots pubescent. This variety appears to be common in the western and southern parts of the United States, and does not appear to have been introduced into cultivation. In the absence of fruit, it would be difficult to distinguish this variety from *Ostrya carpinifolia*.

(A. H.)

The tree grows, according to Sargent, on dry gravelly slopes and ridges, often in the shade of oaks and other large trees ; and is a native of Canada and the United States, occurring on the northern shores of Lake Huron in western Ontario, eastward through the valley of the St. Lawrence to Chaleur Bay and Cape Breton Island ; extending southward to Northern Florida and Eastern Texas, and westward to Northern Minnesota, the Black Hills of Dakota, Nebraska, and Kansas. It is most abundant and of its largest size in Southern Arkansas and Texas.

I saw it at Mt. Carmel in Illinois, and in the Arnold Arboretum, where it was a finer tree in size and habit than *Carpinus caroliniana*. It is known in America as

---

[1] *Ann. Sc. Nat.* sér. 2, xvi. 246 (1841), and *Hist. Vég.* xi. 218 (1842).

Ironwood, and is used for levers and tool handles, the wood being very tough and strong.   Michaux states that on the estate of Duhamel du Monceau, in France, there were trees 20 feet high, from which self-sown plants had sprung up.

It was introduced into England by Bishop Compton in 1692, but is rarely met with except in botanic gardens.   At Kew there are four trees, 20 to 30 feet in height.   Others are growing at Eastnor Castle and at Grayswood, near Haslemere, where, though not planted above twenty years, it is growing vigorously, and looks as if it would make a handsome tree.   A tree in the Edinburgh Botanic Garden was, in 1905, 39 feet high by 3 feet 3 inches in girth.   Seedlings raised in my garden grow more freely than those of the common hornbeam ; but not so fast as those of *Ostrya carpinifolia*.   (H. J. E.)

OSTRYA JAPONICA, Japanese Hop Hornbeam

Ostrya japonica, Sargent, *Garden and Forest*, vi. 383, f. 58 (1893), *Forest Flora Japan*, 66, t. 22 (1894), and *Silva N. Amer.* ix. 32 (1896); Shirasawa, *Icon. Ess. Forest. Japon*, text 49, t. 25, ff. 1-14 (1900).

Ostrya virginica, Maximowicz, *Mél. Biol.* xi. 317 (1881).

Ostrya italica, sub-species *virginiana*, Winkler, *Betulaceæ*, 22 (1904).

A tree attaining in Japan a height of 80 feet, with a tall straight stem, 5 feet in girth, but usually smaller.   This species is considered by Maximowicz and Winkler to be identical with the American species, and there is said to be little or no difference in the fruit, which I have not seen.   In cultivation, the Japanese tree is readily distinguished as follows :—Leaves (Plate 201, Fig. 10) velvety to the touch on the upper surface, which is covered with a dense erect pubescence ; nerves, ten to twelve pairs, fewer than in the other species ; base slightly cordate.   Young branchlets densely white pubescent, without glandular hairs, which are also absent from the petiole and midrib of the leaf.

According to Sargent, this species is nowhere abundant in Japan, occurring only as scattered individuals in the forests of deciduous trees which cover Central and Southern Yezo, and growing also in the province of Nambu in Northern Hondo. Shirasawa, however, gives a more extensive distribution, stating that it is found also throughout the central chain of Hondo, in the provinces of Musahi, Kaï, and Totomi, and also at Nikko ; and farther south, in the island of Shikoku.   *Ostrya iaponica* is also a native of China, being an exceedingly rare tree in the mountain forests of Eastern Szechwan and Western Hupeh, where it was discovered by Père Farges and by myself.   *Ostrya mandschurica*, Budischtschew,[1] recorded from Manchuria, is probably identical with this species.

The Japanese Hop Hornbeam was introduced in 1888 into the Arnold Arboretum by seed sent from Japan by Dr. Mayr, and has proved hardy in the climate of Eastern Massachusetts.   There are two trees at Kew, sent by Prof. Sargent in 1897, which are now about 15 feet high and growing vigorously.   There is also a healthy young tree at Grayswood, Haslemere.   (A. H.)

---

[1] In Trautvetter, *Act. Hort. Petrop.* ix. 166 (1884).   I have seen no specimens of this.

NOTHOFAGUS

Nothofagus, Blume, *Mus. Bot. Lugd. Bat.* i. 307 (1850); Oerstedt, *Vidensk. Selsk. Skrift.* V. ix. 331
   (1873); Solereder, *System. Werth Holzstructur*, 253 (1885); Krasser, *Ann. K.-K. Naturhist.
   Hofmuseums, Wien*, xi. 149 (1896).
Calucechinus and *Calusparassus*, Hombron et Jacquinot, *Voy. Pôle. Sud.* Atlas, tt. 6-8 (1853).
Lophozonia, Turczaninow, *Bull. Soc. Imp. Nat. Mosc.* xxxi. 396 (1858).
Fagus, section *Nothofagus*, Bentham et Hooker, *Gen. Pl.* iii. 410 (1880).

THIS genus comprises the beeches inhabiting extra-tropical South America, Australia,
Tasmania, and New Zealand, and was formerly considered to be a section of the
genus Fagus, which, however, as now limited, includes only the species of the
northern hemisphere. The two genera are distinguished as follows :—

NOTHOFAGUS.—Trees or shrubs, with deciduous or evergreen foliage.[1] Flowers
monœcious or rarely diœcious, either solitary or in groups of threes. Fruit : involucre,
two-, three- or four-valved, usually bearing externally transverse entire, toothed or
lobed lamellæ, with or without gland-tipped processes; or in rare cases the valves
are smooth and without appendages; nuts, solitary or three in each involucre.

FAGUS.—Trees with deciduous foliage. Flowers monœcious; the staminate
numerous in globose heads, the pistillate in pairs. Fruit: involucre, covered
externally with bristly, deltoid or foliaceous processes; nuts, two in each involucre.

About seventeen[2] distinct species of Nothofagus are known, constituting three
natural sections, based on the characters of the foliage :—

I. Leaves deciduous, soft in texture, folded in bud along the lateral nerves,
crenate or serrate in margin.

1. *Nothofagus antarctica*, Oerstedt. Large tree, S. America. Introduced into
cultivation. Leaves ovate, $\frac{3}{4}$ to 1 inch long; lateral nerves three to five pairs;
margin slightly lobed, unequally crenate, with three to five teeth between the ends
of each adjacent pair of nerves.

2. *Nothofagus Montagnei*,[3] Reiche. Shrub or low tree. Chonos Archipelago.
Not introduced. A little-known species, of which I have seen no specimen; leaves
$\frac{1}{2}$ inch long, firmer in texture and more conspicuously veined above than those of
the preceding species, from which it is also distinguished by the yellow-coloured
pubescence on the branchlets.

---

[1] Bunbury, in *Bot. Fragments*, 322 (1883), writes an interesting article on the different types of foliage which are met
with in this genus.

[2] *N. alpina*, Reiche (*Fagus alpina*, Poeppig et Endlicher), is a doubtful species.

[3] *Calucechinus Montagnei*, Hombron et Jacquinot, *loc. cit.* t. 7 (1853). *Fagus Montagnei*, Philippi, *Linnæa*, xxix. 45
(1857).

3. *Nothofagus pumilio*,[1] Krasser.   Shrub or low tree, S. America.   Not introduced.   Leaves ovate, 1 to 1¼ inch long; lateral nerves five to seven pairs; regularly bicrenate in margin, with two teeth between the ends of each adjacent pair of nerves.   Young branchlets clothed with dense yellow pubescence.

4. *Nothofagus obliqua*, Blume.   Large tree, S. America.   Introduced.   Leaves, 1 to 2½ inches long, ovate-oblong, glaucous beneath; lateral nerves seven to eleven pairs; lobulate and serrate in margin.

5. *Nothofagus procera*,[2] Oerstedt.   Large tree, S. America.   Not introduced.   Leaves, 2 to 3½ inches long, oblong; lateral nerves fifteen to sixteen pairs; doubly serrate in margin.

6. *Nothofagus Gunnii*,[3] Oerstedt.   Shrub, Tasmania.   Not introduced.   Leaves, ½ inch long, ovate; lateral nerves six pairs, prominent beneath; margin regularly crenate, with one tooth between the ends of each adjacent pair of nerves.

II. Leaves evergreen, margin not entire.   In this section, the leaves are usually very coriaceous, and glabrous, or with only slight pubescence.

A. *Leaves minutely crenulate in margin.*

7. *Nothofagus apiculata*,[4] Krasser.   Tree attaining 40 feet in height, New Zealand.   Not introduced.   Leaves 1 inch long, elliptical, conspicuously mucronate at the apex.

B. *Leaves crenate in margin.*

8. *Nothofagus Menziesii*, Oerstedt.   Large tree, New Zealand.   Introduced.   Leaves, ½ inch long, ovate, rhomboid or orbicular, doubly crenate, with two pubescent pits on each side of the midrib near the base on the lower surface.

9. *Nothofagus Cunninghami*, Oerstedt.   Large tree, Australia, Tasmania.   Introduced.   Leaves, ½ inch long, ovate, deltoid or rhomboid, simply crenate; lower surface without pits or resinous papillæ.

10. *Nothofagus betuloides*, Blume.   Large tree, S. America.   Introduced.   Leaves, ¾ to 1 inch long, ovate, crenate or obtusely dentate; lower surface dotted with resinous papillæ.

C. *Leaves serrate in margin.*

11. *Nothofagus Dombeyi*,[5] Blume.   Large tree, S. America.   Not introduced.   Leaves, 1 inch long, lanceolate; nerves six pairs; lower surface dotted with resinous papillæ.

12. *Nothofagus nitida*,[6] Krasser.   Large tree, S. America.   Not introduced.   Leaves, 1 to 1½ inch long, trapezoid-ovate, acuminate; nerves six pairs; lower surface not dotted.

13. *Nothofagus fusca*, Oerstedt.   Large tree, New Zealand.   Introduced.

---

[1] *Fagus pumilio*, Poeppig et Endlicher, *Nov. Gen. et Sp.* ii. 68, t. 195 (1835).   *Fagus antarctica*, var. *bicrenata*, DC. *Prod.* xvi. 2, p. 120 (1864).

[2] *Fagus procera*, Poeppig et Endlicher, *Nov. Gen. et Sp.* ii. 69, t. 197 (1835).

[3] *Fagus Gunnii*, Hooker, *Icon. Plant.* t. 881 (1852).

[4] *Fagus apiculata*, Colenso, *Trans. N. Zeal. Instit.* xvi. 335 (1884).

[5] *Fagus Dombeyi*, Mirbel, *Mém. Mus. Paris.* xiv. 467, t. 24 (1827).

[6] *Fagus nitida*, Philippi, *Linnæa*, xxix. 44 (1857).

Leaves, 1 to 1½ inch long, thin in texture, ovate, rounded at the apex; serrations large, few, irregular; nerves four to seven pairs.

14. *Nothofagus Moorei*, Krasser. Large tree, Australia. Introduced. Leaves, 2 to 3 inches long, ovate-lanceolate, acuminate; sharply and finely serrate; nerves ten to fifteen pairs.

III. Leaves evergreen, entire in margin. In this section, the leaves on young trees are glabrous; but on older trees they become densely tomentose on the under surface. They resemble considerably the leaves of certain species of *Vaccinium*.

15. *Nothofagus cliffortioides*, Oerstedt. Large tree, New Zealand. Introduced. Leaves, ¼ to ½ inch long, ovate, rounded at the base; tomentum whitish.

16. *Nothofagus Solandri*,[1] Oerstedt. Large tree, New Zealand. Not introduced. Leaves ¼ to ½ inch long, oblong, cuneate at the base; tomentum whitish.

17. *Nothofagus Blairii*,[2] Krasser. Large tree, New Zealand. Not introduced. Leaves ¾ inch long, ovate, rounded at the base, apiculate at the apex; tomentum yellowish. (A. H.)

NOTHOFAGUS CLIFFORTIOIDES

Nothofagus cliffortioides, Oerstedt, *Vidensk. Selsk. Skrift.* V. ix. 355 (1873).
Fagus cliffortioides, J. D. Hooker, in *Hook. Icon. Plant.* tt. 673 (1844) and 816 B (1852), *Flora New Zealand*, i. 230 (1854), and *Handb. New Zealand Flora*, 250 (1864); Kirk, *Forest Flora New Zealand*, 201, tt. 101, 101 A (1889); Cheeseman, *New Zealand Flora*, 643 (1906).

An evergreen tree, attaining, in New Zealand, about 50 feet in height and 6 feet in girth. Young branchlets pubescent; buds minute, ovoid, shining, brown. Leaves (Plate 202, Fig. 4) persistent for two or three years, distichous and crowded on the branchlets, coriaceous, minute, ¼ to ½ inch in length, entire in margin; on young plants ovate, rounded at both base and apex, green and glabrous on both surfaces; on adult trees, ovate or ovate-oblong, rounded and unequal at the base, subacute at the apex, minutely punctate above, greyish-white with dense appressed pubescence beneath; petioles short and pubescent. Male flowers solitary; stamens eight to twelve. Fruit: involucre ⅛ to ¼ inch long, three-lobed, each lobe with two or three entire transverse lamellæ; nuts one to three, winged, one or two triquetrous, the third flattened.

This tree is known in New Zealand as the "mountain birch," and is confined to mountainous regions except in the south-western corner of the South Island, where it descends to sea-level. It is not found in the northern part of the North Island; but elsewhere is very common in the forests[3] between 2000 and 4000 feet elevation,

---

[1] *Fagus Solandri*, Hooker, *Icon. Plant.* t. 639 (1844).
[2] *Fagus Blairii*, Kirk, *Trans. N. Zeal. Inst.* xvii. 297, 306 (1885).
[3] A view of a forest of this species in the South Island at 3000 feet, showing a dense undergrowth of young beech and tall smooth stems of older trees, is given in Schimper, *Plant Geography*, 760, f. 460 (1904).

often forming the timber line, when it becomes a mere bush. The wood, though of no great size, is used for telegraph-poles, fencing-posts, railway sleepers, and wharf-piles, the heartwood being very durable in situations where it is exposed to alternations of dryness and moisture.

N. cliffortioides is extremely rare in cultivation. There are two specimens at Enys, Cornwall, which, according to Mr. John D. Enys, were 35 feet and 28½ feet high respectively in 1905; but when Elwes saw them in that year they were very slender and not thriving. These trees are semi-deciduous, most of the leaves, after turning brilliant red in autumn, falling off during winter; whereas, in New Zealand, the foliage is strictly evergreen. Another tree is growing at Messrs. Veitch's nursery at Coombe Wood, where it stands out of doors without any protection. It is very slow, however, in growth, and is only about 12 feet in height.   (A. H.)

NOTHOFAGUS MENZIESII

Nothofagus Menziesii, Oerstedt, *Vidensk. Selsk. Skrift.* V. ix. 355 (1873).
Fagus Menziesii, J. D. Hooker, *Icon. Plant.* t. 652 (1844), and *Flora New Zealand*, i. 229 (1854); Kirk, *Forest Flora New Zealand*, 175, t. 89 (1889); Cheeseman, *New Zealand Flora*, 640 (1906).

An evergreen tree, attaining in New Zealand 100 feet in height and 15 to 25 feet in girth. Bark silvery-white, resembling that of the common English birch. Young branchlets covered with dense erect brown pubescence. Leaves (Plate 202, Fig. 9) persistent for two or three years, distichous on the branchlets, about ½ inch long, coriaceous, deltoid, ovate or rhombic; cuneate at the base, obtuse at the apex; glabrous; upper surface dark-green, shining; lower surface pale-green, with usually two (occasionally only one or none) small pits fringed with brownish hairs near the base of the midrib; lateral veins about three pairs; margin irregularly and doubly crenate; petioles short, pubescent. Male flowers solitary; calyx four- to six-lobed; stamens six to twelve. Fruit: involucre ¼ to ⅓ inch long, cleft into four narrow lobes, each with five transverse scales, cut to the base into recurved linear gland-tipped processes; nuts three, one two-winged, two three-winged, the wings produced upwards into sharp points.

This species, which is known in New Zealand as the "silver birch" or "red birch," is common in the mountain forests of both the North and South Islands, ascending from sea-level to 3500 feet. The wood is dark-red, strong, and compact, and being easily worked, is suitable for making furniture.

A small tree of this species is growing in the Temperate House at Kew; and we are not aware that it has ever been tried in the open air. The tree has handsome foliage, and should be hardy in Cornwall and the south of Ireland.   (A. H.)

NOTHOFAGUS FUSCA

Nothofagus fusca, Oerstedt, *Vidensk. Selsk. Skrift.* V. ix. 355 (1873).
Fagus fusca, J. D. Hooker, *Icon. Plant.* tt. 630, 631 (1844), and *Flora New Zealand*, i. 229 (1854);
  Kirk, *Forest Flora New Zealand*, 179, t. 90 (1889); Cheeseman, *New Zealand Flora*, 641
  (1906).

An evergeen tree, attaining in New Zealand 100 feet in height and 25 feet in girth. Bark of young trees white and smooth, becoming on old trees furrowed longitudinally and brown in colour. Young branchlets minutely pubescent. Leaves (Plate 202, Fig. 8) persistent for two years, distichous on the branchlets, $\frac{3}{4}$ to $1\frac{1}{2}$ inch long, thin in texture, glabrous, broadly ovate, cuneate at the base, rounded at the apex, dark-green and shining above, pale-green beneath; sharply serrate with a few large teeth in the upper two-thirds of the leaf; lateral nerves four to five pairs; petioles about $\frac{1}{8}$ inch, pubescent. Male flowers solitary or in threes; calyx five-toothed; stamens eight to sixteen. Fruit: involucre nearly $\frac{1}{2}$ inch long, viscid-pubescent, four-lobed, each lobe with three to five transverse entire or fringed scales; nuts three, as in *N. Menziesii*.

This species is a native of New Zealand, where it is known as the "black birch" or "red birch." It grows in forests at elevations between sea-level and 3500 feet, being a splendid tree. Its distribution is North Island from Monguni and Kartaia southwards, but local to the north of the East Cape; South Island from Nelson to Foveaux Straits, but rare in Canterbury and Eastern Otago. The wood is dark-red, strong and compact, and more durable than that of the other species; it is frequently used for wharves, bridges, and fencing posts.

A small tree of this species about 10 feet high, and said to be thirty years old, is growing in the Coombe Wood Nursery. At Castlewellan, Co. Down, there is a tree[1] about 18 feet high, which was imported from New Zealand some years ago as a small plant in a Wardian case. It is growing rapidly, making an annual growth of a foot. The old leaves remain on the branches till the new ones appear, changing before they fall to a brilliant red, which contrasts well with the light green of the young growths. (A. H.)

NOTHOFAGUS MOOREI

Nothofagus Moorei, Krasser, *Ann. K.-K. Naturhist. Hofmuseums, Wien*, xi. 163 (1896).
Fagus Moorei, F. v. Mueller, *Fragm. Phyt. Austral.* v. 109 (1865); Bentham, *Fl. Australiensis*, vi.
  211 (1873).

An evergreen tree, attaining in Australia 150 feet in height. Young branchlets pubescent; buds ovoid, acute, brown. Leaves (Plate 202, Fig. 7) persistent for two or three years, distichous on the branchlets, coriaceous, glabrous; shining, dark-

---

[1] This is figured as *Fagus cliffortioides* in Earl Annesley's *Beautiful and Rare Trees*, 71 (1903).

green above; pale-green beneath; ovate or ovate-lanceolate; 2 to 3 inches long on barren branchlets, about 1 inch long on flowering shoots; unequally cuneate at the base, acuminate at the apex, finely and sharply serrate; lateral nerves 8 to 12 pairs, prominent on the upper surface; petioles very short, pubescent. Flowers unknown. Fruit: involucre about ⅝ inch long, four-lobed; lobes lanceolate, acute with pubescent scales terminating in glandular processes; nuts three, two three-winged, the other two-winged.

This tree was discovered by Mr. C. Moore, Curator of the Botanic Gardens, Sydney, in New South Wales, and is the only southern beech occurring in a subtropical region. It forms dense forests at the head of Bellinger River and Bealsdown Creek, at about 4000 feet altitude; and a few trees have also been seen near the source of the Macleay River.

It was introduced into cultivation at Kew about fifteen years ago, and there is a small tree now growing there in the Temperate House. The only specimen living in the open air, so far as we know, is growing in the garden of Mr. Thomas Acton at Kilmacurragh, Co. Wicklow. It was 18 feet high in 1906, and had bark resembling that of *Prunus avium*. (A. H.)

NOTHOFAGUS CUNNINGHAMI

Nothofagus Cunninghami, Oerstedt, *Vidensk. Selsk. Skrift*, V. ix. 355 (1873).
Fagus Cunninghami, J. D. Hooker, *Journ. Bot.* ii. 152, t. 7 (1840), and *Flora Tasmaniæ*, i. 346 (1860); F. v. Mueller, *Fragm. Phyt. Austral.* v. 110 (1865); Bentham, *Flora Australiensis*, vi. 210 (1873); Rodway, *Tasmanian Flora*, 182 (1903).

An evergreen tree, said to attain in Tasmania 200 feet in height and 40 feet in girth. Bark, as seen in cultivated trees, roughened by small scales and fissuring longitudinally. Young branchlets densely and minutely pubescent. Buds conical, sharp-pointed and curved at the apex, shining, brown. Leaves (Plate 202, Fig. 5) persistent for two or three years, distichous and crowded on the branchlets, coriaceous, about ½ inch in length; broadly ovate, deltoid or rhombic; cuneate or cordate at the base, acute at the apex; unequally crenate in margin; both surfaces glabrous, veins inconspicuous and scarcely prominent beneath; petioles short, pubescent. Male flowers solitary; stamens eight. Fruit: involucre ¼ inch long, four-lobed, each lobe with five or six rows of dorsal transverse scales, split up into gland-tipped processes; nuts usually three, two lateral triquetrous and three-winged, the other flattened and two-winged.

This species is very common in Tasmania, where it is known as the "Tasmanian myrtle," and forms a large proportion of the forests in the mountainous and western humid districts. It ascends to 4000 feet, becoming at this elevation a mere shrub, a few feet in height. It also occurs on the mainland of Australia, in Victoria, in a few scattered localities, being most common according to F. v. Mueller on the Bawbaw mountains, and less common at Dandenong, Mount Juliet, Wilson's Promontory,

La Trobe, Tyre's and Thomson's Rivers, and in a few places in the cooler and moister parts of Gipp's Land. The wood[1] apparently varies: one kind, "red myrtle," being of a bright pink colour, with a grain like that of the English beech, and considered to be suitable for cabinetwork; another kind, "white myrtle," brownish-grey in tint, is not so attractive in appearance.

N. Cunninghami is very rare in cultivation. The finest tree (Plate 154), said by Lord Barrymore to have been planted about fifty years ago, is growing at Fota, Co. Cork, and measured, in 1904, 48 feet high by 3 feet 3 inches in girth. This tree has numerous branches, many of them ascending from near the base of the trunk. A tree at Kilmacurragh, Co. Wicklow, with branches ascending and curving at the tips, was 40 feet high by 3 feet 4 inches in 1906. This tree has excrescences on the trunk, similar to the so-called "wood-balls," which are often seen on the common beech. It flowered in 1906. At Osborne, Isle of Wight, there is a tree, 30 feet by 2 feet 2 inches, which when Elwes saw it in 1906 seemed thriving.

It seems to be as hardy as any of the genus, and might be planted with good prospects of success in the extreme south-west of England near the sea.

(A. H.)

NOTHOFAGUS BETULOIDES

Nothofagus betuloides, Blume, *Mus. Bot. Lugd. Bat.* i. 307 (1850); Reiche, *Chil. Buch.* 15 (1897); Wildeman, *Voy. Belgica*, 74 (1905); Macloskie, *Princeton Univ. Expedit. Patagonia, Botany*, 329 (1903-6).

Fagus betuloides, Mirbel, *Mém. Mus. Paris*, xiv. 465, t. 4 (1827); Loudon, *Arb. et Frut. Brit*. iii. 1982 (1838); Hooker, *Journ. Bot.* ii. 155 (1840), and *Fl. Antarct.* ii. 349, t. 124 (1847).

Fagus Forsteri, Hooker, *Journ. Bot.* ii. 156, t. 8 (1840).

Betula antarctica, Forster, *Comm. Goett.* ix. 45 (1789).

Calusparassus betuloides and *C. Forsteri*, Hombron et Jacquinot, *Voy. Pôle Sud*, Atlas t. 7 (1853).

A large evergreen tree. Bark smooth, grey; scaling near the base in old trees. Young branchlets slender, viscid, covered with short pubescence. Buds minute, brown, ovoid. Leaves (Plate 202, Fig. 3) persistent for two or three years, distichous and crowded on the branchlets, rigid, coriaceous, $\frac{3}{4}$ to 1 inch long by $\frac{1}{2}$ inch or slightly more in breadth, ovate, rounded at the base, acute at the apex, crenate or serrate in margin; upper surface shining, glabrous, often viscid; lower surface finely reticulate, glabrous, dotted with resinous papillæ; petioles short. Male flowers solitary; calyx funnel-shaped, four- to seven-lobed; stamens ten to sixteen, with long and slender filaments. Fruit: involucre four-lobed, with erect filiform glandular processes.

According to Loudon, both *N. betuloides* and *N. antarctica* were introduced in 1830, but he had not seen a specimen of either. Sir W. J. Hooker[2] states that healthy young trees of both species, the first, as far as he knew, that ever had reached Europe, were sent in Wardian cases to Kew from Cape Horn in 1843, being

---

[1] Report on Tasmanian Timbers by Mr. R. A. Ransome, of the Stanley Works, Chelsea, in *Kew Bull.* 1889, pp. 114, 115.
[2] *Notes Bot. Antarctic Voyage*, 64 (1843); cf. also Loudon, *Gard. Mag.* 1843, p. 442.

a consignment from his son, who was attached as botanist to the Antarctic expedition of the *Erebus* and *Terror*. (A. H.)

The largest and finest specimen which I have seen in cultivation is at Bicton (Plate 155). It measured in 1906 about 50 feet in height and 6½ feet in girth; but bore no fruit on two occasions when I saw it. At Pencarrow, Cornwall, a tree,[1] reported by Mr. Bartlett to have been obtained from Messrs. Veitch in 1847, was 36 feet by 4 feet ·3 inches in 1903. One at Coldrinick in the same county was measured in 1905 by Mr. Bartlett, who gives its dimensions as 45 feet by 5 feet 5 inches. There is another specimen,[2] about sixty years old, growing on Sir John Llewellyn's property at Caswell Bay, near Swansea, which he tells me measured 25 feet by 3 feet 2 inches in 1907. It is close to the sea and in consequence has been shorn off by the sea wind to the same height as the Portugal laurels and poplars which grow beside it.

At Grayswood, Haslemere, at 600 feet elevation, a tree, said by Mr. B. C. Chambers to have been planted in 1882, measured 34 feet by 2 feet 3 inches in 1906. At Hafodunos, Denbighshire, a tree, reported by Col. Sandbach to have been planted in 1855, was in 1904 36 feet high by 5 feet 2 inches in girth at 3 feet from the ground, dividing at 5 feet up into two stems. There is no tree growing now at Kew, one, a healthy specimen, having been killed[3] by frost in January 1867. There is also a tree growing at Ashridge Park, Herts, which is about 30 feet by 3 feet, on which Miss Woolward has observed fruit, and a smaller one is in the Knap Hill Nursery, near Woking. A tree at Powerscourt, Co. Wicklow, was, in 1906, 33 feet high by 2 feet 10 inches in girth.

I have a sample board from a tree of this species, which grew on the rockery at Lucombe and Pince's Nursery, Exeter. This was cut down when the nursery was cleared for building in March 1903. In 1886, it was 35 feet high and 2 feet 8 inches in girth at 3 feet from the ground,[4] and had not grown much in the succeeding years. The timber was of poor quality, and had begun to decay.

(H. J. E.)

NOTHOFAGUS OBLIQUA

Nothofagus obliqua, Blume, *Mus. Bot. Lugd. Bat.* i. 307 (1850); Reiche, *Chil. Buch.* 8 (1897); Wildeman, *Voy. Belgica*, 75 (1905).

Fagus obliqua, Mirbel, *Mém. Mus. Paris*, xiv. 465, t. 23 (1827); Hooker, *Journ. Bot.* ii. 153 (1840).

Fagus glauca, Philippi, *Linnæa*, xxix. 43 (1857).

Lophozonia heterocarpa, Turczaninow, *Bull. Soc. Imp. Nat. Mosc.* xxxi. 396 (1858).

A deciduous tree, attaining in Chile a height of over 100 feet. Bark, according to Reiche, dark in colour and fissured. Young branchlets glabrous; buds small,

---

[1] Figured in *Gard. Chron.* xxxiii. 10, f. 5 (1903), where it is erroneously stated to have been introduced from New Zealand.

[2] Figured in *Gard. Chron.* 1872, p. 466, f. 136, and 1886, xxv. 104, f. 18.

[3] J. Smith, *Records of Kew Gardens*, 277 (1880).   [4] *Gard. Chron.* xxv. 104 (1886).

conical, sharp-pointed, glabrous, brown, few-scaled, and appressed to the branchlets. Leaves (Plate 202, Fig. 2) variable in size, 1 to 2½ inches in length, ½ to 1 inch in breadth, thin in texture, ovate-oblong, unequal at the rounded or cuneate base, sub-acute or obtuse at the apex; dark-green above, very pale beneath, both surfaces glabrous except for slight pubescence on the midrib and nerves; margin shallowly lobulate in the lower half, the lobules and upper part of the leaf serrate with minute triangular acute teeth; nerves 8 to 11 pairs, prominent on the lower surface, running obliquely to the margin; petiole ⅛ to ¼ inch long. Male flowers solitary; calyx irregularly lobed, stamens thirty to forty. Fruit: involucre four-valved, valves pubescent on the back with lobed appendages bearing stalked glands; nuts three, two trigonous and three-winged, one flattened and two-winged.

This species is very variable, especially as regards the size and pubescence of the leaf, and De Candolle[1] distinguished three varieties:—Var. *valdiviana*: leaves small, glabrous, with cuneate base; var. *macranthera*: stamens long, leaves pubescent beneath; and var. *macrocarpa*, with the nuts longer than the valves of the involucre.

N. obliqua was introduced[2] into England by Lobb in 1849, and in the following year it was said to have been growing freely in the open air in Messrs. Veitch's nursery at Exeter. None of the original plants appear, however, to have survived.

Plants raised from seed, brought from Chile by Elwes in 1902, have grown with great vigour at Kew,[3] being now about 8 feet in height. At Monreith, Sir Herbert Maxwell, who received a plant from Kew, reports that it has borne without injury 20° of frost, and may be assumed to be perfectly hardy. In Lord Ducie's garden at Tortworth, this tree has grown with astonishing vigour, being now 12 feet high and 8 inches in girth; it endured the severe frost of May 1905 without any apparent injury. The seedlings which were raised at Colesborne, however, never throve, and died before attaining any size, which is possibly due to the presence of lime in the soil.   (A. H.)

NOTHOFAGUS ANTARCTICA

Nothofagus antarctica, Oerstedt, *Vidensk. Selsk. Skrift.* V. ix. 354 (1873); Reiche, *Chil. Buch.* 11 (1897); Wildeman, *Voy. Belgica*, 73 (1904); Macloskie, *Princeton Univ. Exped. Patagonia*, Botany, 326 (1903-1906).

Fagus antarctica, Foster, *Comm. Goett.* ix. 24 (1789); Loudon, *Arb. et Frut. Brit.* iii. 1982 (1838); Hooker, *Journ. Bot.* ii. 149, t. 6 (1840), and *Fl. Antarct.* ii. 345, t. 123 (1847).

Calucechinus antarctica, Hombron et Jacquinot, *Voy. Pôle Sud*, Atlas, tt. 6, 7 (1853).

A deciduous tree, attaining in Terra del Fuego at low elevations a very large size. Young branchlets covered with dense erect pubescence, persistent in the second year. Buds, ⅛ inch long, ovoid, slightly compressed, glabrous, few-scaled. Leaves (Plate 202, Fig. 1) ½ to 1 inch long, crumpled and uneven in surface, oblong-

---

[1] *Prod.* xvi. 2, p. 119 (1864).

[2] *Gard. Chron.* 1849, p. 563; Lindley, *Journ. Hort. Soc.* vi. 265 (1851); Lindley and Paxton, *Flower Garden*, ii. 166 (1852).   [3] Cf. *Kew Bull.* 1906, p. 379.

ovate, unequal and usually cordate at the base, rounded at the apex, variable in pubescence, dark-green above, light-green beneath; nerves usually four pairs; margin with three or four pairs of shallow lobes, which are minutely and irregularly dentate, the teeth being rounded or acute; petiole ⅛ to ¼ inch, pubescent. Male flowers solitary; calyx five-partite; stamens ten, as long as the calyx. Fruit: involucre four-partite, each lobe with three to four transverse reddish scales; nuts three, the central one two-winged, the lateral pair three-winged.

Two distinct forms occur :—

1. Var. *sublobata*, DC.[1]—Petiole and upper surface of the leaf glabrescent; lower surface glabrous except on the nerves, which are clothed with long appressed hairs.

2. Var. *uliginosa*, DC.[1]—Leaves pubescent on both surfaces with minute erect hairs.

This species was introduced in 1843, as mentioned in our account of *N. betuloides*, but it is doubtful if any of the original plants are still living. The only specimen which we have discovered is a bushy tree, about 15 feet high, which is growing alongside a fine tree of *N. betuloides* at Hafodunos, Denbighshire. Colonel Sandbach believes it to be about thirty years old.

Plants raised from seed, collected by Elwes in Chile in 1902, are now in cultivation at Kew,[2] in a peat-bed, and have attained about 6 feet in height. They are vigorous in growth, and have passed through the severe frosts of 1906-1907 without injury, and look as if they might grow to be trees of considerable size.   (A. H.)

DISTRIBUTION OF THE SOUTH AMERICAN BEECHES

In extra-tropical South America, the beeches are the dominant trees, extending from a point on the west coast of Chile about lat. 33°, southward to Patagonia and Tierra del Fuego, and crossing the Andes into Argentina. The best account of the Chilean beeches is given by Reiche,[3] from whom, supplemented by my own observations in 1901-1902, I take the following particulars.

The most northerly species is *N. obliqua*, which extends on the coast up to about lat. 33°, but in the extreme north does not form forest except in the interior valleys. About lat. 35° it is the principal tree in the forest which formerly clothed the lower slopes of the Andes, but which is now fast vanishing before the attacks of man. The tree is called Roble Pellin by the Spaniards, and grows to a large size with a tall straight trunk, attaining a height of 120 feet or more, and a girth of 20 to 30 feet. In the forest country, which commences south of the Maule River, it is mixed with *N. Dombeyi*; and these two species form the principal timber supplies of Chile, and are largely cut for house-building, railway sleepers, and other purposes. Some cargoes of this timber have lately been imported into England, under the name of Chilean Oak; and by the courtesy of the Great Western Railway Company, I have received one of these sleepers, which has a dense reddish wood, not at all resembling

---

[1] *Prod.* xvi. 2, 120 (1864). De Candolle's var. *bicrenata* is *Nothofagus pumilio*.

[2] Cf. *Kew Bull.* 1906, p. 381.

[3] *Beiträge Kenntniss Chilen. Buchen* (Valparaiso, 1897).

that of the European beech, and apparently suitable to take the place of the lower grades of mahogany in the manufacture of furniture. It cracks, however, badly in drying, and will require very careful seasoning. In the forests of Chillan (lat. 36°), *N. obliqua* grows up to 4000 or 5000 feet, being replaced at higher elevations by *N. antarctica* and *N. pumilio*; and reaches its southern limit in the region of Lake Llanquihue. A photograph by Mr. Bartlett Calvert, who accompanied me on my journey, shows the appearance of this tree in the forest at about 3000 feet, near the source of the Renaico River (Plate 156). Here the undergrowth is usually composed of a dense thicket of the Chilean bamboo (*Chusquea sp.*); and in the more open places the ground is often carpeted with a dense bed of *Alstrœmeria aurantiaca*, whose brilliant orange flowers produce a most lovely effect. In the wetter places it is associated with *Drimys Winteri* and many beautiful shrubs and herbaceous plants, of which *Eucryphia pinnatifida*, *Embothrium coccineum*, *Tropæolum speciosum*, and several species of Fuchsia and Calceolaria are the choicest ornaments of our gardens in the warmer and damper parts of the south-west of England and Ireland. In many parts of the Chilean forests it is often covered with a lovely parasitic plant, *Myzodendron linearifolium*, DC.,[1] which hangs in silvery masses from the branches.

Nothofagus Dombeyi is known to the Chileans by its Indian name of *Coigue*, and is a large and common tree in Chile. It is usually associated with *N. obliqua*, but does not extend so far to the northward, not being found to the north of the Maule river. It is widely spread in Araucania, Valdivia, and Llanquihue, and occurs also on the Argentine side of the frontier. It grows on the island of Chiloe, and has been collected on the river Aysen (lat. 45°); but its extreme southern limit is not accurately known.

Nothofagus nitida, which has been much confused with *N. Dombeyi*, is a common forest tree in the coast mountains of Valdivia, and grows on Chiloe and the Guaitecas Islands. The distribution of this species has not yet been satisfactorily determined.

Nothofagus procera, known as *Rauli*, is less common than *N. obliqua*, to which it is allied, and usually grows scattered in the forest. Its northerly limit lies between 35° and 36° lat., and it does not occur farther south than the province of Valdivia, where it becomes a stately tree. It does not cross the frontier into Argentina.

Nothofagus antarctica is widely distributed, extending from about lat. 38° to Tierra del Fuego. It is the commonest species which I found on my tour at high elevations, both on the Chilean and Argentine sides of the frontier. It is associated with Araucaria at 4000 feet, and is common also in the plain of Valdivia in marshy situations. In the mountains around the great lake of Nahuelhuapi, the leaves of this species had already assumed their autumnal tint in February. *N. antarctica* and *N. betuloides* are the dominant trees in Patagonia and Tierra del Fuego; and

---

[1] I found this species in the low country about Temuco in Chile, and also on the Argentine side of the frontier in two or three localities. Two other species also occur :—*M. oblongifolium*, DC., which I found on *Nothofagus antarctica*, near the baths of Chillan at 5000 to 7000 feet elevation ; and *M. punctulatum*, DC., which I gathered on *Nothofagus Dombeyi* at Lake Meliquina, and in the dense evergreen forest which skirts the glaciers of the great Tronador mountain at 2000 feet in lat. 40°.

according to Dusén,[1] their distribution is regulated by the amount of rainfall.   In the western parts of Tierra del Fuego, where the rainfall is heavy, the coast forest is evergreen and is mainly composed of *N. betuloides* and *Drimys Winteri*; and *N. antarctica* is only met with in the mountains.   In the eastern part of Tierra del Fuego, where the rainfall is slight, the latter species descends to sea-level and grows in mixture with *N. betuloides* and *N. pumilio*.

In Western Patagonia, the evergreen forest predominates in the Archipelago and on the western side of the mountain range, where much rain falls and the prevailing winds are south-westerly; whereas, on the eastern slopes of the mountains, where the climate is comparatively dry, the forests are composed of deciduous trees. According to Dusén, the deciduous-leaved forest is well seen at a point 30 miles up the River Aysen.   In the inland region the ground is covered by a thin park-like forest, which is almost exclusively composed of one species, *N. antarctica*.   This tree does not grow in such close masses as the European beech, and, owing to the absence of dense shade, there is a luxuriant undergrowth of herbs and shrubs. These park-like forests prevail up to 2300 feet.   Above this elevation steppes occur, which are studded with small groves of *N. pumilio*, the ground being covered with mosses.   At 3000 feet *N. pumilio* is only a low tree, which gradually becomes smaller as it ascends, until at 4300 feet it forms a stunted forest of dwarfed trees, with their branches interlaced together.

An earlier account of the Antarctic beeches is given by Sir J. D. Hooker,[2] who states that *N. antarctica* strongly resembles the European beech in its deciduous leaves, form of trunk, and smooth bark.   It ascends much higher at Cape Horn than *N. betuloides*, and is much the larger tree of the two when it is found growing at sea-level.   *N. betuloides*, however, grows to a very large size about the Straits of Magellan, and being evergreen, is a marked feature of the scenery in winter, as its upper limit is sharply defined, and contrasts with the dazzling snow that covers the matted and naked branches of *N. antarctica*.   Captain King[3] observed many trees of *N. betuloides* 3 to 4 feet in diameter, one being as large as 7 feet.   He describes the wood as heavy and far too brittle for masts or even boat-hooks, but cutting up into tolerable planks.   Hooker considered the timber of the deciduous species to be superior.

N. betuloides, while much commoner in the south, extends along the coast range as far north as Valdivia.[4]   It is replaced in the Guaitecas Islands by *N. Dombeyi* and *N. nitida*.[5]

Nothofagus pumilio has been much confused with *N. antarctica*, of which it was made var. *bicrenata* by De Candolle.   It is very distinct in both foliage and fruit.   It extends from Chillan and Nahuelbuta in Chile southward to the Straits of Magellan, and is usually a shrub, constituting the scrubby growth which prevails

[1] *Princeton Univ. Exped. Patagonia, Botany* 2, 10, 26 (1903-1906); and Engler, *Bot. Jahrbüch.* xxiv. 179 (1897).

[2] *Fl. Antarct.* ii. 345.

[3] *Voyage of the "Adventure" and "Beagle,"* i. 576 (1839).   Ball, *Notes of a Naturalist in S. America*, 225, says that *N. betuloides* "has a thick trunk, commonly three or four feet in diameter, but nowhere attains any great height.   Forty feet appeared to me the outside limit attained by any that I saw at Eden Harbour or elsewhere."

[4] Reiche, *loc. cit.*   [5] Dusén, *loc. cit.*

above timber-line in the mountains. I saw this species near the baths of Chillan, lat. 37°, where it grew as a bush at 6000 to 8000 feet on the crests of the ridges in volcanic soil. According to Reiche, it occurs as undergrowth in the Araucaria forests at 4000 to 5000 feet elevation; but in sheltered situations in the mountains of Valdivia and Llanquihue it occasionally becomes a tree 60 feet in height. (H. J. E.)

ARBUTUS

Arbutus, Linnæus, *Gen. Pl.* 123 (1737); Bentham et Hooker, *Gen. Pl.* ii. 581 (1876). *Unedo*, Hoffmannsegg et Link, *Fl. Port.* i. 415 (1809).

EVERGREEN trees or shrubs, belonging to the order Ericaceæ. Leaves simple, alternate, spirally arranged on the branchlets, coriaceous, persistent, stalked, pinnately-veined, entire or serrate, without stipules. Buds with spirally imbricated scales, within which the young leaves lie flat and are not rolled or folded. Flowers perfect, regular, in terminal compound racemes or panicles. Pedicel with two bracteoles, in the axil of an ovate bract; bracts and bracteoles scarious, persistent. Calyx five-lobed, free, persistent, unaltered at the base of the fruit. Corolla gamopetalous, hypogynous, urceolate or globose, with five obtuse, recurved, imbricated teeth. Stamens ten, included; filaments free, inserted on the base of the corolla, dilated and pilose at the base; anthers deflexed, dorsifixed, two-celled, opening by two pores, each anther with two awns on the back, against which insects knock in their search for honey and scatter the pollen through the pores. Pollen-grains united in tetrahedral masses of four grains each. Disc annular. Ovary superior, five- or occasionally four-celled; style columnar, stigmatose and obscurely five-lobed at the apex; ovules numerous. Fruit a berry or drupe, the endocarp often being imperfectly developed. Seeds numerous, small, angled, with a coriaceous testa and a horny albumen.

About twenty species are known, inhabiting the western and south-western parts of North America, Central America, Ireland, the countries in Europe bordering upon the Mediterranean, the Canary Islands, Morocco, Algeria, Asia Minor, the Crimea, and the Caucasus. Many of the species are only shrubs or very small trees, and others are not hardy or have not been introduced. Only four species[1] attaining a considerable size in cultivation in the open air in England, one of which is a hybrid, will be dealt with :—

A. *Leaves serrate. Young branchlets glandular-pubescent.*

1. *Arbutus Unedo*, Linnæus. Ireland, Southern Europe, Asia Minor, Morocco, and Algeria.

Leaves green beneath; petiole ¼ inch. Older branchlets dark brown, rough, and fissuring.

---

[1] *Arbutus canariensis*, Lamarck, growing in the open air, is five feet high at Mount Usher in Wicklow; but at Newry this species requires protection in winter.

Arbutus arizonica, Sargent, a native of the high mountains of Southern Arizona, if introduced, might be hardy.

2. *Arbutus hybrida*, Ker-Gawler. A hybrid.

Leaves slightly glaucous beneath; petiole ½ inch. Older branchlets fawn-coloured, smooth.

B. *Leaves entire. Young branchlets glabrous.*

3. *Arbutus Andrachne*, Linnæus. Albania, Greece, Asia Minor, Crimea, Caucasus.

Leaves slightly glaucous beneath, contracted into short broad points at the apex, tapering at the base in cultivated trees; petiole ½ inch.

4. *Arbutus Menziesii*, Pursh. Western N. America, from British Columbia to California.

Leaves glaucous, almost white, beneath; rounded or with a minute sharp point at the apex; sub-cordate or rounded at the base; petiole 1 inch.

ARBUTUS UNEDO, Strawberry Tree

Arbutus Unedo, Linnæus, *Sp. Pl.* 395 (1753); Loudon, *Arb. et. Frut. Brit.* ii. 1117 (1838); Boswell-Syme, *Eng. Bot.* vi. 28, t. 882 (1866); Hooker, *Stud. Fl. Brit. Islands*, 243 (1878); Mathieu, *Flore Forestière*, 225 (1897).
Unedo edulis, Hoffmannsegg et Link, *Fl. Port.* i. 415 (1809).

A small tree, attaining in Ireland 40 feet in height and 10 feet or more in girth, usually a shrub in the Mediterranean region. Bark rough, brownish-red, more or less fissured, and only rarely scaling off in part. Young branchlets reddish or green, covered with gland-tipped hairs, which persist in the second year; older branchlets brown, rough, slightly fissuring on the surface. Buds minute, reddish; scales imbricated, ovate, acute, ciliate. Leaves 2 to 4 inches long by 1 to 2 inches broad, very variable in shape, oblong, oblong-lanceolate, elliptic, or ovate, acute at the apex, tapering at the base; upper surface dark-green, glabrous and shining; lower surface pale-green, glabrous, with prominent midrib and inconspicuous pinnate-reticulate venation; margin serrate or biserrate, the serrations acute or rounded. Petioles short, about ¼ inch long, glandular-pubescent.

Flowers appearing in autumn, inodorous, in short drooping glabrous terminal panicles. Calyx-lobes minute, triangular. Corolla usually white, rarely pinkish, urceolate, with rounded ciliated teeth; ovary glabrous. Fruit ripening in the following autumn, at the same time as the appearance of the flowers of the succeeding year; a stalked berry, pendulous, sub-globose, ¾ inch in diameter, orange-scarlet, densely covered with minute pyramidal spine-like excrescences, edible, superficially resembling a strawberry, but entirely different in structure.

Seedling.—Cotyledons two, raised above ground on a short caulicle, oval, rounded at the apex, abruptly narrowed at the base into a flat petiole, entire, ¼ inch long, dull-green above, pale-green beneath. Young stem reddish, with short glandular hairs; primary leaves alternate, minute, oval or obovate, serrate and minutely glandular-pubescent in margin; tap-root about 2 or 3 inches long.

Varieties

In the wild state there is considerable variation in the size and shape of the leaves, dependent upon conditions of soil, shade, and climate. Fliche[1] describes two distinct forms in France. In the hot and dry region of the Esterel, the leaves are small in size, not exceeding 2½ inches in length by ¾ inch in breadth, and are very coriaceous, spathulate, with feebly serrated and revolute margins. In the forest of La Pinouse, near Quillan in Aude, which is mainly composed of *Pinus sylvestris* with a slight mixture of beech and silver fir, the climate being cool and the altitude considerable, the Arbutus has very large leaves, often 5 inches long by 2 inches broad, which are lanceolate with sharply serrate and non-revolute margins.

The following varieties are often cultivated :—

1. Var. *rubra*, Aiton, *Hort. Kew*, ii. 71 (1789). (Var. *Croomei*,[2] Hort.).—Flowers pink or reddish. Mackay[3] noticed a single plant of this variety, growing on red slate near Glengariff.

2. Var. *integerrima*, Sims, *Bot. Mag.* t. 2319 (1822) (vars. *integrifolia* and *rotundifolia*, Hort.).—Leaves entire and smaller than in the type. This is said to have been raised by Loddiges from seed of the ordinary form. The leaves vary in shape, often being obovate or almost orbicular.

3. Var. *quercifolia*, Hort.—Leaves obovate-lanceolate, with a few irregular teeth in the upper half, about 2 inches long by ¾ inch broad. In cultivation at Kew.

4. Var. *turbinata*, Persoon.—This variety occurs wild in Greece, and is remarkable for its large top-shaped fruit, more than an inch in length.

5. Other varieties have been noted, which I have not seen, as *salicifolia* with narrow leaves, *crispa* with crumpled leaves, and *plena*, with semi-double flowers.

Distribution

This species is widely spread throughout the maritime regions of the countries bordering on the Mediterranean, occurring in Spain, France, Corsica, Sardinia, Italy, Istria, Herzegovina, Dalmatia, Greece, Turkey, Syria, Algeria, and Morocco. It is also met with in the maritime belt along the Atlantic from Portugal to Kerry in Ireland. It occurs either as undergrowth in the forests, where in favoured situations it reaches the dimensions of a small tree, or is one of the shrubs composing the *maquis* or heaths, which spread over large tracts of siliceous soil that have been denuded of trees in past ages. It is apparently only in Ireland that the Arbutus grows to be a forest tree, moderate in size, but equalling in height and girth the trees of other species, with which it is associated.

In France, the Arbutus is common in the departments whose shores are bathed by the Mediterranean and extends inland as far as Drôme and Lozère ; it is not unfrequent along the west coast from Bayonne to La Rochelle, and is recorded[4] from

---

[1] *Bull. Soc. des Sciences*, 1886, p. 26.   [2] Figured in *Garden*, xxxiii. 320 (1888).   [3] *Fl. Hibernica*, 182 (1836).

[4] Arbutus is very abundant, in company with oak and mountain ash, in a wood, about 1½ mile in length, on the abrupt and rocky slope of the cliff of Trieux, near Paimpol, in Côtes-du-Nord. Cf. Dr. Avice, in *Bull. Soc. Bot. France*, xliii. 123 (1896), and Coste, *Flore de la France*, ii. 506 (1903). In a note on the occurrence of this species in the Landes, in *Bull. Soc. Bot. France*, xlix. p. lvii. (1902), it is stated that wherever the Arbutus grows, in that region, holly is absent, the two species seeming to exclude each other.

an outlying station in Brittany. In Corsica, it is very common as a shrub in the *maquis*; but in some of the forests grows to be a considerable size, as in that of Bonifatte near Calvi, where I measured trees 25 feet in height and 1 foot in diameter, which were growing at 2000 feet altitude. In Corsica, a liqueur, called *acqua vida de bagui*, is made from the berries. In Spain and Algeria, I noticed it as a shrub, growing in ravines in the forests; but in Italy it sometimes attains a considerable size.

The Arbutus is unquestionably wild[1] in the south-west of Ireland, where it is associated with other plants, which like it are Mediterranean in type and not indigenous to other parts of the British Isles. It has been known to the Irish since early times, and is called *caithne* (pronounced *cahney*) in Kerry and *cuince* in Clare. The former name occurs in several place-names in Kerry, as Derrynacahney, the "oak-wood of the Arbutus," two miles south-east of Crusheen; Cahnicaun wood, near the Eagle's nest, Killarney, which is *coill caithneacan*, the "wood of the little Arbutus," in Irish; Ishnagahiny Lake, five miles south-east of Waterville, which is *uisge-na-geaithne*, "Arbutus water," in Irish. The Clare name, *cuince*, is supposed to occur in several place-names, anglicised as *quin*, which, however, often represents a family name of another signification. Cappoquin, in Waterford, means the field of the Arbutus, and Feaquin, in Clare, the wood of the Arbutus. The occurrence of names like Quin, a parish in Clare, and Quinsheen, one of the islands in Clew Bay, Mayo, may point to an extension of the distribution of this plant far to the north in ancient times.

At present, *Arbutus Unedo* is restricted to Co. Kerry and the extreme south-western part of Co. Cork. In the latter county it is thinly scattered through the woods in the vicinity of Glengariff, growing in company with oak, birch, holly, hazel, and mountain ash, and attaining about 25 feet in height and 3 feet in girth. It is said to grow here and there among the mountains to the west of Glengariff, and was seen by R. A. Phillips at Adrigole, ten miles to the west, high up in the mountains amongst rocks, and without the shelter of other trees. Phillips believes that it does not now grow to the eastward of Glengariff; and he could not find it in its former station, Ballyrizzard, near Crookhaven.

. The Arbutus has its head-quarters in Co. Kerry, in the Killarney district, being particularly abundant and luxuriant on the islands and shores of the lakes generally, where it forms a considerable part of the natural forest. At the base of Cromaglaun mountain, near the tunnel on the Kenmare road, there is a wood composed almost exclusively of Arbutus; and it is also met with on the Cloonee lakes south of the Kenmare River.[2]

About Killarney the tree is indifferent as regards soil, as it grows on limestone on Ross island, on sandstone on Dinis island, and on slate, grit, and conglomerate

[1] Its right to be considered an indigenous plant was contested by Smith, who, in his *History of the County of Kerry* (1756), states that it was introduced by the monks of St. Finnian, who founded the Abbey of that name on the banks of the lake, in the sixth century. Babington, in *Mag. Nat. Hist.* ix. 245 (1836), says this idea is inconceivable as the tree grows in isolated spots far up in the mountains, and is truly an aboriginal. All Irish botanists, and they are supported by authorities like Sir J. D. Hooker and Prof. Fliche of Nancy, are agreed as to the tree being an undoubted native of the south-west of Ireland.

[2] There are six trees on the islands in Glenmore Lake near Dereen, and a few on the mountains beside the lake, according to information I received when visiting Dereen in July 1907.

elsewhere. It is much more affected by climate and aspect than by soil, and seeks the most humid and mildest situations. In the Killarney basin it occupies practically the whole northern shore of the northern lake, but does not grow on the exposed islands of this lake. It is absent from the shore itself, when this is marshy or composed of shingle or sand, and grows on the rocky headlands, where it forms a natural wood with oak, holly, and mountain ash. It is very common on the long indented promontory of Muckross, and reaches its greatest dimensions on Dinis Island, which is perhaps the dampest and most sheltered spot in the whole district, protected by high mountains on the east and west, but open to the south. It usually does not extend far from the lake shore, but in the very humid and shaded Torc ravine it recedes into the general woodland along the rocky banks of the torrent, and ascends to an elevation of several hundred feet. It flourishes also on the rocky and sheltered islands of the southern lake.

In dense woods it has a fairly straight and single trunk; but in the open it usually divides at a short distance from the ground into two or more stems, which tend to be spirally twisted and are often curved, each of them terminating in a much-branched wide crown of foliage. The bark of old trees scales off in longitudinal strips and becomes purplish-grey in colour, assuming in the sunlight a reddish tinge, resembling in this respect the branchlets, which are pale-green on the shaded side and crimson on the sunny side.

The largest trees seen by me were about 40 feet in height; one had three stems 4 feet 10 inches, 4 feet 3 inches, and 3 feet 2 inches in girth respectively, the butt measuring close to the ground 17 feet round. Plate 157 represents one of the finest of these trees on Dinis island. Major Waldron has recently found trees up to 5 feet 7 inches in girth. Much larger trees existed formerly, as one measured by Mackay in 1805, which was $9\frac{1}{2}$ feet in girth. The Arbutus woods, like those of Kerry generally, suffered much from the ironworks, which were established in the eighteenth century, and the largest trees were cut down at this period.

INTRODUCTION

The date of the introduction of the Arbutus into English gardens is unknown; but Mrs. J. R. Green has kindly sent me the following extract from the State Papers,[1] showing that its existence in Kerry attracted in the sixteenth century the attention of the English settlers, who called it *wollaghan*, a corruption of *ubhla caithne* (pronounced *oolacahney*), or "arbutus apples," a name used for the edible fruit :—

"You shall receive herewith a bundle of trees called wollaghan tree, whereof my Lord of Leicester and Mr. Secretary Walsingham are both very desirous to have some, as well for the fruit as the rareness of the manner of bearing, which is after the kind of the orange, to have blossoms and fruit green or ripe all the year long, and the same of a very pleasant taste, and growing nowhere else but in one part of Munster, from whence I have caused them to be transported immediately unto you,

---

[1] *Cal. State Papers, Ireland*, A.D. 1586, p. 240.

praying you to see them safely delivered and divided between my said Lord and Mr. Secretary, directing that they may be planted near some ponds or with a great deal of black moory earth, which kind of soil I take will best like them, for that they grow best in Munster about loughs and prove to the bigness of cherry trees or more and continue long." (A. H.)

CULTIVATION

Though the Arbutus can hardly be called a tree in most parts of England, because it is rarely planted in situations which will enable it to assume a tree-like habit, yet it is so beautiful as a shrub, that no garden should be without it in districts which are warm enough in winter and damp enough in summer to allow it to thrive. It is easily raised from seed, and I have found little difference between the growth of seedlings raised from English and from French seed. Both suffer severely from frosts exceeding about twenty degrees, and from cold dry winds, and should therefore be kept under glass in winter till they are 2 or 3 feet high, when they should be planted out in a well-drained but not dry or heavy soil, in a place well sheltered from the north-east, but not overhung by other trees. Severe winters injure and often kill Arbutus in the eastern and midland counties, and large specimens are rarely seen except on the west and south-west coasts. Even there I have never seen one rivalling what Henry describes in Ireland, and it does not seem to be a long-lived tree in England. The best I have seen, perhaps, is on Sir E. Loder's beautiful grounds at Leonardslee, which is about 30 feet high, with a clean stem 8 or 10 feet high and 3 feet 4 inches in girth. The largest tree on record[1] was one growing at Mount Kennedy, Wicklow, which in 1773 was 13 feet 9 inches in girth. It was supposed then to be somewhat more than 100 years old. In 1794 it was still living, though it had been split by the wind, and torn up by the roots; and fresh healthy shoots were springing up from some branches which had layered.

The wood, which is of a reddish-brown colour, is hard and takes a good polish, but is very liable to split in drying, and so far as I know is not used for anything but small ornamental work, though it seems very suitable for inlaying or parquet.

(H. J. E.)

ARBUTUS HYBRIDA[2]

Arbutus hybrida, Ker-Gawler, *Bot. Reg.* t. 619 (1822); Loudon, *Arb. et Frut. Brit.* 1119 (1838); Gard. Chron. ix. 211, f. 37 (1878).
Arbutus andrachnoides, Link, *Enum. Hort. Berol.* i. 395 (1821).
Arbutus serratifolia, Loddiges, *Bot. Cab.* t. 580 (1821).
Arbutus intermedia, Heldreich, *Flora*, 1844, p. 14.
Arbutus Unedo-Andrachne, Boissier, *Fl. Orient.* iii. 966 (1875).

Arbutus hybrida, being a cross between *A. Unedo* and *A. Andrachne*, is variable in the wild state, sometimes being exactly intermediate between the two

---

[1] Hayes, *Practical Treatise on Planting*, 128 (1794).
[2] This name, though not the oldest, is the one by which the species has been usually known, and is adopted by us.

parents, and sometimes more closely resembling one of them. As seen in cultivation, the bark is smooth, like that of *A. Andrachne*. The branchlets have the glandular pubescence of *A. Unedo*, and the leaves are serrate, as in that species; but have the slightly glaucous tint and conspicuous veins of the other species; petioles glandular-pubescent. The flowers are borne in spring in large drooping panicles, which are usually glandular-pubescent. The fruit is of moderate size, and slightly tubercular on the surface.

According to Loudon, var. *Milleri*, with large leaves and pink flowers, was raised in the Bristol nursery, being a cross between the red-flowered variety of *A. Unedo* and *A. Andrachne*. This seems to be rare in cultivation.

Arbutus hybrida originated in the Fulham nursery early in the nineteenth century. It is, however, known in the wild state, being recorded by Heldreich and Halacsy for several localities in Greece. It is also reported to have been found by Albow[1] at Pizunda, on the north-eastern shore of the Black Sea, which is remarkable, as *A. Unedo* does not occur wild in this district, and the identification was possibly erroneous.

A tree growing at Sedbury Park, near Chepstow, the residence of Colonel Marling, V.C., is by far the finest we have seen of this hybrid. It measures 39 feet high by 5 feet 10 inches at 5 feet, and 7 feet 4 inches at 3 feet from the ground. It is grafted on a stock of *A. Unedo*, but shows more of the character of *A. Andrachne* in its habit and bark. It has been propagated by inarching, and seems to be a hardier tree than *A. Unedo* (Plate 158).

There are fair-sized trees at Kew.   (A. H.)

ARBUTUS ANDRACHNE

Arbutus Andrachne, Linnæus, *Sp. Pl.* 566 (1762); *Bot. Reg.* ii. t. 113 (1813); *Bot. Mag.* t. 2024
    (1819); Loudon, *Arb. et Frut. Brit.* ii. 1120 (1838).
Arbutus integrifolia, Salisbury, *Prod.* 288 (1796).
Arbutus Sieberi, Klotzch, *Linnæa*, xxiv. 71 (1851).

A large shrub or small tree, attaining 30 to 40 feet in height. Bark peeling off in thin papery layers, smooth, thin, and reddish brown. Young branchlets reddish or green, glabrous; older branchlets olive-green or brownish, smooth. Buds minute, reddish. Leaves, larger usually in cultivated trees than those of *Arbutus Unedo*, oval-oblong, contracted into short blunt points at the apex, tapering at the base; upper surface dark green, glabrous, shining; lower surface glaucescent, glabrous, with prominent midrib and distinct lateral veins; margin entire. Petiole glabrous, about ½ inch long.

Flowers in erect viscid glandular-pubescent panicles, yellowish white, appearing in spring. Calyx-lobes deep, ovate, acute. Corolla contracted at the apex, with five reflexed short rounded ciliate lobes. Ovary pubescent. Fruit small, about ⅓ inch,

[1] Radde, *Pflanzenverb. Kaukasusländ.*, 127, note (1899).

rarely ⅛ inch in diameter, globose, orange coloured, smooth, hard, glandular on the surface.

Arbutus Andrachne is a small tree or large shrub, resembling *A. Unedo* in habit, and like it occurring often in heaths and occasionally in the forests; and only rarely forming small pure woods. It occurs in Albania, Greece, Cephalonia in the Ionian Islands, Crete, Rhodes, Cyprus, in the maritime regions of Asia Minor, Syria, and Palestine, in the Crimea and in the district of the Caucasus bordering upon the Black Sea. (A. H.)

It was introduced into England from Smyrna in 1724, and cultivated at Eltham by Dr. Sherard.

This tree though rarely planted in modern gardens[1] is, on account of its superior hardiness and its extremely beautiful bark, a more ornamental tree than the native species. Though I have never seen or heard of its producing ripe fruit in England, seedlings may be obtained from Continental nurseries, and some that I brought from Pallanza, in October 1906, have survived the journey without injury. The tree seems to enjoy lime in the soil. The bark is like smooth reddish-brown leather, covered with a thin silvery paper-like skin which peels off annually, and for this alone it is well worth growing. There was a very fine though not tall tree of this species on the lawn at Williamstrip Park, Gloucestershire, on rather heavy soil, which endured the inclement season and severe winters of 1879-80-81 without much injury, but is now dead. I saw in 1903 another which was 36 feet high and 4 feet in girth lying on the ground at Haldon near Exeter, which had been blown down some years before but was still living. The best that I know now living is in the Botanic Garden at Bath, and measures 27 feet by 6 feet 3 inches at 1 foot from the ground, shortly above which it divides into several stems. There is also a handsome tree about 25 feet high at Westonbirt, and one at Mamhead, 30 feet high, which is decaying at the butt. (H. J. E.)

ARBUTUS MENZIESII, Madroña

Arbutus Menziesii, Pursh, *Fl. Amer. Sept.* i. 282 (1814); Sargent, *Silva N. Amer.* v. 123, t. 231 (1893), and *Trees N. Amer.* 728 (1905).

Arbutus procera, Lindley, *Bot. Reg.* xxi. t. 1753 (1836); Loudon, *Arb. et Frut. Brit.* ii. 1121 (1838).

Arbutus laurifolia, Hooker, *Fl. Bor. Amer.* ii. 36 (1840). (Not Lindley.)

A tree attaining in America 100 feet in height and 20 feet in girth, but usually much smaller. Bark of branches and young stems thin, smooth, reddish, peeling off in large thin scales; of older trunks dark reddish brown and covered with small thick scales. Young branchlets glabrous; older branchlets reddish brown, smooth. Buds stouter than in *A. Unedo*, ⅛ inch long; scales ovate, acute, apiculate. Leaves oval or oblong, larger than in *A. Andrachne* or *A. Unedo*, up to 5 inches long by 3 inches broad, rounded or contracted into minute sharp points at the apex,

---

[1] A tree in Kew Gardens, 20 feet high, is figured in *Gard. Chron.* iv. 724, f. 100 (1888).

subcordate or rounded (rarely tapering) at the base; upper surface dark green, shining, glabrous; lower surface glaucous, almost white in colour, glabrous, with prominent midrib and conspicuous lateral veins; margin entire, occasionally serrate on young plants. Petiole stout, ½ to 1 inch long, glabrous, usually winged on one or both sides for some distance by the decurrent base of the leaf.

Flowers appearing in spring, in erect pubescent panicles, about 5 or 6 inches long and broad. Calyx-lobes scarious, white. Corolla white, urceolate. Ovary glabrous. Fruit ripening in autumn, sub-globose, ½ inch in diameter, bright orange-red, glandular on the surface, with a thin flesh and a five-celled thin-walled cartilaginous stone.

Arbutus Menziesii occurs in the Pacific coast region from Southern British Columbia, where it grows on Vancouver Island and the islands at Seymour Narrows, through Washington and Oregon to California, reaching its most southerly point in the Santa Lucia Mountains. In Washington it is not uncommon on the cliffs along Puget Sound, and on high slopes, where it receives plenty of light. It usually grows on rich soil and, according to Sargent, is common and attains its largest size in the redwood forest of Northern California, becoming smaller to the north and south, and only growing as a shrub to the south of the bay of San Francisco. I did not observe it in the dense redwood forest near Crescent City; but found it common inland to the east of the coast range in South-Western Oregon. Here it grew on dry hills at 2000 to 3000 feet altitude, in mixture with *Pinus ponderosa*, *Libocedrus decurrens*, and oak, in thinly forested country; and resembled very much in habit, with its short trunk and broad branching crown, the Arbutus of Killarney. In a ravine near Kerby I measured a tree, 99 feet high by 5 feet 1 inch in girth, with a straight stem, clear of branches to 40 feet; but this grew in exceptionally good soil, and was crowded by other trees—Lawson Cypress, Sugar Pine, *Quercus densiflora*, *Acer macrophyllum*, etc.   (A. H.)

The largest tree of this species known, which has been figured by Sargent,[1] is growing in the grounds of the reservoir at San Rafael in Marin county, California. It measures 100 feet in height and 23 feet in girth at 3 feet from the ground.

Arbutus Menziesii was introduced by Douglas in 1827; but is rather a rare tree in cultivation in England. It appears to be less hardy than the other species now described, and at Kew makes slow growth and often has its leaves and shoots injured by frost. It is found in gardens usually under the name of *A. procera*, and commonly attains the size of 20 to 30 feet. The largest that I have seen is at Bassetwood, near Southampton, the residence of J. R. Anderson, Esq. This tree is no less than 50 feet high, with a stem clear for about 20 feet, and 3 feet 2 inches in girth. A tree at Tortworth is 35 feet high by 4 feet 4 inches.[2] In Scotland, at Castle Menzies, I measured one in 1907 which was 37 feet by 5 feet 2 inches, and did not seem to have suffered much from the severe frost of the previous winter, though the flower buds were killed.   (H. J. E.)

---

[1] *Garden and Forest*, v. 146, f. 23 (1892). In the same journal, iii. 509, f. 515 (1890), the tree is figured in its native forest.

[2] Mr. Clinton Baker informs me that there is a tree at Bayfordbury which was sent to his grandfather about twenty-five years ago from America, as a very small plant. It is now 30 feet high, and bears fruit every year.

SCIADOPITYS

Sciadopitys, Siebold et Zuccarini, *Fl. Jap.* ii. 1, tt. 101, 102 (1844); Bentham et Hooker, *Gen. Pl.* iii. 437 (1880); Masters, *Journ. Linn. Soc. (Bot.)* xviii. 502 (1881), xxvii. 276, 320 (1889), xxx. 21 (1893), and *Journ. Bot.* xxii. 97 (1884).

AN evergreen tree, belonging to the tribe Taxodineæ of the order Coniferæ, attaining in Japan a height of 120 feet and a girth of 12 feet. Bark reddish brown, scaling off in long strips. Branches sub-verticillate. Branchlets brown, glabrous, bearing minute scales, which represent true leaves, and cladodes, which are long, green, and leaf-like, performing the functions of true leaves, but differing from them in structure. The scales are borne spirally on the internodes, and are dry, brown, membranous, ovate-lanceolate, and decurrent. At the apex of the shoot there is a ring of similar scales, deltoid in shape, and densely pubescent on their inner surface, out of the axils of which arise a whorl of cladodes, ten to thirty in number, spreading all round the branchlet. These are 2 to 5 inches long, averaging $\frac{1}{8}$ inch in width, linear, rigid, narrowed towards the base, obtuse and minutely rigid at the apex; upper surface dark green, shining, with a median groove; lower surface green on each side of a deep white stomatiferous central furrow. Buds globose, composed of numerous spirally imbricated greenish scales; terminal bud, at the apex of the shoot, in the centre of the whorl of cladodes, continuing the growth of the main axis in the following year; a smaller bud, often present at the side of the terminal bud, developing into a lateral branch in the next season. As a rule, the main axis is bare, except for the scales, below the apex, which bears the whorl of cladodes and the buds; but on strong-growing shoots a lateral branch is occasionally developed half-way up the internode. The cladodes are leaf-like shoots, and not true leaves, each representing an axillary branch with two coherent leaves; but their true nature has given rise to a great deal of discussion; and the elaborate papers of Dr. Masters cited above may be consulted on this subject.

Male flowers in a terminal compact raceme, about an inch in length; each flower $\frac{3}{8}$ inch long, subsessile; anthers numerous, spirally arranged, short-stalked, with an acute and reflexed crest and two pendulous cells, opening by a vertical slit; pollen-grains globular, minutely tuberculate. Female flowers, terminal small cones composed of spirally arranged lanceolate bracts, which are serially continuous with the true leaves, empty at the base of the cone, higher up with fleshy semi-lunar ovular scales in their axils, half the size of the bracts and bearing one to nine ovules in a transverse series on their inner surface. As the cones increase in size, the ovular scales outgrow the bracts, and in the mature cone are much larger than and almost entirely coalesced with them.

The cones, which are borne on short stout stalks, clothed with a few membranous bracts, either remain terminal and erect or are pushed aside by the growth of a lateral branch. They take two years to ripen, and remain persistent on the tree for some months after the dehiscence of the seeds. Ripe cones, about 3 inches long by 1½ inch in diameter, oblong-ovoid, obtuse at the apex, composed of woody scales, which result from the coalescence of the ovular scales and bracts of the flower. The scales are fan-shaped, about ¾ inch wide; upper margin rounded and reflexed; outer surface convex, marked by a transverse rugged irregular ridge; inner surface concave, with slight depressions for the seeds. Seeds, five to nine on each scale, reversed, oval, compressed, dark brown, surrounded by a narrow membranous reddish-brown wing, notched at the base and marked at the apex by the white hilum; seed with wing, about ¾ inch long by ¼ inch wide. The seedling has a long slender tap root, and a terete green glabrous caulicle about an inch in length, which bears two cotyledons. These are sessile, linear, tapering to an obtuse apex, a little more than ½ inch long, dark green above, paler below with indistinct lines of stomata. Primary leaves like the cotyledons, but longer.

Sciadopitys is a monotypic genus, only one species being known, which is a native of Japan.

SCIADOPITYS VERTICILLATA, Umbrella Pine[1]

Sciadopitys verticillata, Siebold et Zuccarini, *loc. cit.* 1844; Kent, *Veitch's Man. Coniferæ*, 287 (1900); Shirasawa, *Icon. Ess. Forest. Japon*, text 22, t. 8, ff. 15-36 (1900); Thiselton-Dyer, *Bot. Mag.* t. 8050 (1905); Mayr, *Fremdländ. Wald- u. Parkbäume*, 407 (1906).
Taxus verticillata, Thunberg, *Fl. Jap.* 276 (1784).

The species has been described above. The tree is known in Japan as *Koyamaki*, or pine of Mt. Koya, one of the localities where it is found growing wild. Thomas Lobb sent a living plant in 1853 from the Botanic Garden at Buitenzorg in Java to Veitch's nursery at Exeter; but it soon died. It was afterwards introduced by seeds brought from Japan by J. Gould Veitch in 1861, some being also sent about the same time by Fortune to Standish at Ascot.

A variety in which the leaves are striped with yellow was introduced by Fortune; but this seems to be now unknown in cultivation.

Plants only 3 feet high produced cones in 1876 in the nursery of Messrs. Thibaut and Keteleer at Sceaux.[2] The tree appears to have first borne fruit in Scotland[3] at Ardkinglas, in 1878, and in England[4] at Kew and Coombe Wood, in 1884. Proliferous cones,[5] which bear cladodes at their apex, are of frequent occurrence in Japan, and have also been borne by trees cultivated in Europe. In

[1] This is a translation of *Sciadopitys*, a name given on account of the leaf-like cladodes spreading out from the apex of the shoot, like the ribs of an umbrella. [2] *Gard. Chron.* v. 827 (1876).
[3] *Jour. of Forestry*, 1879, p. 508. [4] *Gard. Chron.* i. 80 (1884). [5] Masters, *Journ. Bot. loc. cit.* f. 4.

these the bracts, which are ordinarily completely coalesced with the fruit-scales, become detached from them towards the apex of the cone, and are scale-like in character, producing cladodes in their axils. (A. H.)

Shirasawa states that the tree grows wild in mixture with *Abies firma* and *Cupressus pisifera* at 600 to 5000 feet in the forests of Kiso and Shinaño. Matsumura adds Mt. Hoonokawa in Tosa.

According to Mayr, the tree is similar to the silver fir in its capacity for bearing shade; but is extremely slow in growth, only attaining in Kiso, where the climate is favourable to it, a height of 30 feet in fifty years. Trees 110 feet in height and 2 feet in diameter average about 250 years old. Mayr gives a figure of two old trees, growing at Agematsu, which were nearly 120 feet in height and 4 feet in diameter; and this shows how the tree, even at an advanced age, preserves a narrow pyramidal form with an upright leader, without any sign of flattening of the crown.

I saw this tree in its native forest to the best advantage in the lovely valley of Atera, on the west side of the Kisogawa, below Agematsu, at from 2000 to 3000 feet elevation. Here it was scattered in a forest of mixed conifers and hardwoods, and seemed to grow only on rocky slopes and ridges, where its narrow-pointed top made it conspicuous. The seedlings were numerous in dense shade growing on a bed of humus, and those that I took up had long but scanty roots, running deep, but spreading little. Their growth was very slow, not more than 3 to 6 inches annually for the first twenty years at least. On a steep rocky hill above the forester's house, at the end of the tramway which has been made up this valley, the largest trees were growing, mixed with Thujopsis, the undergrowth being very dense, and composed of Rhododendron, with *Shortia uniflora* spreading over the ground in great sheets. The largest that I was able to measure were 90 to 100 feet high and 9 to 10 feet in girth, one being 11 feet 9 inches at 5 feet from the ground. Plate 159A fairly represents the appearance of the tree here.

In the forest near Koyasan I saw it again, mixed with *Cupressus obtusa*, but not attaining so large a size, though it seemed that in the dense shade the seedlings of Sciadopitys were more numerous and vigorous. Though often planted in parks and temple gardens, I never saw any trees as fine as those figured by Mayr at Agematsu, and it is clear that shade, perfect drainage, and a rich forest soil are essential to this species.

According to Mayr, the wood is white in colour, the sapwood only $\frac{2}{3}$ inch thick being like the heartwood. The wood is comparable to the best kind of spruce, and is soft and elastic. It is used in Japan for boat-building, making bath-tubs and casks, planking, etc.

CULTIVATION

Though this interesting tree has been planted in many places, yet it usually grows very slowly and seems to require a high summer temperature, with a warm and sheltered situation. Ripe seed was produced in Ireland at Castlewellan in 1900, but the seedlings which I raised from this source were always weakly, and notwithstanding every care died after two or three years. Seed sent from Japan in 1906 also germinated weakly, and many of the seedlings damped off in the winter without making roots, though very carefully watered. They seem to require a very light, sandy peat when young, and refuse to grow in soil which contains lime, or where moisture is deficient during the summer. By far the finest tree that I have seen in England is at Hemsted, in Kent, where a tree was in 1905 no less than 38 feet high by 2 feet in girth, and showed its true habit very well. Owing to its being rather crowded by other trees, a photograph of this was difficult to take, but after several attempts had been made, Mr. Edwards was able to get the one reproduced in Plate 159B. The next largest I have seen is at Coombe Royal, in South Devon, where a tree about 25 feet high is growing, with a forked stem.

As usually seen in gardens in England, it forms a shrubby pyramid, and I have seen no others over 15 to 18 feet high. At Castlewellan, however, it seems to thrive very well, and should do well in the south and west of Ireland and Wales.

At the Villa Trubetskoi, near Intra, on Lake Maggiore, I saw a vigorous tree about 45 feet high, which had divided into three stems, and in 1906 bore no cones.

Sciadopitys is perfectly hardy, and bears without injury the severe winter climate of Boston in New England, and of Grafrath in Bavaria, the thermometer descending in the latter locality to − 18° Fahr. ; but in severe winters the foliage turns brown.

(H. J. E.)

PINUS SYLVESTRIS, SCOTS PINE

Pinus sylvestris, Linnæus, *Sp. Pl.* 1000 (*excl. var.*) (1753); Lambert, *Genus Pinus*, i. tab. I. (1803); Loudon, *Arb. et Frut. Brit.* iv. 2153 (1838); Willkomm, *Forstliche Flora*, 193 (1887); Mathieu, *Flore Forestière*, 579 (1897); Kent, in *Veitch's Man. Coniferæ*, 379 (1900); Kirchner, Loew u. Schröter, *Lebengeschich. Blütenpfl. Mitteleuropas*, i. 175 (1904); Mayr, *Fremdländ. Wald- u. Parkbäume*, 347 (1906); Borthwick, in *Trans. R. Eng. Arb. Soc.* vi. 205 (1906).

A TREE commonly 100 feet, rarely attaining 150 feet in height, with a girth of 10 to 15 feet. Stem usually straight and cylindrical, with the branches regularly whorled in young trees, forming a pyramidal crown; in older and isolated trees, branching irregular, with a flattened crown. Bark different in the lower and upper parts of the trunk; towards the base thick, fissured into irregular longitudinal plates, scaly, and reddish brown or greyish brown in colour; on the upper part of the stem,[1] owing to the outer portion continually falling off in thin papery scales, the bark remains very thin, smooth, shining and bright red. Young shoots greenish, smooth and shining; becoming greyish brown in the second year; marked with the pulvini of the scale-leaves, which are early deciduous. Buds long-oval, pointed, usually non-resinous, covered by lanceolate acuminate scales, fimbriated on their edges, the upper ones with their tips free and not recurved. Leaves two in a bundle; sheaths at first white, ⅛ inch long, speedily becoming shrivelled, brown, and short; the pair of leaves close together, but not appressed, usually about 2 inches long but varying under different conditions from 1 to 4 inches, dark green with interrupted lines of stomata on the convex side, glaucous with many well-defined lines of stomata on the flat inner side, plano-convex in cross-section, linear, stiff, acute at the apex, somewhat bent, smooth, finely serrate in margin; resin-canals marginal. The leaves persist usually three years.

Male flowers in dense clusters at the lower part of the current year's shoot, ¼ inch long, oval, short-stalked, surrounded at the base by four yellowish bracts; anther with small rounded upright connective. Female flowers, solitary, opposite or occasionally whorled, apparently terminating the young shoot, erect at first, but becoming pendant immediately after pollination, stalked, globose, reddish, composed of rounded bracts and almost circular ovular scales, the latter having a beak-like process on the upper side and bearing two minute ovules.

Cones shortly stalked, variable in shape, usually ovoid-conic with an acute apex, oblique or nearly symmetrical at the base, greyish or dull brown in colour, 1 to 3

---

[1] According to Shaw of Boston, who is the greatest living authority on the genus *Pinus*, this peculiarity of the bark of the upper part of the tree being thin and reddish, owing to the constant shedding of scales, occurs only in three pines, viz. *P. sylvestris*, *P. densiflora*, and *P. patula*.

inches long.   Scales dark brown on the inner surface, oblong, ending in a rhomboidal apophysis, which is variable in form in different varieties and even in the same cone; flattened, with a transverse keel and an elevated or depressed umbo, or raised and pyramidal, with four to five concave sides; occasionally the apophysis ends in a hooked process.   The cones open in spring to let out the seeds, which may be carried by strong winds to an immense distance; and the empty cones usually remain on the tree till the following autumn.   Seeds long-oval, $\frac{1}{8}$ to $\frac{1}{5}$ inch, some blackish, others grey in colour, surmounted by a wing half-oval in shape, which is three times as long as the body of the seed.

Seedling.—Cotyledons, four to seven, triangular in section, linear, slightly curved upwards, about $\frac{4}{5}$ inch long; stomata absent on the outer surface, present on the inner two surfaces, without wax, so that the cotyledons are green in colour and not glaucous.   Primary leaves elliptic in section with hairs on their edges.   The seedling grows about 2 to 4 inches high in the first year, ordinary needles being produced in the second year; branches usually appear in the third year.   The primary root is long, attaining about 8 inches in the first year, and giving off many lateral fibres.

Varieties

The common pine, spread over an immense geographical area and growing in the most diverse conditions of soil and climate, exhibits considerable variations in most of its characters.   The stem may be straight and cylindrical with a single leader, only branching at the top and giving rise to a flattened crown of foliage in old age; or it may be dwarf, branched from the base and crooked, simulating the smaller forms of *Pinus montana*.   The young cones are usually reversed immediately after flowering; but in certain regions they remain erect.   The adult cones vary in size and shape and in the form of the apophyses, which may be flat or raised, pyramidal or hooked; but all these variations in the apophysis may occur on the same cone. The male flowers may be yellow or reddish in colour.   The leaves vary in length from one to four inches, and may be broad or narrow, stiff and sharp-pointed or soft in texture; and in some cases they are much more glaucous than in others.   They vary in duration from two to five years.

Many of these varieties occur in individual trees in the same forest; and in many cases, when the condition of the soil is changed as by draining, pines which have been small and stunted assume the ordinary tall form, and the shape of the cones probably does not remain constant, when the seedlings are raised in a new locality. It is difficult on this account to establish clearly marked geographical varieties.

The experiments, which have been carried out at Les Barres,[1] over a long term of years, show, however, that there are races of pines, which preserve their characters of straightness of stem, quickness of growth, or the reverse; but these races cannot be distinguished by characters of cones or leaves; and are the result of the selection of seed from vigorous or weak individuals.

---

[1] Cf. Pardé, *Arboret. Nat. des Barres*, 71 (1906), where a full account is given of the plots of Riga, Haguenau, Scotch, and certain French varieties of *Pinus sylvestris*, which were mostly planted between 1823 and 1835.

The following varieties, occurring in the wild state, have been distinguished, though they are not so clearly defined in nature, as they seem to be from their description.

1. Var. *genuina*, Heer. Cones usually solitary, long-stalked, symmetrical, acute at the apex; apophysis flat or convex, not hooked. Needles about 2 inches long, persistent three years.

This is the common pine, growing on good soil in Germany, Southern Scandinavia, Poland, and North-Western Russia. Two races have been distinguished on the Continent in cultivation :—

(*a*) *rigensis* (*Pinus rigensis*, Desf.). Riga pine, raised from seeds collected near Riga. At Les Barres, this is the best race of *P. sylvestris*, the stem being very straight and cylindrical, rising to a great height, and with few lateral branches ; bark very red, stripping off above in very thin papery scales. Von Sievers states that the form native to the Baltic provinces of Russia is superior in growth and timber to that introduced there by seed from Germany. Willkomm, however, is of opinion that the so-called Riga pine is only a fine tall-growing form, and occurs in North Germany and Poland, as well as in Russia.

(*b*) *Haguenensis*, Loudon. Haguenau pine, raised from seed obtained in the forest of Haguenau in Alsace. At Les Barres, this form, though vigorous in growth, is defective, on account of its tendency to form numerous irregular branches, so that the stem is not so clean and does not reach the same height as the Riga variety. The bark is not so red, and is not so fine-scaled as in that variety.

Two trees of the Haguenau variety, raised from seed, procured by Loudon in 1828, are growing at Seggieden in Perthshire, and are now about 65 feet high by 8 feet in girth. According to the forester, they are distinguishable in bark, buds, shoots, and leaves from the Scots pine growing near them.

2. Var. *scotica*.—This variety, which grows wild in the Highlands of Scotland, differs in the redder bark of the stem ; in the shorter more glaucous leaves ($1\frac{1}{2}$ inch long), often persistent four years ; and in the shorter cones ($1\frac{1}{2}$ inch long), which are symmetrical, with apophyses usually flat near the base, tending to be pyramidal in the upper part of the cone.

3. Var. *engadinensis*, Heer.—Bark reddish ; needles short, 1 to $1\frac{1}{2}$ inch long, thick and stiff, persistent for five years ; buds resinous. Cones ovoid-conic, 2 inches long, oblique at the base ; apophyses convex on the outer side of the cone, umbo large and blunt. A small tree, rarely 30 feet high, growing in the Engadine Alps.[1] It is perhaps a hybrid between *P. sylvestris* and *P. montana*.

4. Var. *lapponica*.—*Pinus lapponica*, Mayr, *Fremdländ. Park. u. Waldbäume*, 348 (1906). This variety, which grows in the north of Norway and Sweden and in Finland, is considered by Willkomm and Christ[1] to be identical with var. *engadinensis*, with which it agrees in the short, straight stiff leaves, persistent for

[1] Dr. Christ, in *Flore de la Suisse*, 197, and *Suppl.* 31 (1907), considers the Engadine pine to be precisely the same as specimens he examined, which were collected at Quickjock in Lapland (lat. 67°). He also mentions (p. 285) a curious form of the common pine, slender and tall in habit, with very short green needles, which grows at Flims in Switzerland, and also in one or two places in Silesia.

five years, in the resinous buds, and in the small cones with hook-like apophyses. Mayr, however, considers it to be a distinct species, and gives the characters which distinguish it from the common form of *P. sylvestris*, without pointing out in what respect it differs clearly from var. *engadinensis*.

5. Var. *nevadensis*, Christ.—Needles broad, short and stiff, very white on their flat surfaces. Cones nearly sessile, oblique, with very pyramidal apophyses. Occurs in the Sierra Nevada in the south of Spain.

6. Var. *reflexa*, Heer.—Needles as in the common form. Cones long and slender, conic, with long hooks to the apophyses. This variety has been found growing on high peat-mosses in Switzerland and on poor sandy soil in Prussia, and occurs sporadically elsewhere.

In the Caucasus and Asia Minor, *P. sylvestris* differs from the European form, in having very long and broad needles ($3\frac{1}{2}$ inches long), and very oblique cones with hooks on their outer side directed downwards. Specimens from the Amur have very long leaves (4 inches or more) with cones of the ordinary form. The pine of the Ural and Altai mountains (var. *uralensis*, Fischer) is only distinguished by having short and stiff leaves. In the dry climate of the south of France, in the Cevennes and in Provence, the needles of this species become short and are often disposed in slender tufts at the ends of the branchlets.

In the French Alps near Modane, *P. sylvestris* grows in mixture with *P. montana*, var. *uncinata*; and it is difficult to distinguish between these trees in this locality,—the branches densely covered with short leaves, persistent for four or five years, being alike in both species; and the cones of both have hooked apophyses.[1] However, at Modane, as elsewhere, the reddish bark and the dull colour of the cones will distinguish *P. sylvestris*; while in the other species the bark is never red and cones are shining brown.

Similar forests occur in Switzerland, where *P. sylvestris* and *P. montana* appear to pass one into the other; and the occurrence of these apparently transitional forms has given rise to the belief that they are hybrids between the two species; but this is not established beyond doubt.

Willkomm describes two interesting varieties, due to poverty of soil and exposure. One is the shore-pine of the Baltic provinces of Prussia, which has a short bent stem with an irregular crown of foliage or is a mere bush; the cones are very oblique and hooked. Another form is peculiar to the peat-mosses in Austria and Germany; the stems are rarely more than 6 feet high, very slender, and branched to the base; needles very stiff and short (about 1 inch long), persistent for two years; cones very small, with hooked apophyses.

Several varieties have arisen in nurseries or as sports in the wild state.

Var. *virgata*, Caspary.—Main branches irregularly whorled, arising from the stem at an angle of 30° to 60°, elongated and giving off a few twig-like branchlets, only the outermost of which are furnished with leaves. This curious variety[2] was first

---

[1] The young cones of *P. sylvestris* at Modane remain erect (like those of *P. montana*) and are not reversed immediately after pollination, as is usually the case elsewhere.

[2] Willkomm, *Forstliche Flora*, 199 (1887).

noticed in France; and some years later, in 1881, was found in the forest of Wandsburg in Prussia.

Var. *argentea*, Steven.—Cones and leaves with a silvery tint. Found in the Caucasus.

Var. *monophylla*, Hodgins.—A shrub, with the needles in each sheath attached to each other throughout their length, apparently forming one needle, but easily separated. Originated at Dunganstown, near Wicklow, about 1830.

Var. *microphylla*, von Schwerin.—Needles thin, sharply pointed, only $\frac{1}{2}$ inch long. Originated as a seedling in 1883 at Wendisch-Wilmersdorf.

Var. *aurea*.[1]—A low tree of dense habit, with leaves of a golden yellow colour usually in spring, the foliage becoming green in summer.

Var. *variegata*.—Leaves variegated. This form has arisen several times in cultivation; but was once found wild in Prussia by Caspary.

Var. *pyramidalis*.—Fastigiate in habit. Schübeler says that trees of this kind are common in the forests of Norway and Finland.

Var. *pendula*, Caspary.[2]—A weeping form, found in a wood near Tilsit, in East Prussia.

Various dwarf forms are known, as *pumila*, *nana*, *globosa*.

DISTRIBUTION

The common pine has an extraordinarily wide distribution, occurring in regions of the most diverse climates and on almost all soils, and in the mountains as well as in the plains. It grows in Eastern Siberia, where the temperature falls to − 40° Fahr., and the period of vegetation hardly lasts for three months; and is met with in Southern Spain, where the summer heat reaches 95° Fahr., and the period of vegetation lasts for nine months of the year. It occurs in dry regions like Provence, where there is little humidity in the air, and in the west of Scotland, where the air is laden with moisture all the year round. It is by preference a tree of siliceous soils, but occurs on almost all geological formations; and in Scotland, Norway, and Sweden grows on peat-bogs too wet for the spruce to exist on.

The area of distribution includes almost all Europe and the greater part of Northern Asia. The northerly limit, commencing on the north-west coast of Norway at Alten (70° N. lat.), passes through Lapland, south of the Enara lake (68° 50′), and touches Pasvig Fjord on the Arctic Sea at 69° 30′. Extending through the Kola peninsula from Kola bay, it crosses the White Sea at 66° 45′ and in the Petchora territory goes as far north as 67° 15′; and crosses the Ural at about 64°. In Siberia it never reaches quite as far north as the Arctic circle, though it nearly touches it on the Ob and the Yenisei rivers; east of the Lena river it descends to about 64°. It reaches its extreme easterly point (about 150° E. long.) in the Werchojansk Mountains. The eastern limit descends from there through the Stanovoi Mountains

---

[1] There is a useful note on the propagation of this variety in *Gard. Chron.* xi. 405 (1892).

[2] *Schrift. Phys. Oekonom. Gesell. Königsberg*, 1866, p. 49, fig. 1.

and the Seja territory to the Upper Amur.   According to Komarov,[1] it is a scarce tree on the banks of rivers in Manchuria.   Its southerly limit in Siberia is not well known; but it is known to occur in the mountains of Dahuria, in the territory around Lake Baikal, and in the Altai Mountains.   Its southern limit in European Russia is a very irregular line, which begins in the Ural south of Orenburg at about lat. 52°, is most to the north in the government of Tula (lat. 54° 30'), and descends from there to Kharkof (lat. 49°), passing into Galicia about lat. 50°.   Far south of this line, and separated from it by the Russian Steppes, on which no pine trees grow, occurs an area of distribution, not yet well made out, which includes the Caucasus, the mountains of the Crimea, Asia Minor,[2] and North-Western Persia.   There is also an isolated area, in which the pine is found growing wild, in Macedonia, on Mount Nidjé.   From Galicia the southern limit in Europe (exclusive of the last-mentioned area) passes southwards to the Transylvanian Alps; thence it extends along the mountains to Servia, where the tree grows on the Kopavnik mountain (about lat. 43°), continues through the mountains of Bosnia, Dalmatia, Illyria, Venetia, and through Lombardy to the Ligurian Apennines (about lat. 44°).   It passes into France, across the Maritime Alps, into the Cevennes, and reaches the Eastern Pyrenees; in Spain it descends through the mountains of Catalonia, Aragon, and Valencia to the Sierra Nevada in Andalusia, which is its extreme southerly point in Europe (lat. 37°). The westerly limit beginning here, stretches north-west through the mountains of Avila to those of Leon in North Spain; and is continued through the mountains of Scotland to the north-west coast of Norway.

In this vast area the pine is very irregularly distributed.   The largest forests occur in the Baltic provinces of Russia, in Scandinavia, in Northern Germany, and in Poland.   Towards the south it only occurs in mountains, and rarely forms pure forests of considerable extent.   According to Huffel,[3] it is rare in Roumania, where he saw it at the confluence of the Lotru and Oltu rivers at 1700 feet altitude, and in the valley of Bistritza.[4]

In the British Isles, the common pine is found wild at the present day only in the Highlands of Scotland, where a few forests still remain.   These occur in the valley of the Spey at Rothiemurchus, Duthill, Abernethy, and Glenmore, and in the valley of the Dee at Invercauld, Braemar, and Glen Tanar.   There is also a fine wild forest, the "Black Wood," on the south side of Loch Rannoch in Perthshire.[5]   That of Ballochbuie near Invercauld is probably the finest now existing.

The pine was widely spread over the British Isles in ancient times, as is evidenced by the occurrence of remains of logs, stumps of trees, and cones in the

---

[1] *Flora Manshuriæ*, i. 175 (1901).

[2] *Pinus sylvestris* grows on the Armenian plateau, and has been described in *Linnæa*, xxii. 296 (1849), as *P. armena*, Koch; *P. Kochiana*, Klotzsch; and *P. pontica*, Koch.   Cf. *Moniteur Jardin Botanique Tiflis*, ii. 26 (1906).

[3] *Forêts de la Roumanie*, 6 (1890).

[4] M. B. Golesco, in an article on the forests of Roumania, in *Bull. Soc. Dendr. France*, i. 171 (1907), states that in the Muscel district *P. sylvestris* is only found on calcareous soils; and in a letter to Elwes confirms this statement, adding that it attains a diameter of one metre, and does not grow on the adjoining schist.

[5] Buchanan White, *Flora of Perthshire*, 282 (1898), gives as additional localities for wild trees in Perthshire, Breadalbane, in Glen Lyon and near Killin and Tyndrum; and mentions one or two other places where the pine is doubtfully native. According to the Rev. E. G. Marshall, *Journ. Bot.* xliv. 160 (1906), it is certainly native in the forest of Glenavon, but quite scarce, and the seedlings appear to be destroyed by deer browsing on them.

peat-mosses and submerged forests.[1] In the south of England extensive forests occurred in Neolithic times, when the existing peat-mosses began to form; but in other parts of the three kingdoms it is probable that the pine existed in many places in historic times.[2]

Of its existence in a wild state until lately in England, the evidence is very meagre. Holinshed,[3] writing in 1586, says: "The firre, frankincense, and pine we do not altogether want, especiallie the firre, whereof we haue some store in Chatleie Moore in Darbishire, Shropshire, Andernesse, and a mosse neere Manchester, not far from Leircesters house; although that in time past not onelie all Lancastershire, but a great part of the coast betweene Chester and the Solme were well stored." According to the Rev. Abraham de la Pryme[4] there was a wood of wild pine on a hill at Wareton in Staffordshire in his day, the beginning of the eighteenth century; and, in an old deed, fir trees were mentioned as growing scattered in Hatfield Chase in Yorkshire about the year 1400, the last surviving aboriginal pine here being cut down about 1670. The Wareton pines were described by Ray in a note[5] dated Oct. 14, 1669: "We rode to see the famous fir-trees, some $2\frac{1}{2}$ miles distant from Newport, in a village called Wareton in Shropshire,[6] on the land of Mr. Skrimshaw. There are of them thirty-five in number, very tall and straight, without a bough till towards the top. The greatest, and which seems to be the mother of the rest, we found by measure to be $14\frac{1}{2}$ feet round the body, and they say 56 yards high, which to me seemed incredible. The tenant's name of the house close by these fir-trees is Firchild, whose ancestors have been tenants to it for many generations." These trees, according to Dr. Higgins[5] of Newport, are mentioned in an old book, *Historia Vegetablium Sacra*, published in 1694 by Westmacott, who says there were thirty-six of them, one of them being $47\frac{1}{2}$ yards high. Withering,[7] writing in 1776, states that the trees at Wareton were no longer existing in his time. Pine forests apparently occurred in Roman times in the north of England, and remnants of these may have existed down till a recent period, concerning which the late Professor Newton told me of some very old Scots pines that used to grow about forty-five years ago on Wretham Heath, Norfolk, which local tradition said had never been planted, but grew there wild. They were always spoken of as the "Deal[8] Trees," all other trees of this species that were planted being named Scotch firs. Whether there is any real foundation for this tradition is very hard to say, but it is possible that the seed

---

[1] Cf. Clement Reid, *Origin British Flora*, pp. 16, 152 (1899):—"Remains of this tree are found in Neolithic deposits, in 'submerged forests' and at the base of peat-mosses, nearly throughout Britain and in Ireland. In late Glacial times at Bovey Tracey, Devon, and at Hoxne, Suffolk (in bed C?). Abundantly in the preglacial strata of Norfolk, but not in any of the interglacial deposits in Britain. During the Neolithic period it seems to have been one of our commonest trees; but afterwards disappeared from the southern half of England."

[2] The orchid, *Goodyera repens*, which was formerly supposed to grow only in wild coniferous forests, as in the Highlands of Scotland, has begun to appear, of late years, in various localities, where the Scots Pine has been planted, both in England and in France; and the problem as to how the seeds of the orchid reach these plantations is still unsolved. Cf. *Kew Bulletin*, 1906, p. 293; *Actes Premier Congrès Internat. Bot. Paris*, 382 (1900); and Fliche, in *Mém. Acad. Stanislas*, 1878.

[3] *Holinshed's Chronicles*, i. 358 (1807), reprint of the edition published in 1586.

[4] *Phil. Trans.* No. 275, p. 980 (1701).

[5] Derham, *Memorials of John Ray*, 25 (1846).

[6] Wareton, now usually written Warton, is in Staffordshire, not far from the Shropshire boundary.

[7] *Botany*, ii. 593 (1776).

[8] According to Britten and Holland, *Dict. Eng. Plant-Names*, 146 (1886), *deal-tree* is used for *Pinus sylvestris* in East Anglia and Northamptonshire, the cone being commonly called *deal-apple*.

from which these trees grew might have been brought from Norway in early times; and Sir H. Howorth suggests that the existence of the Capercaillie, whose bones have been found in Tertiary deposits in the eastern counties, would have been impossible unless either pines or spruce existed to feed them in winter.

The Rev. Leonard Blomfield read a paper before the Bath Antiquarian Field Club on December 9, 1885, in which he tried to prove that the numerous Scots pines, now growing in the neighbourhood of Bournemouth,[1] are descended from aboriginal trees; and gave the following list of names of places in England in which the word *fir* occurs, indicating that these localities were in early days probably noted for woods of *Pinus sylvestris*:—Firbank in Westmoreland; Furbecke or Firbeck, and Firbie or Firby, in Yorkshire; Furbie, Firby, or Firsby in Lincolnshire; Furcombe in the parish of Farnborough, Berkshire; Furle or Furleigh in Pevensey Rape, Sussex; Furland, a tithing of Crewkerne in Somerset; and Furland Hill, between Brixham and Dartmouth.

Loudon, p. 2167, says that the tree only began to be planted in Britain about the end of the seventeenth century; but the following extract from a letter[2] of James I. to the Earl of Mar, dated Oct. 30, 1621, shows that the introduction of the Scots pine into England was earlier.

"The Marquis of Buckinghame, being desirous to have firre trees planted aboute his house at Burleigh on the Hille, hath earnestlie requested us to cause him to be furnished as well with the seede as with young trees, which his desire wee willinglie wold have performed with all expedition. And because wee know none who so readilie can give us satisfaction in this pointe as your selfe, we have thoughte good by these presentes to require you with all expedition to cause some store of seede to be gathered eyther in your owne boundes or in those of the Marquis of Huntlie, where it may be soonest had, and so soone as possiblie may be, sende a man of purpos to Burleigh on the Hille with so much of the freshest and fairest thereof as convenientlie may be caried. And that yee cause sette downe in writing at what time and in what kinde of grounde the same is to be sowed, and with the maner of sowing thereof; also when the time of year is fitting for removing and setting of plantes and young trees. Yee shall likewise sende one to Burleigh with four or five thousand of them, with the like instructions of time, place, and maner of setting and preserving."

There is no reference to these trees in the *History of Burley on the Hill*, published in 1901; and enquiries have elicited no information, except that there are now on the estate six or eight Scots firs, which are not more than 25 feet high. A local woodman, about 60 years of age, whose father was woodman before him, never heard of the existence of old pines at Burley.

The common Gaelic word for Pine is *gius*. It occurs in a few Scottish names of places, as Craiggush, Kingussie Altnaguish, Dalguise. This word is commonly used for pine also in Ireland, and *ochtach* occurs in books. In spite of the wide prevalence in ancient times of pine in Ireland, place-names with either of these words

---

[1] The submerged pine forest on the sea-coast at Bournemouth is described by Sir C. Lyell in *Principles of Geology*, ii. 536 (1872).

[2] *Historical MSS. Commission, Report on MSS. of Earl of Mar*, p. 103 (1904).

are rare. Mr. T. P. O'Nowlan, a competent Gaelic scholar, has given me the following list : [1]—

GOOSE ISLAND, *oilean gius*, in Lough Derg, Co. Tipperary, "island of pines."

CLONYGOOSE, *cluain gius*, parish in Carlow, north of Borris, "meadow or plain of pines."

MULLAGHANUISH, *mullach an gius*, near Ashford, Limerick, "hill-top of the pine."

GARROOSE, *gardha gius*, near Bruree, Limerick, "garden of pines."

KNOCKNAGUISH, *cnocan an giuis*, about three miles north of Kenmare, "little hill of pine." KNOCKNAGUSSY, similar in meaning, is situated about three miles south-west of Lough Mask in Co. Galway, "hill of the pine."

KNOCKHOUSE, *cnoc gheainas*, three miles south-west of Mullinavat, Co. Kilkenny, "hill of pines"; the Gaelic word used here being a local variation of the common form.

OGHTY ISLAND, *oilean ochtaigh*, near Roundstone, Co. Galway, "pine island."

DROMOGHTY, *drom ochtaigh*, about three miles north of the tunnel on the Kenmare road, Co. Kerry, "pine ridge."

Apparently, though the pine tree was centuries ago well known in Ireland, there is very scanty evidence as to its existence as an indigenous tree in modern times. Everywhere in Ireland the roots of pine trees are often found *in situ* in the upper layers of the peat-mosses, showing that forests of pine grew in the peat and attained a considerable size. These peat-mosses are probably of late formation.[2]

Ray[3] quotes Mr. Harrison as an authority for pine "growing wild in the mountainous parts of Kerry where the Arbutus grows," about the beginning of the eighteenth century. Smith,[4] writing in 1761, says that "these trees have been much destroyed in recent years; for, except a small shrub here and there among the rocks, there are none standing at present of any large size."

Mackay[5] mentions, in 1825, a solitary pine tree standing near the foot of Mount Nephin in Mayo, which was supposed to be the last remnant of the pine forest of that county. This tree,[6] very large and very old, was living in 1866, the exact locality being an open bog at Deal Castle, near Crossmolina, at the head of Lough Conn, and had been fenced in by the Earl of Arran.

Hayes,[7] writing in 1794, speaks throughout his valuable book of *Pinus sylvestris* as Scots fir; and evidently in his day all the pines in Leinster at least were the product of Scotch seed.

In France the common pine is never met with growing wild in the plains. It is confined in the wild state to the Alps of Savoy, of Dauphiné, and of Provence, the

---

[1] While the above was passing through the press, Mr. O'Nowlan sent me a further list, as follows :—Lough Aguse, name of two lakes, one near Pettigo, Donegal, and another in Fermanagh; Lough Ayoosy and Aghoos, in Mayo; Cappayuse in Roscommon; Meenaguse in Donegal; Drumgoose and Derrynoose in Armagh; and Annagoose Lake in Monaghan.

[2] The evidence for this is too large a subject to be entered upon here. In certain peat-mosses no less than three distinct forests are discernible, occupying different depths; and the uppermost forest, always of *Pinus sylvestris*, probably dates from historic times.

[3] *Synopsis Methodica*, 442 (1724).

[4] *State of the County Kerry*, 372 (1761).

[5] *Catalogue of Plants in Ireland*, 83 (1825).

[6] *Cybele Hibernica*, 277 (1866).

[7] *Practical Treatise on Planting*, 133, 167 (1794).

mountains of Auvergne, the Cevennes and Pyrenees, and is specially noted as occurring chiefly on the slopes with a southerly aspect.[1]   It is common in the eastern part of the Pyrenees, between 4000 and 6600 feet altitude.

In Spain the common pine is also restricted to the mountains, only forming woods on northern slopes; and in the Sierra Nevada, forms the large and splendid forest of La Granja on the north side of the Sierra de Guadarrama, where it ascends to 6500 feet.[2]

In Germany very extensive and pure forests of pine occur in the north-east, always on sandy soil in the plain.   These forests are called heaths, as they contain wide stretches covered only with heather and many swampy areas.   Such forests are common in the provinces of East and West Prussia, Pomerania, Brandenburg, Posen, upper Silesia, Saxony, and in the kingdom of Saxony.   Other large forests of the same kind occur more isolatedly in north Schleswig, Hanover, Jutland, and Holland. In the valley of the Rhine, both in Alsace and Baden, very fine forests of pine are also met with, as at Haguenau, likewise on sandy soil.   In the mountains of Middle and Southern Germany, the pine only grows in small groves or as isolated trees. Similarly throughout the Alps and Carpathians, in Hungary, and on both shores of the Adriatic the pine is rare, only occurring in small woods.

The pine does not occur wild in the islands belonging to Denmark, and is totally absent from the Hungarian plain, the Bakony forest in Hungary, the Central Carpathians, Banat, and Slavonia, and is not met with in the alpine and subalpine regions of the high mountains of Central Europe.   (A. H.)

In Switzerland this tree does not seem to attain such a large size as in Scandinavia.   A tree at Campodials in the Grisons, figured on Plate xi. of *Les Arbres de la Suisse*, is said to be 80 feet high by 10 feet 6 inches in girth at 4 feet from the ground.   It grows on the edge of a forest at about 3000 feet elevation, on crystalline rocks.

In Norway and Sweden the common pine constitutes by far the largest portion of the forest, and flourishes farther north than any other tree except the birch. Though truly virgin forest is now becoming a rarity in Norway, and in the more accessible parts of Sweden and Finland, yet the area of land covered with pine and spruce is still so large and so much better suited than England for the production of commercial timber that we shall, in my opinion, never be able to produce it of such good quality and at so low a price.   I have seen *Pinus sylvestris* at its best in the forests of South-Eastern Norway in the valley of the Glommen; where the bright yellow bark of the upper part of the tall trunks on the banks of the river is a marked feature of the scenery; and in the far north in upper Saltenfjord where the oldest pines known to me now exist; though here, as elsewhere, they are rapidly being felled to supply the great demand for building and mining timber.   In the more central provinces of North and South Trondhjem, and on the coast, the pine does not seem to grow to such a great size, probably because the soil and climate are too wet to suit

---

[1] It is not a native of the French side of the Vosges, but occurs on the German side at 1300 to 3000 feet altitude.   It is not wild in the Ardennes or in the Jura.

[2] Mentioned by Christ in *Flore de la Suisse*, 198 (1907).   Cf. *supra*, p. 574, var. *nevadensis*.

it as well as the spruce, for the pine is a lover of a sandy soil and a dry long winter, with a hot sunny summer.

Dr. Schübeler, in his *Viridarium Norvegicum*, i. 375, gives many details about the pine, from which I gather that its range extends from the south, where it reaches an elevation of 3500 feet above the sea, to the inner valleys of Finmark, where in lat. 70° N. it attains in Alten and Porsanger fjords as much as 60 feet high and 7½ feet in girth. He tells us that formerly there were pines on the Dovrefjeld, near Jerkin, at an elevation of 3200 feet, as much as 1 foot in diameter, where no trees now exist; and that near Roros, now one of the bleakest and coldest towns in Norway, the forest was, in 1773, so dense as to be almost impassable. The tallest pines in Norway that he mentions were near Holden in Lower Thelemarken, where one was measured 104 feet high, with a diameter at the ground of 2 feet 10 inches, and at 70 feet high of 9½ inches. Another at the same place was 105 feet high, and 5 inches in diameter at 96 feet up. At Klosterskogen in Skien, one was measured 108 feet by 6 feet 5 inches at breast height. The greatest girth that he mentions is about 15½ feet.

I have myself measured at Graddis in Junkersdal, within the Arctic Circle, and at an elevation of at least 1200 feet, pines of over 50 feet high and 12 to 13 feet in girth. One of these, which was cut down, was 34 inches in diameter and about 240 years old, but the outer rings were so close that I could not count them accurately, the first 100 years' growth being over 26 inches in diameter, showing that in this latitude at least the increase after this time is very slow. The tallest that I saw in this valley was 84 feet high near the Government Forest Nursery at Storjold. I observed that in Junkersdal the natural regeneration from seed was poor, and that in the upper parts of the valley the young seedlings were very small and stunted, and birch seemed to be taking their place. In this valley on July 10, 1904, vegetation had only just commenced, and the pines had not pushed their young growth, though *Cypripedium Calceolus* was in flower. A severe frost which took place in April, – 14° to – 16° Réaumur, after warm weather in March, had killed most of the young shoots where not protected by snow.

Schübeler gives several illustrations of the curious forms which this tree sometimes assumes. His Fig. 59 shows a tree in which the branches are very short and which has the shape of a northern spruce rather than that of a pine. Fig. 60 shows a branch with a great bunch of forty to fifty closely packed cones surrounding it. Figs. 61 and 62 show the power which the tree possesses of sending out upright stems of considerable size from a fallen trunk whose roots still retain their hold on the ground. Fig. 63 shows an immense witches' broom, forming a dense mass of living twigs in a ball 10 feet in diameter, which surrounds the trunk of a pine growing at Aaseböstäl in Nordfjord.

It is occasionally planted in Iceland,[1] but does not long survive the severe climate, though Hooker was told that a single dwarf tree grew on an island in a lake between the head of Borgarfjord and Reyholt.

As little is known with regard to the so-called Riga pine, which was for long

[1] Babington, in *Journ. Linn. Soc.* (*Bot.*) xi. 50 (1870).

the most celebrated for masts and shipbuilding purposes, and has been found in France to be the best variety in cultivation, we may refer our readers to a recent publication by Von Sivers,[1] with a map of the distribution of pine and spruce, which shows a comparatively small area of the former. The author states that though the pine is everywhere at home, it grows best on sand, especially where that is underlaid by good soil, and that in favourable places it reaches often a height of 150 feet. The area which is occupied by pine plantations in Estland, Livland, and Kurland is estimated at 638 square kilometres. It would, therefore, seem that the production of pine timber is not sufficient to continue the large export upon which in the past reliance could be placed. And though there are still large reserves of pine forest in Northern Sweden and Finland, yet it was stated by Mr. A. Howard at a recent meeting of the Society of Arts, in a discussion on Sir Herbert Maxwell's paper on Forestry, that the size of the deals imported from the Baltic is steadily diminishing, and that a much smaller proportion of 11-inch boards is now sent than was formerly the case.

In the forests of the lower valleys of the Altai Mountains in Siberia I have seen the pine attain a greater size than anywhere in Europe, some trees in the valley of the Biya river, a tributary of the Ob, which I observed in 1899, being estimated at 150 to 160 feet in height, and clean to 100 feet, at which height they looked as if they were 5 or 6 feet in girth.[2]

CULTIVATION[3]

Of all the many species of pine, none is so widely distributed in Europe, so common all over Great Britain, so easy to grow as the Scots pine, or Scotch fir, as it is often incorrectly called. Its vigorous constitution and rapid growth when young enables it to exist and even to thrive in almost all situations, and though the variations which it has produced in a wild as well as in a cultivated state are innumerable, yet the most casual observer can hardly fail to distinguish it from any other species which is likely to be seen in cultivation. I have seen the tree in the greatest perfection on the sandy soils of Surrey, Sussex, Bedfordshire, and Notts, on the rich loams of the south-western and midland counties, on the dry sandy glacial deposits and heath-clad hills of the Highlands, and in many parts of Europe.

Whether the Scots pine was at first principally propagated in England from native Scotch seed or from German seed is doubtful, and probably the earliest planted trees came from various sources; but so far as my experiments have gone, it seems as though the seedlings grown from acclimatised trees are now more flourishing, and grow faster in the south of England than those from German, Highland, or Scandinavian seed. I have tried plants of the same age from all these sources in Gloucestershire, and have found those sent me from the New Forest the most promising in their younger stages. If rapidity of growth at first is any indication

---

[1] *Die Forstlichen Verhältnisse der Baltischen Provinzen*, Riga, 1903.

[2] Farther east, near Krasnoyarsk, a pine has been measured, which at 200 years old was 40 archines (93 feet) high, and 11 verschoks (19¼ inches) in diameter; but this is far surpassed by the pines found near Bélovège, where trees 150 years old are said to measure 60 archines (140 feet) high by 12 verschoks (21 inches) in diameter, and contain as much as 100 sagènes (about 250 cubic feet) of timber. Cf. *Les Forêts de la Russie, Paris Exp.* 1900.

[3] Loudon's excellent account of the culture should also be referred to, pp. 2178-2183.

of vigour, I should prefer them, though I would not plant Scots pine as a forest tree on any soil where I could get larch to grow even fairly well; and on dry chalk and limestone soils it never grows with the vigour that it does on sandy soils.

Large parts of the open heath of the New Forest, though constantly pastured by horses, are becoming overgrown with Scots pine to such an extent that if they escape fire it seems as though they would eventually turn those open wastes into a more or less dense pine wood.[1] But on clay soils, and wherever a rank growth of grass, ferns, or briars is found, natural reproduction is comparatively rare, and over the whole of the Cotswold Hills I only know of a few places where self-sown pines can be seen.

If natural reproduction is desired, the best way of encouraging it is to uncover lines or patches of soil in the winter, on which the seed falling in April can germinate; but the growth of these self-sown plants is, as usual with almost all natural seedlings, at first much slower than that of planted trees. In very old pine woods of 100 to 150 years' growth, such as are found in Strathspey and in a few parts of England, the accumulated carpet of dead pine needles seems to prevent the young seedlings from establishing themselves; and in the Belvidere plantation at Windsor Park, which is one of the finest in England, I saw no self-sown seedlings under the fine old trees, many of which are 100 feet and more in height.

In such cases it is best to burn the heather or to graze it closely with sheep and cattle, and in many cases this is a necessary preliminary to preparing the ground for natural reproduction in Scots pine woods; but if the soil produces grass rather than heather, the regeneration is always less successful and requires more assistance.

I shall not attempt to give any estimate of the financial results of planting Scots pine as an unmixed plantation, because the conditions of soil and climate are so varied that any estimates, such as we see commonly given in books on forestry, are usually misleading. On very sandy, dry soil it will probably pay as well or better than any other tree, because it can be planted so cheaply, and will regenerate itself so easily.[2] But it must be kept thick enough to clean its stem before the branches get large, and in fact it may be better not to thin at all until 20 to 30 years old, when the weaker stems which will hardly pay to cut and carry out will be killed by their stronger neighbours. On high moorlands also it may be, and now often is, as profitable a crop as larch, because it grows well in windy and exposed situations; but I would not plant it, except as a nurse to other trees, on any soil where experience has shown that a more valuable tree will grow to fair timber size, and the plan often adopted of mixing it in larch plantations on calcareous soil has led in many places to absolute failure.

With regard to the possible yield of Scots pine in England, I have heard of nothing better than a part of the Dipton Woods near Hexham, Northumberland, the property of Lord Allendale. This was described in *Trans. Scott. Arb. Soc.* xx.

[1] I was informed during a recent visit to the New Forest that the commoners already complain that the pasture is deteriorating from this cause.

[2] I have seen no better example of natural regeneration than on the Duke of Bedford's property at Old Wavendon heath, near Woburn.

84 (1907), as containing something like 9000 cubic feet per acre, which at 4d. per foot works out at £150 per acre. Mr. Gillanders informs me that the soil is fine fresh sand, the elevation 550 feet, and the aspect north-east.

Some of the best foresters consider it an excellent nurse for oak, but beech is now usually preferred for this purpose, on any soil where the latter will thrive.

From 70 to 100 years, or in the Highlands 120 to 150 years, is about the age at which the tree is usually mature for felling; as, when younger than this, the timber is comparatively soft and inferior, though after creosoting it may be utilised for many purposes where strength is not important. Where pit-props are saleable, it is more profitable to cut the crop as soon as large enough for that purpose.

PROPAGATION

No coniferous tree is more easy to raise from seed, or easier to transplant. In Scotland the cones are usually gathered in autumn and the seeds extracted by kiln drying when required for sowing; but they are better left on the trees till spring, and the seeds may then be easily extracted by damping the cones and exposing them to the sun till they open. The seeds will keep good for several years, and if not wanted to sow at once, are better extracted in the summer after they are ripe and kept until the following March, when they will germinate as readily as, and perhaps produce stronger plants than, those extracted by artificial heat in winter and sown the next spring.

The best nursery practice is to sow them broadcast on slightly raised beds of sandy soil about three feet wide, and cover with about half an inch of fine earth, some of which may be raked off just before they begin to germinate, leaving a fresh surface uncaked by the rain and sun.

If sown too thickly, the plants will be drawn up closely, and will not remain two years in the seed-bed without becoming crowded. Some people advise transplanting at one year old, but in my experience two years is better, and, if carefully handled, the percentage of loss caused by transplanting is very small.

If the plants are to be put out on heath or sandy land, the stronger ones may be permanently planted out from the seed-bed; but in all soils which are grassy and weedy, it is better to keep them one or two years in nursery lines, which should be about 1 to 1½ feet apart, and the plants 3 to 6 inches apart in the rows, according to whether they are intended to remain one or two years in lines.

It is rarely desirable or necessary to allow them to remain more than four years in the nursery; but if plants larger than 1½ to 2 feet are required for special purposes, they must be transplanted when four years old and put in rows about one foot apart and two feet apart in the rows.

The best time for planting out large trees is in the autumn, as soon as the terminal buds become hard; but small plants should not be transplanted till after the period of severe frost has passed, or they will in most soils be lifted by frost. If, however, it is necessary to do so, stones should be put round the collar of the tree, not only to keep them fast in the ground, but also to keep out the drought during the first year.   I have found that this is a very successful method to adopt with all small trees on stony soil liable to drought.

An account of the best way of growing the Scots pine from seed was written by the Earl of Haddington in 1760 to his grandson, and is quoted in the Highland and Agricultural Society's volume on the *Old and Remarkable Trees of Scotland*, published in 1864.   This account is very practical and based on personal experience, and interesting as showing how much care was taken by the planters of those days to ensure good results.

REMARKABLE TREES

As to the height the Scots pine attains in Great Britain, many particulars have been given by Loudon, which in most cases cannot be relied on for accuracy, but we have reliable measurements which show that the tree rarely exceeds 110 feet, and more usually is not over 100 feet.

In the *Victoria County History of Hants*, it is stated in vol. ii. p. 469, that trees 130 feet high were growing at Beaulieu, but Lord Montagu tells me that he has never actually measured one over 116 feet, of which height one was blown down some years ago.   I saw these trees in June 1906, and though many exceed 100 feet, and are clean to 70 or 80 feet, with a girth of about 7 feet, I could find none over 110 feet.   At Rooksbury Park, near Wickham, Hants, there are some which, I think, are taller, growing, mixed with beech and oak, in a dense thicket of rhododendron. The largest I measured here was about 115 feet by 10 feet 4 inches, dividing at about 17 feet into three tall, clean stems.

At Carclew, Cornwall, the seat of Colonel Tremayne, there is a fine avenue of pines, the tallest of which I found to be about 110 feet (Plate 160).   At Pain's Hill, Surrey, Henry measured a tree of 106 feet.   In the Belvidere plantation in Windsor Park, one of the finest old pine woods in England, planted about 1760, there are many trees of 100 feet and some perhaps a little more.   There were some very tall trees at Hursley Park, Hants, of which I have no exact measurement, but I hear that few, if any, of them remain.

At Blickling Hall, Norfolk, the property of the Marchioness of Lothian, there is an immense tree, perhaps one of the oldest in England, which, when described by Grigor in 1841,[1] was 70 feet high and 16 feet in girth at 6 feet from the ground.   He thought it the largest tree of the kind in Norfolk.   When I saw it in April 1907 it was 96 feet by 17 feet 1 inch, dividing at about 10 feet into two main trunks, which were chained together 40 feet up.   It had a large burr at the base.   At Stratton Strawless there are some fine trees planted about 1740 by Robert Marsham, measuring about 100 feet by 9 feet.   A tree was reported by Loudon at Castle Howard, Yorkshire, as being 120 feet high, with a bole 100 feet long, but I could not identify this tree as still living in 1905.   At Cocklode House, in Sherwood Forest, there is a fine avenue of Scots pines about 160 years old, which are 90 to 100 feet high, and 9 to 10 feet in girth, but many of them have been blown down.

The tallest that I have ever seen or heard of is in the grounds at Petworth,

---

[1] *The Eastern Arboretum*, p. 100.

Sussex, the seat of Lord Leçonfield, where the soil seems particularly favourable to very tall trees. A careful measurement of this in 1905 gave the height as 120 feet by 11 feet in girth, and a bole of 35 to 40 feet, where it divides into two stems. Sir Hugh Beevor, who saw it in 1904, did not make it quite so tall. This tree appears in the foreground of Plate 162.

No park in England contains a greater number of fine and picturesque old pines than Bramshill Park, Hants, the seat of Sir Anthony Cope, who tells me that he believes them to have been planted about the year 1600, and to be some of the oldest in England. The soil here is very light and sandy, and the oldest pines are in avenues, which have become rather irregular in course of time. The tallest trees that I measured here were not over about 80 feet high by 10 to 12 feet in girth, but there is one splendid tree in the Gravel Pit drive which is about 80 feet by 16 feet, of which I give an illustration (Plate 161). There are many self-sown seedlings of various ages in this park, but no other trees of remarkable size.[1]

At Dyrham Park, Gloucestershire, the seat of the Rev. W. T. Blathwayt, there is a very fine tree, of which I am indebted to the owner for a photograph, which measures about $73\frac{1}{2}$ feet high by 14 feet 9 inches in girth, dividing at 4 feet into three trunks. There are some very fine clean Scots pines in Stowe Park, near Buckingham, one in the Queen's Quarter being over 100 feet high, with a clean bole over 60 feet long and 11 feet 3 inches in girth. In the Fir Grove, at Bayfordbury, Herts, there is a tree,[2] with a clear stem of over 50 feet, which measured in 1905, 95 feet high by 9 feet 7 inches in girth.

In Wales I have heard of no Scots pine of greater size than one at Penrhyn Castle, which Henry measured in 1904, and found to be 110 feet by $7\frac{1}{2}$ feet, and about 70 feet to the first branch. At Gwydyr Castle he measured one about 85 feet by 11 feet 2 inches on which a mountain ash seedling was growing.

In Scotland there are so many fine old trees that it is impossible to mention more than a few of them. Perhaps the finest, if not the tallest, is a tree at Inveraray, of which a beautiful photograph by the late Vernon Heath is in the museum at Kew. I measured this tree in September 1905, when it seemed to have changed very little in appearance, and though supported by chains above the fork, is very sound and healthy. It measures 110 feet by 14 feet, forking at about 35 feet, and leaning considerably to one side. The Duke of Argyll informs me that it was probably planted about 1620. Plate 162 shows the present appearance of this tree. There are many other very fine Scots pines at Inveraray on the lower slopes of Dun-i-cuach, but none equal to this in height or girth.

On the banks of the Tay, near Dunkeld, there is a very graceful tree of weeping habit though of no great size (Plate 163), which measures 77 feet by 11 feet 6 inches in girth; and at Blair Atholl there are some curious old pines in a row

---

[1] Bunbury, who visited Bramshill Park in 1859, mentions the tradition that these trees were introduced by James I. from Scotland at the same time that he began building Bramshill, and states that there were three magnificent Scots pines at Eversley Rectory, which were coeval with those of Bramshill. Cf. Lyell, *Life of Sir C. J. F. Bunbury*, ii. 138, 139 (1906).

[2] This tree measured, in 1816, 5 feet 8 inches in girth, according to an entry in an old note-book, now in the possession of Mr. H. Clinton Baker.

by the Inverness road, which the Duke of Atholl informs me are probably part of the booty carried off by his ancestors in 1684 from Inveraray, as described in *Chronicles of the Atholl Family*, by the present Duke.[1] Nothing can better illustrate the importance which was paid to trees and planting even at this early period, when the Highlands were hardly civilised; than that so many exotic trees should have existed at Inveraray, and that it should have been thought worth while to carry them to such a distance when wheeled carriages could not have traversed the country.

An immense Scots pine, which I have not yet been able to visit, grows at Guisachan, Inverness-shire, now the property of Lord Portsmouth, whose forester, Mr. Davidson, informs me that in February 1907 it was 53 feet 10 inches high and 16 feet 8 inches in girth at the ground, and 15 feet 7 inches at 5 feet. At 11 feet from the ground, below the first branch, it is 16 feet 10 inches in girth. The trunk has been cut into at the base, which is believed by old people living near to have been done by smugglers, as an illicit whisky-still once existed near it. A drawing of this tree was made for the late Sir Dudley Marjoribanks, of which I have a copy. Mr. E. Ellice tells me that there are a number of very large old pines in the Guisachan Woods, girthing over 14 feet.

Mr. E. Ellice of Invergarry informs me that there are a considerable number of old native Scots pines at that place, among them one which attracted the late Mr. Gladstone's attention, and of which he sends me a sketch, with the following measurements:—Height, 70 to 80 feet; girth at the ground, 20½ feet; at 5 feet, 16 feet 3 inches; at 10 feet, 15 feet 9 inches; at 13 feet, 17 feet. A figure of this will appear in Vol. IV. Other trees near it measure 14 feet 3 inches, 12 feet 10 inches, and 12 feet 9 inches; and these appear to be the parents of many more which may be divided, according to their age, into three classes: those of 120 to 150 years, of which there are some hundreds; those of from 80 to 100, of which there may be 15,000 to 20,000; and younger trees.

The finest forest in this locality is in Glen Malie, on Lochiel's property,

---

[1] " In 1684 or 1685 the Marquis of Atholl did carry out of the orchard enclosures and shrubberies at Inveraray—

	£	s.	d. Scots
600 Silver and Spanish fir trees, 6 years' growth	1800		
500 Pinaster trees, 12 years' growth	500		
500 Pine trees, 10 years' growth	500		
400 Yew trees, 16 years' growth	266	13	4
6000 Holland trees (holly)	1800		
600 Beech trees	600		
2000 Lime trees, 4 years	400		
400 Buckthorn, 8 years	120		
600 Black and White Poplars, 13 years	200		
400 Chestnut	266	13	4
200 Horse Chestnut	200		
300 Walnut	200		
200 Fir trees, 5 years	400		
20,000 Ash, Plane, and Elm trees	2400		
200 Pear and Apple trees	400		
200 Plum trees	200		
300 Cherry	300		
1000 Apple and Pear stocks	3000		
	£13,553	6	8 Scots.

This claim was settled for £13,000 Scots or £1333, 6s. 8d."—*Chronicles of Atholl Family*, i. 265.

running up from the shore of Loch Arkaig for five miles, and in it there is one tree, even larger than that last mentioned, of which Mr. Ellice has sent me a sketch.   At the narrowest part of the trunk, three feet from the ground, this tree measures 18 feet 8 inches in girth, and at about 10 feet divides into three tall trunks, each of which girths between 11 and 12 feet.   Just below this fork it girths about 30 feet, and appears to be sound throughout.

At Novar, in Ross-shire, there is an old plantation containing a number of very fine Scots pines, one of which measures 105 feet by 10 feet 3 inches, and larger ones can probably be found in this district, as well as in the sandy district which extends east from Inverness, where many large plantations of this tree flourish exceedingly.

The finest individual trees and the finest Scots pine plantation that I have seen is in a place called Wishart's Burn, near Gordon Castle, Banffshire, on red sandstone soil.   Though supposed to be about 180 years old, most of the trees are still in good health and quite sound, though wind has made some gaps in the plantation. When I visited them in April 1904 the tallest tree was about 117 feet high by 10 feet 11 inches in girth.   It forks at about 45 feet, but carries its girth so well that the bole would, I think, measure 45 feet by 28¼ inches quarter girth, about 245 cubic feet, and the tops might contain 50 feet each, making a total of 345 cubic feet (Plate 164).   Another tree standing near it was 114 feet by 8 feet 10 inches, and I estimated that the older trees here average over 100 feet high by 8 feet in girth.   Mr. Webster, gardener to the Duke of Richmond, who showed me this beautiful spot, agreed with me that the average number of trees to the acre here was about sixty, and their average contents about 100 cubic feet; but many have been cut and sold at as much as £7 : 10s. each, to make masts for large herring boats.   One of these trees probably is the one figured by Loudon (p. 2162) as a model of a fine Scots pine clear of branches to 50 feet, and containing 260 feet of timber.

There are also very fine plantations of Scots pine in the neighbourhood of Castle Grant, the seat of the Dowager Countess of Seafield, in Inverness-shire, a place celebrated for its good forestry, and where better examples of thickly grown self-sown pine may be seen than anywhere else in Scotland.   Mr. Grant Thomson, who has had charge of the extensive woods here for forty-five years, told me that the oldest planted trees are about 180 years old ; many of these are 80 to 90 feet high and 8 feet in girth, and number sixty to seventy per acre.   Some have already begun to decay at the heart, and it was noticeable that on the thick bed of decayed pine needles under them seedlings would not grow.   This has been referred to by Prof. A. Schwappach in a paper on the "Forests of Scotland" in *Trans. Roy. Scot. Arbor. Soc.* xv. 13 (1898).

In this neighbourhood are the most celebrated and extensive natural forests of Scots pine in Great Britain, which I visited in April 1904.   Glenmore forest, the property of the Duke of Richmond, was perhaps the best of these until 1783, when a great part of the mature timber was sold to an English merchant named Osbourne, who cut it down in twenty-two years and floated the timber to Speymouth, where forty-seven ships of upwards of 19,000 tons burden were built from them at an expense for labour only of £70,000 (cf. Loudon, p. 2161). Glenmore Lodge lies at an elevation of 1050 feet on the shore of Loch Morlich, where some of the finest pines still stand. One of them (Plate 165) is interesting as having for many years been the eyrie of an osprey whose nest is visible in the photograph, which was taken in the interval between two snowstorms; but the birds, though carefully protected, have not bred there since 1900.[1] This tree measures 56 feet by 13 feet, and is very characteristic of the native Scots pine in its habit. But perhaps the most interesting tree in this forest is one from which a plank, now preserved at Gordon Castle, was cut and presented by Mr. Osbourne to the Duke, in 1806, as a memento of the forest. I could not count the rings of wood in it exactly, but the Duke of Richmond informs me that there are about 236. I measured the plank 5 feet 5 inches wide at the butt end and 4 feet 4 inches at the top. The sapwood is worm-eaten, and the colour of the wood has become very dark. I saw still lying on the hillside above Glenmore Lodge, near the upper limit of the Scots pine, at an elevation of about 1400 feet, a huge top, over three feet in diameter where it was cut off, and was assured by Francis M'Pherson, an old woodman, who showed it to me, that it was the top of the identical tree from which the above-mentioned plank was sawn. Though overgrown with moss and heather, much of the wood appeared to be still sound, after lying for nearly a hundred years. In confirmation of this I may state that Mr. J. Michie showed me, in Ballochbuie Forest, the remains of a pine which was sawn up and found sound after lying seventy years on the ground.

In Abernethy Forest there are also many fine old pines, one of which, Mr. Grant Thomson's favourite, is shown in Plate 166. It measures about 60 feet high by 14 feet 3 inches in girth, and, though it divides into five tops, is a most graceful tree. Much of this forest was burnt down many years ago, but has become self-sown with young trees, and is now open wood covered with long heather, and a favourite wintering ground for deer. We measured a group of the best clean self-sown trees supposed to be about 120 years old, and estimated them at about 120 per acre, with an average timber height of 40 to 50 feet and a cubic content of about 25 feet (Plate 167). Such trees, where they stand, are worth about 6d. per foot. Many cones are gathered in this forest for seed, of which about 8 ounces from a bushel is the average produce; and there is a large nursery where they are raised, the growth being very slow as compared to what one sees in England, on account of the cold and damp situation.

I next visited Ballochbuie Forest, by the kind permission of His Majesty, who preserves this beautiful forest with great care. It is now perhaps the largest area of natural forest in Scotland, extending for several miles along the south side of the upper valley of the Dee. The photographs, Plates 168 and 169, give a good idea of the picturesque scenery of this forest and of the fine trees in it, many of which are 80 to 90, and some as much as 100 feet high, by 7 to 8 feet in girth. The

---

[1] Mr. S. R. Clarke has sent me a photograph of a Scots pine at Fasnakyle, which is used annually as an eyrie by the Golden Eagle.

best of those which I saw might measure 60 feet timber height by 18 inches quarter-girth or 135 cubic feet.   In the best stocked areas the trees might average 120 per acre at 120 years old, of which forty trees of the first class would average perhaps 40 feet, forty second class 25 feet, and forty of the third class perhaps 12 feet each, or about 3000 feet to the acre.   But these figures are only a rough estimate, as the King's trees are not cut for sale, and in consequence Mr. Michie could not give me exact figures, but thinks they are worth from 6d. to 8d. per foot standing.

The Black Wood of Rannoch is an ancient natural pine wood extending for about three miles along the south shore of Loch Rannoch, and though there are no trees of exceptional size, there are many very picturesque ones, which are protected by the owner, B. C. Vernon-Wentworth, Esq., whose residence at Dell is near the east end of the wood.   The largest which I measured was 91 feet by 11 feet 7 inches, with a very spreading base 15 feet 5 inches round at one foot from the ground.   The greater part of the wood is open and covered with long heather, among which seedlings were fairly numerous wherever the soil was exposed.   Many of the large trees were blown down thirteen years ago, and their timber, which was of very fine quality, was used by Sir J. Stirling Maxwell in the interior work of his house at Corrour.

There is an excellent account of the Black Wood of Rannoch and its history in chap. xxxiv. of that admirable book, Hunter's *Woods and Forests of Perthshire*, which, though now getting out of date, as it was published in 1883, gives a better account of the great estates and their trees than exists for any other county in Great Britain.   In this work dimensions are given of the finest trees then existing in the Black Wood, which are remarkable more for their great girth, spreading and massive branches, and picturesque appearance than for their height.

On the shores of Loch Hourn, on the west coast of Inverness-shire, there are many native pines scattered among the birches, but none of large size, a few of those near the sea resembling the stone pine of Italy in habit.   Henry observed that many of these trees do not ripen seed.

With regard to the elevation at which the Scots pine grows in Scotland, we have various somewhat conflicting estimates.   Mr. Michie tells me that Craig Doin (1900 feet) is about the highest level he knows it to reach in Ballochbuie.   Mr. Seton P. Gordon, however, says[1] that he has seen a young Scots pine growing at a height of about 2700 feet not far from the source of the river Dee on the south slopes of Brae Riach, though he regards this as very exceptional.   Mr. Hugh Boyd Watt[2] also considers 2700 feet quite an unusual altitude, and says, "From personal observations made on and around the Cairngorm mountains (and in no other district in this country do forest trees attain higher levels) I can say that even at 2000 feet above sea-level the Scots pine has difficulty in holding its ground. . . . On the southern slopes of Beinn a' Bhuird (Glen Quoich) considerable numbers of fairly well-grown Scots pines reach up from 2000 feet to 2100 feet, and I know no other place where what may be called the forest line is so high. . . . In other localities, apparently favourable to their growth, the pines do not in any numbers exceed an

[1] *Country Life*, 17th Aug. 1907, p. 245 ; 31st Aug. 1907, p. 322.   [2] *Op. cit.* 7th Sept. 1907, p. 359.

altitude of 1500 to 1700 feet. This is approximately the level at which they die out in Abernethy, Rothiemurchus, Glen Feshie, Glenavon, Invercauld, Birse, and Glen Tanar." When stalking on Ben Avon I saw with the telescope some pines in the upper part of Glen Derry which I supposed to be at an elevation of about 2000 feet, and Mr. Michie, who has seen these trees, thinks that this estimate is not far from the mark.   (H. J. E.)

In Ireland the common pine grows with great vigour and beauty, the bark becoming bright red in colour and the leaves very glaucous. The tallest trees, which I have seen, are at Curraghmore, the seat of the Marquess of Waterford, where, near a stream, I measured one 110 feet high and 7 feet in girth; some, but difficult to measure accurately on account of their position in a dense wood, were probably 120 to 125 feet in height, the largest of these having a girth of 9 feet.

At Doneraile Court, Co. Cork, there are some fine pines, growing scattered in an oak wood, the largest of which I made 97 feet by 11½ feet, with a clean stem to 50 feet. These trees are supposed by local tradition to be of native origin, and are called Irish pines; but they have evidently been planted, and there is no means of determining whether they originated from seed collected in Kerry from aboriginal pines still existing there in the 18th century, or, as is more probable, from Scotch seed, as they are probably about the same age as the famous larches at this place, which are reputed to have been sent to Doneraile by the Duke of Atholl.

At Emo Park, Portarlington, there are many fine trees, the largest seen measuring 91 feet by 7½ feet and 88 feet by 9 feet 1 inch. There is also a splendid tree, growing near the gate of Mr. Walpole's beautiful garden at Mount Usher, on the Rossanagh property, which is 11 feet 9 inches in girth, and probably 80 feet in height. At Castledawson, Co. Derry, an old tree measures 80 feet in height by 11 feet 4 inches in girth. There are many fine trees scattered through Coollattin in Wicklow. These grow on moist boggy soil; and I measured two clean of branches to 60 feet, which were 87 feet in height, and 9 feet 5 inches and 8 feet 1 inch respectively in girth.

At Luttrelstown, near Dublin, Hayes[1] measured a "Scots fir, eighty-five years' growth from the seed, of 11 feet 6 inches in circumference, and another of very great height 11 feet 10 inches round." He gives several other instances of the rapid growth of the tree in Ireland.[2]

Mr. T. W. Webber, late Deputy Conservator of Forests in India, in the appendix to his book on the *Forests of Upper India*, gives an interesting account of the growth of Scots pine in Ireland, the planting of which he strongly advocates. To the objection that home-grown timber is of inferior quality, he replies that the wood of *Pinus sylvestris* found in bogs in Ireland is often of great length and thickness, sound, fine-grained, solid and straight, and so excellent that it has been used by coach-builders as superior to Memel timber. Where such timber grew ages ago,

---

[1] *Practical Essay on Planting*, 133, 167 (1794).

[2] At Powerscourt an immense Scots pine was blown down by the great gale of February 1903, which I saw on the ground soon afterwards and which measured about 12 feet in girth. Some boards cut from the tree were kindly sent me by the late Lord Powerscourt, which show its growth to have been very rapid.   (H. J. E.)

similar material might be produced to-day, if close planting and slow growth were the rule.   To prove this, he gives the actual dimensions of Scots fir grown under two different conditions in Ireland.

	Grown thirty to the acre, with spreading crowns.	Grown 200 to the acre, with small crowns.
Girth	5 feet	5 feet.
Height	50 „	75 „
Age	40 years	100 years.
Diameter	20 inches	20 inches.
Heartwood	12 „	19 „
Sapwood	8 „	1 inch.
Rings per inch	4, uneven	10, regular.

These trees are quickly grown on deep soft soil, and are liable to be blown over.   Timber, coarse, knotty, light, and perishable; large amount of sapwood.

These trees were slowly grown on a hill-side on poor and stony soil; standing close they resist storms.   Timber fine-grained, hard, heavy, durable, and equal to best Memel.   Scarcely any sapwood.

Mr. Webber has kindly written to me that the trees just mentioned grow on his own property at Kellyville, near Athy, in Co. Kildare.   A beam, made out of the fine pine timber grown on the hill-side, placed in the front of a conservatory twenty-five years ago, is still sound and good.   Mr. Webber has Scots pine thriving on pure rock, where there is little or no soil.   He states that at Emo Park near Portarlington and on the road to Maryborough there are striking instances of pine succeeding on pure black bog, and self-sown seedlings may be seen spreading all over the turf-moss.   He reiterates the conclusions given above, namely, that the pine should be planted densely on poor soils, where it will resist the wind and yield timber without any appreciable sapwood, whereas on deep soft soils it is easily blown over and yields coarse and valueless timber.   In the bog in Emo Park, Mr. Webber found great bases of Scots pines with their roots in the boulder clay, of gigantic size, showing that the tree was indigenous before the bog began to grow ages ago.

In some parts of Ireland, Scots pine may be seen thriving on deep peat-moss, the condition necessary for success being judicious preliminary drainage.   In mosses soaking with water, trees languish and die on account of the lack of air at their roots. On the other hand, if the drainage is too deep, the upper layer of the peat becomes so dry, that the trees suffer from want of water.   Near Castledawson in Co. Derry, a considerable area of undrained peat-moss is covered by healthy and vigorous pine trees, which are natural seedlings, the product of seeds blown from an adjoining plantation.   Here, however, the peat-moss rests on the side of a sloping sandhill and is not waterlogged.   Natural pine seedlings are often seen on peat-mosses, struggling for life in the wettest situations; and doubtless, if cattle and rabbits were excluded, these would in time take possession.   At Churchill in Co. Armagh, the property of Harry Verner, Esq., considerable plantations of Scots pine, intermixed with a small proportion of larch, were made in 1861 on deep peat-moss, which had been thoroughly drained. These trees, planted 3½ feet apart, are now forty-four years old and average 44 feet in height by 3 feet in girth. Two-year-old seedlings, one year transplanted, were used and a system of pitting was adopted. The holes were made about a foot deep, and were filled in with a mixture of clay and peat. The clay was brought from a distance, and no doubt its use added considerably to the cost of planting. Possibly peat-ashes, obtained by burning peat, heather, etc. on the spot would have answered equally well, and been less costly. The Scots pine succeeds better than any other tree on pure peat-moss, though alder and larch may be added in a certain proportion. At Clonbrock, in Co. Mayo, on an overcut bog, where the peat left uncut was 3 to 4 feet deep, Scots pine eighty years old averages only 47 feet in height by 4 feet in girth. Probably the lesser growth in this case is due to insufficient drainage. As there are immense areas of peat-mosses in Ireland, now yielding no return whatever, the possibility of afforesting them with Scots pine, or with a mixture of Scots pine and larch, is an important question; and the success of the Churchill plantations is encouraging.[1]

Throughout Ireland there are extensive mountain tracts of barren land, covered with stones and rocks, which are of merely nominal value for grazing and are impossible to reclaim for agricultural purposes except at a ruinous expenditure. The Scots pine renders excellent service in turning these wastes to account. The late Lord Powerscourt made extensive plantations on the hill-sides of Co. Wicklow at 500 to 900 feet above sea-level, which paid handsomely. These plantations consisted in the main of a mixture as follows :—200 larch, 1500 Scots fir, and 500 spruce per acre, the plants being notched in, as, in Lord Powerscourt's opinion, they came on eventually as well as those which had been pitted at a much greater expense. The Scots fir have been gradually thinned out, the larch being left as the final crop. Lord Powerscourt was favoured by ready access to the sea, and by proximity to Wales, where his thinnings were readily sold as pit-props. He estimated that the initial cost of planting and fencing is £4 to £5 per acre, and that, during the first twenty to twenty-five years, the thinnings pay for the expense of cutting and the interest on the first cost. After that the thinnings should bring in annually eight shillings an acre; the final crop of larch at fifty years being probably worth about £50 an acre.

(A. H.)

In the United States the Scots pine has been planted with more or less success, but does not seem likely to be as valuable for timber as the native pines. The largest I saw was in the Wellesley Arboretum, near Boston, which was 49 feet high in 1904. In Professor Sargent's grounds it seems to be short-lived, only living for thirty to forty years. Ten miles from Boston, however, near Ponkapoag, it succeeds better on dry sandy soil, and I found some self-sown seedlings. At the Central Experimental Farm, near Ottawa, trees planted in 1888 were about 30 feet high in 1906, but Mr. W. T. Macoun[2] reports that it suffers much from shade, and does not grow so fast as Norway spruce or European larch; though he recommends it for nurses to other trees, and for producing fuel.

---

[1] The plantations on bog land at Knockboy, Co. Galway, were badly made, and turned out a disastrous failure. Cf. Dr. Schlich's report in *Kew Bull.* 1903, p. 22 ; and in *Trans. Roy. Scot. Arbor. Soc.* xvi. pt. ii. 249 (1901).

[2] *Canadian Forestry Journal*, iii. 77 (1907).

Cases of inosculation are rare among pines, but a remarkable instance of this was pointed out to me by Mr. Savile Foljambe in the Catwhins, near the lodge leading out of Thoresby park into the Retford road. It seems probable that when the trees were young they had come in contact, and eventually fused; the iron bands were put on afterwards, but the trees are now dead.

Another case, somewhat similar, occurs in a pine tree growing on the estate of Chenevières, near Montbour (Loiret), France, which a photograph kindly sent me by M. Maurice de Vilmorin illustrates. Here it seems that one tree had forked at or close to the ground, and become connected by a thick branch at a much later period. A third instance of natural inarching in the Scots pine is described and figured by Count von Schwerin[1] from a tree near Teltow in Germany. In this instance a branch of one tree grew into the bark of another and broke off, eventually forming just such a living connection between the two trees as is shown in Vol. I. Plate 4, but much thicker in proportion. The sap of the left hand tree appears to pass through this branch to the other, as the stem is thicker above the junction, and the branch has assumed the yellow bark of the upper part of the trunk.

The large, usually globular masses of dense shoots which sometimes appear on this species, and more rarely on larch and spruce, are not caused by a parasitic fungus. Prof. von Tubeuf[2] says that their origin is unknown, no insect or fungus having yet been discovered which might have caused the growth, which is composed of a mass of small buds, producing densely crowded tufts of short leaves. A specimen which was found at Schwarzenraben in Germany measured 53 centimetres in height and about the same in diameter, the weight of this mass being over eleven pounds.[3] Such growths are not uncommon[4] in England, and I have a photograph of one on a tree at Colesborne, which was about a foot in diameter.

TIMBER

On the timber of the Scots pine so much has been written that I will refer specialists to Laslett,[5] who gives a long account, mostly from a shipbuilder's point of view, of the various foreign varieties known to him as Dantzig, Memel, Riga, and Swedish fir; but makes no reference to the quality of native-grown timber, which, though men-of-war were built from it by Osbourne in the last century, seems to have been unused by the Admiralty since then, as it is now by the Post Office authorities in England, and by architects and builders generally. The reason of this is, no doubt, that the rapid growth of the tree in this country, in our mild climate, causes the wood to be much softer,

---

[1] *Mitt. D. D. Gesell.* 194 (1906).　　　　　　　　　　[2] *Ibid.* 222, fig. 13 (1905).

[3] Count von Schwerin, *Mitt. D. D. Gesell.* 222 (1905), says that in Bavarian Allgau, between Oberstaufen and Weiler, he has seen a forest of sixty-year-old spruce in which almost every tree was more or less affected by these growths, and supposes that the cause, whatever it is, must be contagious. He has seen similar growths on *Picea orientalis* and suggests that some of the horticultural monstrosities such as *Picea excelsa echiniformis* and *C. Lawsoniana forsteckensis* have originated from a similar cause.

[4] A specimen from a tree growing at Hunstanton Hall, Norfolk, was shown at a meeting of the Scientific Committee of the Royal Horticultural Society in April 1899. Cf. *Gard. Chron.* xxv. 270 (1899).

[5] *Timber and Timber Trees,* ed. 2 (London, 1894).

coarser, and less durable than that from North Germany, Russia, and Scandinavia; whilst over-thinning causes it to be much more knotty. It is hardly possible to believe that the same tree can produce timber so different as examples which I showed at a lecture on English timber at the Surveyors' Institute, on 22nd February 1904, taken from an immense tree grown at Powerscourt, of about 12 feet in girth; and the beautiful fine-grained wood which I brought from Northern Norway, and which when well planed shines with a silky gloss. Every intermediate form may be found in this country; but, as a rule, it is little valued in England except for mining timber, for cheap fencing, packing cases, and other uses. In Scotland, where it is as a rule slower grown, and better in quality, its value is kept down by foreign importations, though it is very largely used in making staves for herring barrels and many other purposes.

But when old enough to have produced a large proportion of red heartwood, and free from knots, I have sold Scots pine for as much as 8d. per foot, and have found it very useful and durable timber for roofing and many other estate purposes; and I am informed by Mr. Mitchell, forester to the Duke of Bedford at Woburn, that in that neighbourhood, where it is of good quality, it is largely used for rending lath, and that the buyers will give a high price for it when suitable for that purpose.

Loudon gives many interesting particulars of the uses and quality of pine timber at home and abroad, and quotes[1] Mr. T. Davis who, in 1798, was the Marquess of Bath's forester at Longleat, to the effect that a cart-house, built from it on that estate, remained perfectly sound after eighty years' use. And Pontey, in his *Forest Primer*, published in 1805, also defends the Scots pine against the "almost universally prevalent prejudice against it, which is no doubt based partly on ignorance and partly on the fact that it is often used when too young and unseasoned."

But, whatever may be said against the wood when grown in the south of England, there can be no question that the Highland native pine timber, when clean, is a valuable, and in some cases also a very ornamental timber. I have seen at Castle Grant a very beautiful sideboard made on the estate, which showed the curiously twisted red and yellow grain which Mr. Grant Thomson tells me is only produced by the self-sown native trees; and I am indebted to the kindness of Lady Seafield for some of the same wood, which was cut in the Forest of Abernethy. The entrance hall and a room at Balmoral Castle have been recently panelled with the same sort of wood, which has a very ornamental effect, and Mr. Michie tells me that it has also been largely used for internal decoration in Mar Lodge. I saw in the house of the postmaster of Bodö, in Norway, a very handsome table made from a variety of the wood, which is there called "Rie," and which seems to be caused by a disease in the tree producing excrescences and distortion in a part of its trunk, and I possess a piece of this wood which is so unlike pine wood that no one could recognise it as such. But these trees seem to be as rare in Norway as in Scotland, and command a high price locally.

The oldest example of this wood in the form of panelling that I have seen or heard of is in the room known as Queen Mary's room in Castle Menzies, which was

---

[1] *Op. cit.* 2168.

occupied by Mary Queen of Scots in 1577. This is now perfectly sound, and much darkened in colour by age. The width and somewhat knotty character of the boards tend to confirm what the late Sir R. Menzies told me, that it was made from locally grown Scots pine, which may have come from the Black Wood of Rannoch, or from other native forests that have now disappeared.

It is remarkable that, notwithstanding all that has been written since the Earl of Haddington first raised the question as to the existence of different varieties of this tree in his *Treatise on Forest Trees*, in 1760, there seems to be as yet no exact knowledge as to whether the different kinds of timber produced by different trees are, as I believe, individual variations, largely due to soil, or whether, by sowing seeds from trees possessing superior qualities, they may be reproduced in other soils and situations. The best and most exact records of such experiments that I know of are given by M. M. de Vilmorin in his account of the varieties of *Pinus sylvestris* collected by his grandfather at Les Barres in France, and published in the *Catalogue des vegetaux Ligneux sur le Domaine des Barres* (Paris, 1878), which show how much one family have done for the better knowledge of the economic value of trees, and for the benefit of their country.

Briefly, this trial, extending over a period of over sixty years, shows that, in the soil and climate of Central France, the Riga variety of *P. sylvestris* has, on its first introduction as well as in the second generation, preserved its superiority over other varieties of the same tree—from the various parts of France, from Haguenau on the Rhine, from Switzerland, and from Scotland—by its good growth, freedom from branches, quality of timber, and facility of transplantation. Though this superiority might not be as marked in England, it points to the necessity of careful trials of seed from Riga which, so far as I know, have not yet been made in this country.

In *Scott. Arb. Trans.* ix. 176 (1881), there is a very valuable paper by J. M'Laren and W. M'Corquodale, on " The Supposed Deterioration of the Scots Pine "; it having been stated by George Don and other writers that there were two or more varieties, one of which was very inferior to the other. They conclude, after reviewing the experience of many competent foresters, that the quality of the timber depends on the subsoil and the climate more than on the variety, and that the rich red resinous timber, for which the Highland pine is distinguished, is not to be expected in the south. They say that, since the days of the fine old Memel pine, there is no pine timber imported to our country equal to the old Highland pine, and that what has caused it to fall into disrepute is that it is grown too fast and cut too young, coupled with the fact that it is more difficult to manufacture and dress than foreign timber.   (H. J. E.)

CARYA

Carya,[1] Nuttall, *Gen. Am.* ii. 220 (1818); Bentham et Hooker, *Gen. Pl.* iii. 398 (1880).
Scoria, Rafinesque, *Med. Repos. New York*, v. 352 (1808).
Hicorius, Rafinesque, *Fl. Ludov.* 109 (1817).
Hicoria, Rafinesque, *Alsog. Am.* 65 (1838).

DECIDUOUS trees belonging to the order Juglandaceæ. Branchlets with solid continuous pith, not chambered as in Juglans and Pterocarya. Leaves alternate, compound, unequally pinnate, without stipules; leaflets sessile or sub-sessile, serrate, penninerved.

Flowers monœcious, without petals. Staminate catkins slender, drooping, in threes on a common penduncle or clustered and sessile or subsessile, arising either from buds in the axils of the leaf-scars of the previous year's branchlets or from the base of the current year's shoot, and appearing after the unfolding of the leaves; flowers numerous in the catkin; calyx two- to three-lobed, subtended by an ovate bract; stamens three to ten, filaments short. Pistillate flowers, two to ten, in a cluster or spike, terminal on a leafy branchlet of the year; ovary superior, one-celled, surrounded by a four-lobed cup-shaped involucre, formed by the union of a bract and two bracteoles; calyx one-lobed; stigmas two, sessile; ovule solitary.

Fruit, a nut, enclosed in a four-valved, thickened, hard and woody involucre, four-celled at the base, two-celled at the apex, tipped by the remains of the style; seed solitary, without albumen, filling the cavity of the nut.

The cotyledons remain underground in germination, the plumule being carried by the lengthening of their petioles out of the nut, which splits into two valves. The germination resembles that of the oak, the young stem bearing at first three to eight alternate minute lanceolate scales, above which the leaves are developed. The first leaf is simple, tri-lobed, or trifoliolate; those succeeding, about five or six in the first year, being trifoliolate; all are serrate and stalked. The difference observed in the length of the stems, in two or three seedlings, seen at Colesborne, may not be constant for each species.[2]

Twelve species of hickory are distinguished by Sargent,[3] all natives of North

---

[1] The generic name, *Carya*, though not the first one published, has always been used in England, and is now sanctioned by the regulations drawn up by the International Congress of Botany at Vienna in 1905. Cf. *Verhand. Internat. Bot. Kongress. Wien*, 1905, p. 239. With regard to the specific names, I have not altered those of Nuttall, which have been long in use.

[2] Rowlee and Hastings, *Bot. Gaz.* xxvi. 349, pl. xxix. figs. 9, 10, 12 (1898), give figures of the seedlings of *C. alba* and *C. porcina*.

[3] In *Trees N. Amer.* 132 (1905). Ashe, in *Flora South-Eastern United States*, 333 (1903), raised two varieties to the rank of species, making fourteen species in all.

America, four species extending as far north as Canada, and one species confined to the high lands of Mexico.   Six species[1] (Plate 203) are in cultivation in the British Isles, and will be dealt with in the following account.

The genus is divided into two sections :—

I. *Apocarya*, De Candolle, *Prod.* xvi. 2, p. 144.

   Buds, with four to six valvate scales, which are often obscurely pinnatifid at the apex ; axillary buds, often two to three superposed, the uppermost one stalked.   Husk of the fruit thin, and prominently ridged at the sutures.

   1. *Carya olivæformis*, Nuttall.

   Buds greyish, densely pubescent, without glands.   Leaflets, eleven to thirteen, rarely nine ; margin densely ciliate.

   2. *Carya amara*, Nuttall.

   Buds yellowish, slightly pubescent, glandular.   Leaflets, seven to nine ; margin irregularly ciliate.

II. *Eucarya*, De Candolle, *Prod.* xvi. 2, p. 142.

   Buds with ten to twelve imbricated scales, the outer falling early, the inner accrescent and becoming much enlarged and reflexed at maturity.   Husk of the fruit thick, not ridged at the sutures.

   * *Leaflets five, pubescent.*

   3. *Carya alba*, Nuttall.

   Young branchlets with brown stellate hairs ; base of the shoot marked with a dense pubescent ring.   Leaflets, stellate-pubescent beneath, ciliate in margin.

   ** *Leaflets five or seven, glabrous.*

   4. *Carya porcina*, Nuttall.

   Young branchlets glabrous or with only an occasional hair ; base of shoot without pubescent ring.   Leaflets glabrous beneath, except for axil-tufts, non-ciliate in margin.

   *** *Leaflets, seven or nine, pubescent.*

   5. *Carya sulcata*, Nuttall.

   Branchlets reddish-brown, glabrous towards the tip.   Leaves not fragrant ; rachis nearly glabrous ; nerves in upper lateral pair of leaflets more than twenty pairs.

   6. *Carya tomentosa*, Nuttall.

   Branchlets purplish-grey, pubescent and glandular.   Leaves fragrant ; rachis pubescent and glandular ; nerves in upper lateral pair of leaflets less than twenty pairs.

(A. H.)

---

[1] *Carya aquatica*, Nuttall, the water hickory, a native of river swamps in the southern parts of the United States, is not likely to succeed in any part of the British Isles.   Loudon, *op. cit.* 1444, mentions a tree of this species 40 feet high growing at Milford near Godalming ; but his identification was probably incorrect.

CARYA OLIVÆFORMIS, Pecan Nut

Carya olivæformis, Nuttall, *Gen. Am.* ii. 221 (1818); Loudon, *Arb. et Frut. Brit.* iii. 1441 (1838).

Carya angustifolia, Sweet, *Hort. Brit.* 97 (1827).

Carya illinoinensis, Koch, *Dendrol.* i. 593 (1869).

Carya Pecan, Schneider, ex Sargent in *Bot. Gazette*, xliv. 226 (1907).

Juglans Pecan, Marshall, *Arbust. Am.* 69 (1785).

Juglans illinoinensis, Wangenheim, *Nordam. Holz.* 54 (1787).

Juglans angustifolia, Aiton, *Hort. Kew*, iii. 361 (1789).

Juglans cylindrica, Poiret, in Lamarck, *Dict.* iv. 505 (1797).

Juglans olivæformis, Michaux, *Fl. Bor. Am.* ii. 192 (1803).

Hicoria Pecan, Britton, *Bull. Torrey Bot. Club*, xv. 282 (1888); Sargent, *Silva N. Amer.* vii. 137, tt. 338, 339 (1895), and *Trees N. Amer.* 133 (1905).

A tree, attaining in America 170 feet in height and 18 feet in girth. Bark, brownish, deeply and irregularly divided into narrow forked ridges. Buds, similar to those of *C. amara*, but greyish-white in colour, densely pubescent and without glands; lateral buds ovoid, pointed. Young branchlets densely pubescent, especially towards the tip, where no glands are present, but with minute glands at the base of the shoot. Leaves (Plate 203, Fig. 6), 12 to 20 inches long. Leaflets, eleven to thirteen or more, rarely nine, lanceolate, falcate, subsessile, acuminate, very unequally divided by the midrib; upper surface with stellate pubescence on the midrib and nerves; lower surface covered throughout with fine scattered pubescence and numerous glands; margin plainly ciliate; rachis densely pubescent.

Staminate flowers in sessile or subsessile clustered catkins, usually on the previous year's branchlets. Fruit in clusters of three to eleven, pointed, four-winged and -angled, 1 to $2\frac{1}{2}$ inches long, $\frac{1}{2}$ to 1 inch broad; husk thin, brittle, dark-brown, coated with yellow pubescence, splitting when ripe nearly to the base; nut, thin-shelled, reddish-brown with irregular black markings; seed sweet, reddish-brown.

(A. H.)

The Pecan is a native of the Middle States, occurring from Indiana, Southern Illinois and Iowa, southward through Western Kentucky and Tennessee to Alabama and Mississippi, and extending westward through Missouri, South-Eastern Kansas, Arkansas, Indian Territory and Louisiana to the valley of the Concho River in Texas. It is also met with in the mountains of Mexico.

It chiefly grows on rich alluvial soil, along the banks of streams, and attains a greater size than any other hickory. Ridgway says it is one of the very largest trees of the forest, being only exceeded in height by the tulip tree and the scarlet oak. He records one tree, measured by Dr. Schneck, which was 175 feet high by 16 feet in girth, and another 30 feet in girth at the ground. It is largely cultivated in the Southern States for its fruit, which has been improved by selection and grafting, and is considered the best of the nuts of North America.

It requires a much longer and hotter summer than any part of Great Britain affords; and even in the south of Europe we have not heard of its being successfully grown.   Pardé[1] has not heard of its having ever produced fruit in France, where it is hardy as far north as Grignon,[2] but grows very slowly.

It was introduced, according to Loudon, in 1766 into England, and he mentions, with some doubt as to the species, trees growing in the Horticultural Society's garden and other places near London; but it is probable that this tree has never attained a considerable size in England.   At Tortworth, a tree with the bark beginning to scale was, in 1905, 24 feet high and 1½ foot in girth.   Dr. Warre, the late headmaster at Eton, raised, from seed sent to him from New York, a tree, which is growing in the garden of Mr. E. C. Austen Leigh, at Eton, who informs us that it is about 11 feet high.   Miss Woolward raised at Belton in Lincolnshire a plant, which is now about 4 feet high; but it suffers severely from frost, the young growths being killed back every year.   None of the seedlings which I have raised at Colesborne have thriven, as the summer here is evidently much too short and cold.

(H. J. E.)

CARYA AMARA, Bitternut

Carya amara, Nuttall, *Gen. Am.* ii. 222 (1818); Loudon, *Arb. et Frut. Brit.* iii. 1443 (1838).
Carya cordifolia, Schneider, ex Sargent in *Bot. Gazette*, xliv. 226 (1907).
Juglans alba minima, Marshall, *Arbust. Am.* 68 (1785).
Juglans cordiformis, Wangenheim, *Nordam. Holz.* 25 (1787).
Juglans angustifolia, Poiret, in Lamarck, *Dict.* iv. 504 (1797).
Juglans sulcata, Willdenow, *Berl. Baumz.* 154 (1796).
Juglans minima, Borkhausen, *Handb. Forstbot.* i. 760 (1800).
Juglans mucronata, Michaux, *Fl. Bor. Am.* ii. 192 (1803).
Juglans amara, Michaux f., *Hist. Arb. Am.* i. 177 (1810).
Hicoria minima, Britton, *Bull. Torrey Bot. Club*, xv. 284 (1888); Sargent, *Silva N. Amer.* vii. 141, tt. 340, 341 (1895), and *Trees N. Amer.* 135 (1905).

A tree, attaining in America 100 feet in height and 9 feet in girth.   Bark grey, smooth, ultimately separating on the surface into small thin scales.   Buds bright-yellow, glandular; terminal buds elongated, pointed and oblique at the apex, about ½ inch long, with two pairs of valvate scales, often obscurely pinnatifid at the tip; lateral buds, often two superposed, the uppermost stalked, four-angled, and pointed, the lowermost minute.   Young branchlets, with scattered short pubescence, glandular towards the tip.   Leaves (Plate 203, Fig. 5) 6 to 10 inches long.   Leaflets, five to nine, variable in shape, lanceolate, ovate-lanceolate, or obovate-oblong, long-acuminate; margin with occasional cilia; lower surface with stellate hairs, especially along the midrib; rachis minutely pubescent.

Staminate catkins, pubescent, in threes on a slender peduncle, usually on the previous year's branchlet.   Fruit, one, two, or three at the top of the branchlet, about an inch long; husk thin, glandular, four-winged from the apex to about the middle; nut thin-shelled; seed bitter, reddish-brown.   (A. H.)

This species has a wide distribution, extending to the northward as far as Southern Maine, Ontario, Central Michigan, and Minnesota.   It was the only hickory which I found near Ottawa, where it was common in the Gatineau Valley,

---

[1] *Arbor. Nat. des Barres*, 253 (1906).   [2] Seen by Henry in 1906.

but not there a large tree. It extends westward to South-Eastern Nebraska, Eastern Kansas, Indian Territory, and Eastern Texas, and reaches its southern limit in North-Western Florida and Northern Alabama. It is usually found in lower and moister situations than the other species, and is one of the largest and commonest hickories in Southern New England, where it attains 70 to 80 feet in height and 10 to 12 feet in girth. It grows to its largest size on the alluvial lands of the lower Ohio basin; and in Southern Indiana, Ridgway records a tree 113 feet in height and 6 feet 3 inches in girth.

The bitternut is the commonest species of hickory in England, and grows to a considerable size. The finest tree is perhaps the one at Bute House, Petersham, which was, in 1903, 76 feet in height by 7 feet 5 inches in girth (Plate 170).

At Barton, near Bury St. Edmund's, there are three trees, two in the arboretum, one of which I found in 1905 to be 80 feet by 5 feet 4 inches, and the third on the lawn, 74 feet by 7 feet 6 inches, but forked near the ground. These trees ripened fruit[1] in 1864; but of late years do not appear to have borne any.

At Syston Park, Lincolnshire, there are four trees of this species, one of which was flowering freely on 16th May 1905, when it measured 71 feet by 5 feet 3 inches. Two of these are figured on Plate 171 from photographs taken by Miss F. Woolward.

At Arley Castle there are five trees, the tallest of which measures 72 feet by 4 feet, the others being about 60 feet high with girths ranging from 3 feet 7 inches to 4 feet 7 inches. These are supposed to have been planted about 1820.

At Bicton, a tree measured, in August 1906, 65 feet by 4 feet; and another growing at the Wilderness, White Knights, near Reading, is exactly the same size. At Devonshurst, Chiswick, a tree, now cut down, measured, in 1903, 68 feet by 5 feet.

There is a tree, 58 feet by 6 feet 2 inches, in the Botanic Garden, Glasnevin; but we have heard of no others in Ireland, where hickories seem to have been very little planted. (H. J. E.)

CARYA ALBA, Shagbark Hickory

Carya alba, Nuttall, *Gen. Am.* ii. 221 (1818); Loudon, *Arb. et Frut. Brit.*. iii. 1446 (1838).
Carya ovata, Schneider, ex Sargent in *Bot. Gazette*, xliv. 226 (1907).
Juglans ovata, Miller, *Dict.* ed. 8, No. 6 (1768).
Juglans ovalis, Wangenheim, *Nordam. Holz.* 24 (1787).
Juglans compressa, Gaertner, *Fruct.* ii. 51 (1791).
Juglans alba, Michaux, *Fl. Bor. Am.* ii. 193 (1803).
Juglans obcordata, Poiret, in Lamarck, *Dict.* iv. 504 (1797).
Juglans squamosa, Michaux f., *Hist. Arb. Amer.* i. 190 (1810).
Hicoria ovata, Britton, *Bull. Torrey Bot. Club*, xv. 283 (1888); Sargent, *Silva N. Amer.* vii. 153, tt. 346, 347 (1895), and *Trees N. Amer.* 139 (1905).

A tree, attaining in America usually 70 to 90 feet, rarely 150 feet in height and 15 feet in girth. Bark grey, ultimately separating into long strips, attached by the

[1] Bunbury, *Arboretum Notes*, 100 (1889).

middle and free at one or both ends, giving the trunk a characteristic appearance. Terminal buds ovoid, obtuse, $\frac{1}{2}$ to $\frac{3}{4}$ inch long, with ten to twelve imbricated scales; outer scales persistent during winter, falling a little before the unfolding of the leaves, triangular, keeled, apiculate or contracted into long points, dark-brown, pubescent; inner scales downy, enlarging to 2 or 3 inches long, as the bud opens. Lateral buds about $\frac{1}{3}$ inch long, with four to five scales visible externally. Branchlets purplish-grey, covered with brown stellate hairs, scattered below, denser nearer the tip, and with a few yellow glands; base of the shoot girt with a dense ring of pubescence. Leaves (Plate 203, Fig. 1) 8 to 14 inches long. Leaflets, five, upper three obovate, lower pair ovate, all shortly acuminate; margin ciliate with irregular tufted hairs, densest near the points of the serrations; upper surface glabrous, except for stellate hairs on the midrib and nerves; lower surface with a fine stellate pubescence, densest on the midrib and nerves; rachis stellate-pubescent.

Staminate catkins, glandular-hirsute, pedunculate in threes at the base of the year's shoot. Fruit solitary or in pairs, sub-globose, 1 to $2\frac{1}{2}$ inches long, splitting freely to the base; husk dark reddish-brown, glabrous, $\frac{1}{8}$ to $\frac{1}{2}$ inch thick; nut four-angled and -ridged, white; seed large, lustrous, light-brown, sweet with an aromatic flavour.

The above description is drawn from trees cultivated in England, which resemble in all essential characters, except the imperfect development of the fruit, specimens obtained by Elwes at the Arnold Arboretum. The size and shape of the leaflets, which are always five in number, and the amount of pubescence on the branchlets and on the rachis and surface of the leaves are very variable. Two or three trees[1] of this species at Kew, which are about 25 feet high, have very large leaves which turn a brilliant yellow in autumn. Another tree at Kew, which was labelled *Carya alba*, differs from all other specimens which I have seen, as follows:—Branchlets and leaf-rachis almost glabrous, only showing when young a few stellate hairs. Leaflets, five, lanceolate, narrow, long-acuminate, nearly glabrous, with only a few stellate hairs, confined on the upper surface to the midrib, and scattered over the lower surface; margin non-ciliate. Buds, as in the typical form, but pointed and smaller. This tree, which is about 30 feet high, has very smooth bark, and is growing very vigorously. It is probably a glabrous form of *Carya alba*, and may possibly be *Hicoria borealis*, Ashe,[2] which grows in Michigan and Southern Ontario.

(A. H.)

The Shagbark, according to Sargent, is widely distributed from the St. Lawrence valley near Montreal, along the northern shores of Lakes Erie and Ontario, west to South-Eastern Nebraska, and south throughout the Middle States and along the Appalachian mountains to Western Florida, Northern Alabama, Eastern Texas, and Central Kansas.

It attains its largest size in the Southern Alleghanies and in the lower Ohio basin, the largest trees recorded, one of which is figured in *Garden and Forest*, ii.

---

[1] These have been labelled *C. tomentosa*, from which they differ in having five-foliolate non-fragrant leaves; and are the trees referred to in *Gard. Chron.* xxviii. 295 (1900).

[2] Ashe, *Notes on Hickories*, 1896; Britton and Brown, *Illust. Flora N. United States*, iii. 512, fig. 1156 b (1898). This form is not recognised by Sargent as a distinct species.

463, being in Southern Indiana, where Ridgway[1] says that trees of 130 feet high by 3 to 4 feet in diameter were not rare, and that some were certainly 150 feet, many trunks which seemed less than half the total height being 70 or 80 feet to the first limb. One of these in Wabash Co. measured 78 feet to the first limb, and was 14⅜ feet in girth.

Such giants, however, hardly exist now; and during my travels in America I never saw a hickory more than about 100 feet high; whilst those in New England and Canada are usually from 60 to 80 feet in height, and seldom exceed 6 feet in girth. The bark, which separates externally into long loose flakes, serves to distinguish this species readily from the others. The nuts, which vary much in size and shape and thickness, are superior in quality to any other native nut (except that of the Pecan), and are largely eaten. Some of the better varieties, which have thin shells and larger kernels, have been selected and propagated, so that the improvement of this fruit in cultivation is likely to be as great as that of the walnut.

This species was introduced in 1629, according to Loudon, who mentions large trees growing in 1838 at Syon, Fulham Palace, and other places near London. None of these are now living, and apparently the tree does not attain a great age in England. It succeeds about as well as the Bitternut.

The largest tree that I have seen in England is hidden in a thick shrubbery at Botley Hill, Hants, the residence of Lady Jenkyns, and has an historic interest from the fact that it is almost certainly one of the trees planted by Cobbett, who lived there for some years about 1820. The old brew-house and oven, which, in his opinion, were two of the most necessary parts of an Englishman's house, still remain, as well as some rather stunted black walnuts. This hickory is about 75 feet high by 5 feet 4 inches in girth, with a bole of 30 feet.

At the Wilderness, White Knights, near Reading, a tree measures 55 feet by 5 feet 8 inches. At Castle Howard, Yorkshire, there is a healthy symmetrical tree, growing on the site of an old nursery near the timber yard, which may have been sown *in situ*. I found it in 1905 to be 50 feet high by 3 feet 3 inches in girth.

In the Pinetum, Brocklesby Park, Lincolnshire, there is a fine tree, which Mr. W. B. Havelock informs us is 79 feet high by 4 feet 10 inches in girth, with a straight stem running nearly to the top. It is growing in strong loam, sheltered by surrounding belts of trees, and is supposed to be about sixty years old. The bark is shaggy, and hangs in strips from the trunk (Plate 173).

A fine tree of this species is growing in Syston Park, near Grantham, the seat of Sir John Thorold, who told me that his father planted it and the four bitternut trees mentioned above about fifty years ago. It measured in 1905 62 feet by 4 feet 4 inches, and has the characteristic scaly bark.

At Boynton Hall, Bridlington, Yorkshire, there are three trees of this species, which Sir Charles Strickland informs us are respectively 50 feet by 7 feet, 40 feet by 4 feet, and 25 feet by 6 feet, the last being very bushy in habit, and growing in a very exposed position. These trees were raised from seed brought from America by

[1] *Proc. U.S. Nat. Mus.* 1882, p. 77.

Sir William Strickland, grandfather of the present owner; and several other hickories planted at the same time have disappeared, one being blown down.

At Golden Grove, Carmarthenshire, a seat of Earl Cawdor, a tree, which is supposed to have been planted in 1865, is 42 feet by 3 feet 1 inch. At Fonthill Abbey, Wilts, a tree, with very scaly bark, is 62 feet by 4 feet 1 inch. At Althorp there is a tree 75 feet by 3 feet 6 inches, growing in a dense thicket of laurels near the house.

A hickory which was perhaps of this species grew close to the house at Moncreiffe near Perth, and is mentioned by Hunter as being the finest in Scotland. It had a bole 20 feet long by 5 feet 9 inches, and was cut down about six years ago. The timber was used for making gates.

At Kinblethmont, Arbroath, H. Lindsay Carnegie, Esq., reports a tree 46 feet high by 2 feet 11 inches in girth. It was planted by his father about 1825.

In Ireland, the only specimens which Henry has seen are two small trees about a foot in diameter, growing in the Botanic Gardens at Glasnevin, and two at Kilmacurragh, in Wicklow, about 35 feet high. (H. J. E.)

CARYA PORCINA, Pignut

Carya porcina, Nuttall, *Gen. Am.* ii. 222 (1818); Loudon, *Arb. et Frut. Brit.* iii. 1449 (1838).
Carya obcordata, Sweet, *Hort. Brit.* 97 (1827).
Carya glabra,[1] Spach, *Hist. Vég.* ii. 179 (1834).
Juglans glabra, Miller, *Dict.* ed. 8, No. 5 (1768).
Juglans squamosa, Poiret, in Lamarck, *Dict.* iv. 504 (1797).
Juglans obcordata, Muehlenberg u. Willdenow, *Neue Schrift. Ges. Nat. Fr. Berlin*, iii. 392 (1801).
Juglans porcina, Michaux f., *Hist. Arb. Am.* i. 206 (1810).
Hicoria glabra, Britton, *Bull. Torrey Bot. Club*, xv. 284 (1888); Sargent, *Silva N. Amer.* vii. 165, tt. 352-355 (1895), and *Trees N. Amer.* 144 (1905).
Hicorius glaber, Sargent, *Garden and Forest*, ii. 460 (1889).

A tree attaining in America 90 feet in height and 12 feet in girth. Bark greyish, ultimately fissuring into narrow longitudinal ridges, occasionally on old trees breaking on the surface into loose thick scales. Terminal buds, $\frac{1}{4}$ to $\frac{1}{3}$ inch long, globose or ellipsoidal, with ten to twelve imbricated scales; outer scales usually deciduous in winter, keeled, acute, or pointed, glabrous, ciliate, often glandular; inner scales pubescent, enlarging to 2 inches long as the bud unfolds. Lateral buds small, ovoid, often glandular, with two scales visible externally. Branchlets glabrous, or rarely with a minute pubescence speedily disappearing. Leaves (Plate 203, Fig. 2) 8 to 12 inches long. Leaflets, five to seven, upper three obovate, lower one or two pairs oblong lanceolate, all acuminate; margin without cilia; upper surface glabrous, with numerous minute glands; lower surface glabrous, except for slight tufts of pubescence in the axils of the midrib and lateral nerves, covered with numerous glands; rachis glabrous.

[1] This is given as *Carya glabra*, Schneider, by Sargent in *Bot. Gazette*, xliv. 226 (1907).

Staminate catkins, scurfy-pubescent, pedunculate in threes at the base of the current year's shoot. Fruit, single or in pairs, very variable in size and shape; husk thin, reddish brown, glandular, opening only at the apex or splitting to the middle or to near the base; nut obscurely four-angled; seed small, light brown, poor in flavour.

Carya microcarpa, Nuttall [1] (*Hicoria microcarpa*, Britton [2]), appears to be only distinguished by the nuts, which are very small, compressed and globular. It is impossible to say whether it is in cultivation, as all the hickories in England produce smaller nuts than they do in America. This form is called by Sargent,[3] var. *odorata*, a misleading name, as the tree has apparently no marked odour either in the foliage or in the fruit. (A. H.)

The pignut, which grows usually on dry ridges and hillsides, has a most extensive distribution, ranging from Southern Ontario and Southern Maine in the north to Florida and Mississippi in the south, and extending westward to Nebraska, Kansas, and Texas.

This species of hickory [4] seems to succeed fairly well in England, though we know of very few trees. The best is at Kew, where there is a fine tree near the Temperate House (Plate 172), which is, however, becoming stag-headed. It measured, in 1907, 77 feet by 6 feet. It produces fruit abundantly but the seed, so far as we have observed, is infertile.

Another tree is growing at Leny, near Callander, Perthshire. Though in a somewhat exposed situation and at an elevation of several hundred feet, it had a stem 3 feet 4 inches in girth and might have been 50 feet high before the leading shoot was broken. (H. J. E.)

CARYA SULCATA, BIG SHELLBARK, KINGNUT

Carya sulcata, Nuttall, *Gen. Am.* ii. 221 (1818); Loudon, *Arb. et Frut. Brit.* iii. 1448 (1838).
Carya pubescens, Sweet, *Hort. Brit.* 97 (1827).
Carya cordiformis, Koch, *Dendr.* i. 597 (1869).
Carya laciniosa, Schneider, ex Sargent in *Bot. Gazette*, xliv. 226 (1907).
Juglans laciniosa, Michaux, f., *Hist. Arb. Amer.* i. 199 (1810).
Juglans sulcata, Pursh, *Fl. Am. Sept.* ii. 637 (1814).
Hicoria laciniosa, Sargent, *Silva N. Amer.* vii. 157, tt. 348, 349 (1895), and *Trees N. Amer.* 141 (1905).

A tree attaining in America 120 feet in height and 9 feet in girth. Bark grey, ultimately separating in plates, which sometimes remain for many years hanging on the trunk. Terminal buds, ovoid, obtuse, ¾ to 1 inch long, composed of ten to twelve imbricated scales; outer scales not deciduous in winter, dark brown, ovate, keeled, pointed, tomentose, with scattered glands; inner scales silky pubescent,

[1] *Gen. Am.* ii. 221 (1818). [2] *Bull. Torrey Bot. Club*, xv. 283 (1888).
[3] *Silva, loc. cit.* This varietal name is adopted by Sargent, as the tree appears to have been first described by Marshall, in *Arb. Am.* 68 (1785) as *Juglans alba odorata*.
[4] The date of introduction is uncertain. Loudon states that in 1838 there were plants in the Hackney Arboretum.

606 The Trees of Great Britain and Ireland

becoming 2 to 3 inches long at the opening of the bud. Axillary buds, $\frac{1}{2}$ inch long, with four to five scales visible externally. Branchlets light reddish brown, glabrous towards the tip, covered below with a short dense pubescence. Leaves (Plate 203, Fig. 4) not fragrant, 15 to 22 inches long. Leaflets seven to nine, upper three or five obovate-oblong, lower one or two pairs lanceolate; margin ciliate, the cilia more numerous towards the base; upper surface glabrous, shining; lower surface with scattered short stellate tomentum; nerves in upper pair of leaflets more than twenty pairs; rachis with slight pubescence near the insertion of the leaflets, elsewhere glabrous.

Staminate catkins, glabrous, pedunculate in threes at the base of the current year's shoot. Fruit, solitary or in pairs, 2 inches long; husk downy or glabrous, hard and woody, $\frac{1}{4}$ to $\frac{1}{3}$ inch thick; nut compressed, four- to six-ridged, with a hard bony shell; seed light brown, sweet. (A. H.)

This species usually grows on deep rich alluvial soil, which is inundated for several weeks in the year, and is one of the commonest trees in the great river swamps of Central Missouri and of the Ohio basin. It is rare and local east of the Alleghany Mountains, being occasionally met with in New York, Pennsylvania, and North Carolina; and extends through the central states from South - Eastern Nebraska, Iowa, Illinois, and Indiana, southward to Kansas, Indian Territory, Arkansas, and Tennessee.

It is exceedingly rare in cultivation[1] in England, the largest tree we are acquainted with being one in Tortworth Churchyard, which bore a few nuts in 1905, when it was about 30 feet high by 1 foot 8 inches in girth. There are small plants growing at Hildenley Hall, Yorkshire, and at Grayswood, Haslemere.

(H. J. E.)

CARYA TOMENTOSA, MOCKERNUT

Carya tomentosa, Nuttall, *Gen. Am.* ii. 221 (1818); Loudon, *Arb. et Frut. Brit.* iii. 1444 (1838).
Carya alba, Koch, *Dendr.* i. 596 (1869); Sargent, *Bot. Gazette*, xliv. 226 (1907).
Juglans alba, Linnæus, *Sp. Pl.* 997 (in part) (1753).
Juglans rubra, Gaertner, *Fruct.* ii. 51 (1791).
Juglans tomentosa, Poiret, in Lamarck, *Dict.* iv. 504 (1797).
Hicoria alba, Britton, *Bull. Torrey Bot. Club*, xv. 283 (1888); Sargent, *Silva N. Amer.* vii. tt. 350, 351 (1895), and *Trees N. Amer.* 143 (1905).
Hicorius albus, Sargent, *Garden and Forest*, ii. 460 (1889).

A tree attaining in America 100 feet in height and 9 feet in girth, but usually much smaller. Bark grey, slightly ridged by shallow irregular interrupted fissures, and ultimately covered by closely appressed scales. Terminal buds broadly ovoid, $\frac{1}{2}$ to $\frac{3}{4}$ inch long, of 10 to 12 imbricated scales; outer scales, usually deciduous in winter, dark brown, keeled, apiculate, pubescent, glandular; inner scales silky

---

[1] Loudon states that it was introduced in 1804; but he appears to have been only acquainted in 1838 with small plants growing at the Horticultural Society's Garden, Loddiges' Nursery, and White Knights. None of these now survive.

pubescent, enlarging to 1 to 1½ inch long, as the bud unfolds. Lateral buds much smaller, with 4 scales visible externally. Branchlets densely covered with stellate hairs and shining glands, without any definite ring of pubescence at the base of the shoot. Leaves (Plate 203, Fig. 3) very fragrant, eight to twenty inches long. Leaflets seven or nine, upper three obovate, lower pairs ovate; acuminate; margin regularly ciliate; upper surface glabrous, except for stellate hairs on the midrib, and with minute shining glands; lower surface covered with scattered stellate pubescence and numerous glands; nerves in the upper pair of lateral leaflets usually less than 20 pairs; rachis with stellate pubescence and shining glands.

Staminate catkins, pubescent, pedunculate in threes at the base of the current year's shoot. Fruit, single or in pairs, globose, 1 to 2 inches long; husk hard, thick, glandular, splitting to the middle or nearly to the base; nut four-ridged near the top, thick-shelled; seed small, dark brown, lustrous, sweet.   (A. H.)

This species has a distribution similar to that of the shag-bark and bitternut; but is comparatively rare in the north, though it is found in Southern Ontario. It commonly grows on poor and sandy soil, and is the only hickory in the maritime pine-belt of the Southern States. It does not usually exceed 60 feet in height, except in the rich valleys of the Southern Alleghany Mountains and in Missouri and Arkansas. Ridgway records a tree in Southern Indiana 112 feet by 10½ feet, and says it is a common species in upland woods, being known as the black- or white-heart hickory. Its name of mockernut is derived from the thickness of the shell and the smallness of the kernel of the fruit, which makes it hardly worth eating. The fruit and leaves have a strong fragrant resinous smell, not present in the other species.

Loudon states that this species was "introduced in ? 1766"; but he does not seem to have known of any trees in cultivation in 1838. This hickory is extremely rare, the finest specimen that we know of being one at Golden Grove, Carmarthenshire, which is 50 feet high by 4 feet in girth. It is supposed to have been planted in 1865. Near the Azalea Garden at Kew there is a fine healthy young tree, procured in 1872 from Booth's Nursery, near Hamburg, and now 46 feet high by 2½ feet in girth, remarkable for its fine large foliage, the fragrance of which is especially strong and can be perceived at a distance in the early morning. There are small trees growing at Tortworth and Hildenley.   (H. J. E.)

CULTIVATION OF THE HICKORIES

Though so long introduced into cultivation, and at one time much more commonly planted than at present, no species of hickory has as yet established a reputation which justifies the hope that it may become a tree of real economic importance in this country. As ornamental trees the hickories are not equal either in size or in beauty to others which can be more easily grown, and though at least three of the species may be planted with good hopes of success, as an interesting addition to parks and pleasure grounds, yet their cultural peculiarities must be carefully studied before doing so.

Hickories though easily raised from imported nuts, require special treatment on account of their long thick tap-roots, which make them so difficult to transplant; and as they grow slowly for several years and do not ripen their wood when young, in most places, their ultimate success must always be to some extent a matter of chance.   Though they are found in America, in places where the soil is not specially deep or good, they require a hotter summer than we get to enable them to ripen fruit, and when a tree will not ripen seed, it can hardly be called acclimatised.

I have made many experiments in raising them from seed, which at present have not given very good results, principally, I think, because of unfavourable local conditions; but believe that if the following points, which are based on those adopted in the Arnold Arboretum,[1] are attended to, the trouble will not be thrown away.

The nuts must be procured from America as soon as ripe; and if there is any influence in heredity, as I believe there is, from the Northern, rather than from the Southern or Western States; but it is only fair to say that the seed which I collected myself in Canada and near Boston, did not produce such strong seedlings the first season as those which I procured from Philadelphia.

The nuts should be sown at once in boxes of about 18 to 24 inches deep in rich sandy loam, about 2 inches apart, and covered with an inch or so of light soil.   The boxes may be stored for the winter in a shed, and in spring brought into a frame or greenhouse to induce earlier germination.   They should be kept under glass until all risk of spring frost has gone by, and perhaps are better kept in a frame the whole summer lightly shaded, and watered when necessary.   The leaves will remain on throughout the autumn, when the box should be exposed to the full sun; and as soon as the shoot, which does not exceed 4 to 8 inches in height the first year, is ripe, may again be put away for the winter in a dry place covered with leaves, and protected from mice.

In the following May the seedlings may either be turned out and planted in a deep rich nursery bed, after cutting off the tap-root at about a foot, or if a warm sheltered spot can be found in a wood, where they can be cultivated and sheltered for some years, they may be planted out permanently without cutting the tap-root. But as the danger from vermin and early or late frosts will continue for some years, it may be better to keep them in the nursery till they are 3 to 5 feet high, provided that when transplanted a deep trench is first made on one side, so as to get up the whole of the root with as little injury as possible.   Woods being their natural home they are more likely to grow into good trees when drawn up with others than when exposed in the open; but we cannot point to an instance in Great Britain where they have been so treated, though some of the best trees we know are in dense shrubberies.

As regards soil, it cannot be too deep, rich, or well drained, and a southern or western aspect is to be preferred.   Under such conditions they may attain 50 to 60 feet in height in as many years, and in some parts of England even more.   A certain amount of lime in the soil does not seem to be harmful.

The hickories are not either in America or in England very long-lived trees,

[1] Cf. *Garden and Forest*, x. 116 (1897).

and none of great age are recorded here. When cut down, or when killed to the ground by frost in their young state, they push shoots freely from the stool, though they do not produce suckers.

Dawson in *Garden and Forest*, ix. 77 (1896), gives an account of his method of grafting the cultivated varieties of hickory, and says that the best stock is the bitternut, which grows twice as fast at Boston as the common shagbark. He performs the operation under glass in the month of January, by side-grafting close to the collar of the stock, and plunging the pots into sphagnum moss up to the top bud of the graft.

TIMBER OF THE HICKORIES

The best account of the wood is given by Michaux and Emerson. The timber of the different species is so similar in appearance, that I doubt if any one could identify without the names, the six species illustrated by Hough; and as this author rarely mentions the age or origin of the trees from which his specimens were taken, or shows much personal knowledge of their peculiarities, his work is not of so much practical value as it might have been.

Michaux specially commends the timber of the shellbark and the pignut. Emerson does not say which is best, but says that the most valuable is that which has been grown most rapidly, and places the pignut and shellbark first for weight.

As fuel hickory is, or rather was in days when it was abundant, preferred to all other woods. But its greatest value is for carriage building, axe and tool handles, and especially for cask hoops, of which in Michaux's time large quantities were exported, as well as used at home. Now, however, it is superseded to a great extent for this purpose by iron.

An article on hickory by Mr. J. F. Brown in *Arboriculture*, vi. No. 4, states that the great demand for hoops in the apple-growing districts of Virginia, is rapidly exhausting the local supply of young trees, which is now being filled from Southern Indiana, and that in consequence the supply of second-growth timber fit for wheels and carriage work is likely to become diminished, and in well-settled regions is already exhausted. He states that when the trees are cut and put on the market, no discrimination is made between the different species, though second-growth hickory is always preferred to the timber of old trees, because it is more elastic, tougher, and stronger. He quotes a report of a meeting of over 200 representatives of the carriage-building industry at Chicago, at which it was stated that the hickory trees have recently been attacked by insects to such an extent, that unless some means can be taken to check their ravages, there will be no more hickory available in ten years; and though ash, maple, and other woods have been tried as a substitute, there is no other wood so suitable for this industry as hickory. It is imported to some extent to Europe, usually in the form of second-growth poles, which are produced from the stool and are used by carriage builders.

Cobbett,[1] with his usual enthusiasm for everything from America, urged that the hickory should be planted for coppice wood on account of the value of the hoops

---

[1] *Woodlands*, arts. 295, 296 (1825).

which are made from it, and stated that he had a young tree at Kensington which was seven feet high when only five years old from the seed.   Mayr also thinks that the hickory might have some economic value in the warmer parts of Germany, as it has stood the hardest frosts at Munich without injury.   But so far as we know, none of the trials which have been made in France, where the tree was introduced on a large scale by Michaux 100 years ago, have been successful, and I could not hear that any of the trees which he planted near Paris are now alive.   (H. J. E.)

PLATANUS

Platanus, Linnæus, *Gen. Pl.* 358 (1737); Bentham et Hooker, *Gen. Pl.* iii. 396 (1880); Jankó, in Engler, *Bot. Jahrb.* xi. 412 (1890); Usteri, *Mém. Herb. Boissier*, No. 20, p. 53 (1900); Schneider, *Laubholzkunde*, 435 (1905).

TREES belonging to the order Platanaceæ, which consists of the single genus Platanus. Bark, at the base of old trunks, dark-coloured and scaly; above on the stem and on the branches smooth, thin, light-grey or greenish, separating in large thin scales, which on falling expose large irregular surfaces of pale-yellow or whitish inner bark. Branchlets rounded, zigzag; the apex withering and falling off in summer (no true terminal bud being formed), leaving an elevated orbicular scar close to the uppermost axillary bud, which prolongs the shoot in the following year. Buds in summer concealed in the funnel-like base of the leaf-stalk, after the fall of the leaf surrounded at the base by an incomplete narrow ring-like scar, sinuous in margin and divided into five parts, each marked by a group of bundle dots; a line extending round the twig from the sides of the leaf-scar indicating where the connate stipule fell off. Buds all axillary, uniform in size, conic, covered by a scale in the form of a cap, which contains immediately within two similar scales, and more internally, several scales open at the apex and each with a young leaf at its base. Flower-buds similar, but larger. The three outer cap-like scales split longitudinally as the bud expands, the second and third continuing to grow after the bud is unfolded, ultimately falling and marking the base of the branchlet with ring-like scars.

Leaves deciduous, alternate, simple, stalked, palmately three- to seven-lobed; lobes entire or dentate with minute or coarsely sinuate teeth; venation pseudo-palmate, two strong lateral nerves diverging from the midrib a little above its base, and each often giving off on the outer side a basal nerve, thus forming with the midrib three to five main nerves, each of which ends in the apex of a lobe. Stipules two, lateral, united below into a tube embracing the branchlet above the insertion of the leaf, dilated above and more or less free, thin and scarious on flowering shoots, broad and leafy on vigorous barren branchlets, caducous or occasionally persistent.

Flowers monœcious, fertilised by the wind, appearing with the leaves, the females developing first, minute, densely aggregated in unisexual heads, which are solitary or several in spikes or racemes; the staminate heads on axillary peduncles, the pistillate heads on long terminal peduncles. Staminate flowers; sepals, three to six, scale-like, half as long as the scarious, cuneiform, acute, three to six petals; stamens, three to six, with very short filaments and clavate two-celled anthers, crowned by a pilose capitate connective. Pistillate flowers; sepals, four (three to six), rounded, short; petals, four (three to six), long, acute; staminodes pilose at the apex; ovaries as numerous as the sepals, superior, sessile, surrounded at the base by long hairs, gradually narrowing above into long simple styles; ovules one, rarely two. Head of fruit composed of numerous elongated obpyramidate achenes, surmounted by the persistent style, surrounded at the base by long rigid hairs. Seed solitary, oblong, suspended, containing a thin fleshy albumen and an axile erect embryo. The fruiting heads remain hanging on the tree during winter, the component achenes being ultimately dispersed by the wind.

The dispersal of the pollen in the flowers of plane trees is effected by a peculiar mechanism, which bears some resemblance to that of the yew, and is well described by Kerner.[1]

The planes are readily distinguished by the simple alternate palmately-lobed leaves, the base of the stalks enclosing and concealing the buds. In winter, the conical buds, all lateral, with stipule-lines around the twig and the peculiar narrow sinuous leaf-scars are diagnostic.

The genus is a very ancient one, fossil species[2] having been found in North America in Cretaceous, Eocene, and Oligocene strata. In the Miocene and Tertiary epochs numerous species were spread throughout all Europe, Northern Asia, and North America as far north as the Arctic Circle. In the glacial period those became extinct in the northern parts of their area, and the existing species are confined to Canada, the United States, and Mexico in the New World, and to the Eastern Mediterranean region in the Old World. Their entire absence from Eastern Asia is remarkable, as tertiary plants of circumpolar distribution, which have survived to the present time, are usually found existing both in Eastern North America and in China and Japan.

Six species[3] are now living, which may be conveniently arranged as follows :—

I. Adult leaves glabrous or nearly so, conspicuously toothed in margin.
  1. *Platanus orientalis*, Linnæus. Albania, Macedonia, Thrace, Greece, Crete, Cyprus, Rhodes, and Asia Minor.

    Leaves distinctly lobed, the sinuses extending at least one-third the length of the leaf. Fruiting heads bristly, several on the peduncle. Achenes, with long hairs arising not only at the base, but along the body of the achene; apex pyramidal or conic, acute, passing gradually into the long style.
  2. *Platanus occidentalis*, Linnæus. Eastern North America from Ontario to Texas.

    Leaves indistinctly lobed, the sinuses not extending one-third the length of the leaf. Fruiting heads smooth, solitary, and terminal on the peduncle. Achenes with basal ring of long hairs, elsewhere glabrous; apex truncate or rounded, with a depression, from which arises a very short style.

[1] *Nat. Hist. Plants*, Eng. Transl. ii. 146 (1898).

[2] Cf. L. F. Ward, in *Proc. U.S. Nat. Museum*, 1888, p. 39, who states that a prominent characteristic of these archaic forms is the presence of basal lobes on the leaves. These basal lobes are occasionally met with on the young shoots of the species now living.

[3] *Platanus glabrata*, Fernald, *Proc. Am. Acad.* xxxvi. 493 (1901), is an imperfectly known species from Coahuila in Mexico.

II. Adult leaves densely tomentose beneath, margin entire or minutely and remotely toothed.

* *Lobes elongated.*

3. *Platanus racemosa*, Nuttall, *Sylva*, i. 47 (1842). A tree, attaining 120 feet in height and 18 feet in girth, common on the banks of streams in California. Introduced on the Continent, unknown in cultivation in Britain.

Leaves three- to five-lobed to below the middle; lobes acuminate, tomentose on both surfaces; occasionally with remote minute teeth; base slightly cordate or truncate. Fruiting heads bristly, two to seven on the peduncle. Achenes with ring of basal hairs, elsewhere glabrous; apex pyramidal or rounded, without a central depression; style long.

4. *Platanus Wrightii*, Watson, *Proc. Am. Acad.* x. 349 (1875). A tree, attaining 80 feet in height and 15 feet in girth, common in mountain cañons up to 6000 feet altitude in Southern New Mexico and Southern Arizona in the United States; and in Sonora in Mexico. Not introduced.

Leaves three- to five-lobed to below the middle or near the centre; lobes acuminate, usually quite entire, tomentose on both surfaces, deeply cordate at the base. Fruiting heads smooth, two to four, racemose on the peduncle. Achenes with ring of basal hairs, elsewhere glabrous; apex rounded, with a central depression, from which arises a short style.

** *Lobes short and broadly triangular.*

5. *Platanus mexicana*, Moricand, *Bull. Ferr. Bot.* 79 (1830). A large tree in Mexico, in Nuovo Leon and the provinces to the south of it; frequently planted in the cities of North-Eastern Mexico; the handsomest of all the plane trees.[1] Not introduced.

Leaves three- to five-lobed, densely white tomentose beneath, base truncate or cuneate. Fruiting heads bristly, solitary. Achenes with ring of basal hairs; upper part of the body pubescent; apex pyramidal, continued directly into the long style.

6. *Platanus Lindeniana*, Martens et Galeotti, *Bull. Acad. Brux.* x. 2, p. 343 (1843). Tree, 100 to 150 feet in height. South Mexico, near Jalapa, at 4000 feet altitude. Not introduced.

Leaves with usually three very short lobes, ending in long bristle-like points, densely rusty tomentose beneath; base truncate. Fruiting heads bristly, several on the peduncle. Achenes as in *P. orientalis*.

[1] *Garden and Forest*, ix. 51 (1896).

PLATANUS ORIENTALIS, Oriental Plane

Platanus orientalis, Linnæus, *Sp. Pl.* 999 (1753); Loudon, *Arb. et Frut. Brit.* iv. 2033 (1838); Boissier, *Flora Orientalis*, iv. 1161 (1879); Gamble, *Manual Indian Timbers*, 661 (1902); Schneider, *Laubholzkunde*, 436 (1905).
Platanus palmata, Moench, *Meth.* 358 (1794).
Platanus cuneata, Willdenow, *Sp. Pl.* iv. 473 (1805).
Platanus acerifolia, Willdenow, *Sp. Pl.* iv. 474 (1805).
Platanus laciniata, Du Mont de Courset, *Bot. Cult.* vi. 436 (1811).
Platanus vulgaris, Spach, *Ann. Sci. Nat.* xv. 291 (excl. ε *angulosa*) (1841).
Platanus hispanica, Tenore, *Cat. Ort. Nap.* 1845, p. 91.
Platanus digitata, Gordon, *Garden*, 1872, p. 572.

A tree, with several forms distinct in foliage and habit, most of which appear to have arisen in cultivation. Bark scaling off in thin plates; furrowed and thick at the base of old trunks, variable in the different varieties.[1]

Young branchlets at first densely stellate-pubescent, becoming green and glabrous in summer; brown in the second year. Leaves at first covered with dense loose white stellate pubescence on both surfaces, later nearly glabrous, the pubescence being only retained here and there, mainly on the veins and midrib of the lower surface, which is paler than the dark-green shining upper surface. Petioles at first densely stellate, white pubescent, ultimately glabrescent.

Fruiting heads several (two to seven) on the peduncle, bristly. Achenes with basal ring of rigid long hairs, similar hairs arising also along the body of the achene, the apex of which is more or less acute and ends in a long persistent style.

Seedling:[2] caulicle slender, about ½ inch long, surmounted by two narrow spathulate cotyledons, obtuse at the apex, tapering to a very narrow base, sessile, one-nerved, entire, glabrous, dark-green, about ⅓ inch long. First leaf resembling a petiole in shape, minute and glandular-pubescent. Second leaf spathulate-cuneate, with three teeth at the apex, alternately penninerved, two of the stronger nerves running into the teeth. Third leaf like the second, but larger. Succeeding leaves palmately five-lobed.

In summer, the oriental plane and its varieties are readily recognisable by the leaves, bark, and habit. In winter, the twigs are rounded, striate, glabrous, with numerous inconspicuous lenticels, the apex ending in a short stump, bearing an orbicular scar, marking where the tip of the branchlet fell off in summer. Leaf-scars, on prominent pulvini, almost but not completely surrounding the bud as a narrow ring, sinuous in margin, and with five groups of bundle-dots. Stipular line surrounding the twig at the level of each leaf-scar. Base of the shoot ringed with scars,

---

[1] Boissier states that the bark of the wild tree is rugose, and does not exfoliate, as is usually the case in cultivated trees, especially in var. *acerifolia*, of the origin of which he knew nothing. As there is considerable variation in the size of the scales of the bark on plane trees, it is probable that the difference noted by Boissier is individual and not varietal or specific. Var. *acerifolia* grows usually very fast, and scales off in much larger plates as a rule than is the case in the other varieties, which are slower in growth.

[2] Cf. Lubbock, *Seedlings*, ii. 505, f. 653 (1892).

marking the fall of the previous season's bud-scales. Buds distichous and alternate on the long shoots, arising at an angle of 45°, uniform in size, conical, smooth, lustrous, covered by a glabrous cap-like scale.

VARIETIES

1. Var. *typica*.—The form in cultivation, known generally as *P. orientalis*, slightly different in foliage from the wild form, known as var. *insularis*.

A tree, attaining enormous dimensions in South-Eastern Europe and Western Asia, with a short trunk, dividing into many wide-spreading branches. Leaves (Plate 204, Fig. 4) large, generally exceeding 6 inches broad by 5 inches long, usually five-lobed; lobes extending about half-way to the base of the blade, oblong-triangular with an acuminate apex, entire[1] in margin or with a few sinuate entire teeth; sinuses deep, variable in shape. Each of the two basal lobes often gives off a short lobe below, making the leaf seven-lobed. Base of the leaf truncate or widely cordate, but usually extending along the midrib $\frac{1}{8}$ to $\frac{1}{4}$ inch below where the two main lateral nerves are given off. Upper surface dark-green, glabrous, shining; lower surface paler, glabrous except along the nerves and midrib, and in their axils.

2. Var *insularis*, DC. *Prod.* xvi. 2, p. 159 (1864). The wild form, occurring in Albania, Greece, Cyprus, Crete, Rhodes, and Asia Minor.

A tree, not reported to be very large in size, and said to have rough bark, with small scales, which fall off less readily than in the typical form. Leaves smaller than in the type, scarcely exceeding 5 inches broad and $4\frac{1}{2}$ inches long, very variable in shape, usually five-lobed; lobes oblong-triangular or triangular, coarsely three- to four-toothed, extending about halfway to the base of the blade, which is always cuneate, the lamina descending along the midrib $\frac{1}{8}$ to $\frac{1}{4}$ inch below the insertion of the first pair of main nerves; sinuses deep, variable in form.

Mouillefert[2] distinguishes two varieties of the wild form, one with narrowly lanceolate entire lobes and wide sinuses, the other with lanceolate sinuately-toothed lobes and very deep narrow sinuses. The range of variation in the shape of the leaf in the wild form is considerable.

3. Var. *cuneata*,[3] Loudon, *Arb. et Frut. Brit.* iv. 2034 (1838).

Platanus cuneata, Willdenow, *Sp. Pl.* iv. 473 (1805).

A tree, moderate in size, with bark resembling that of var. *acerifolia*. Leaves (Plate 204, Fig. 3) usually about 5 inches wide by $4\frac{1}{2}$ inches long, occasionally 8 inches by 7 inches; three- to five-lobed, the lobes as in var. *typica*; base of the leaf broadly cuneate, the lamina extending along the petiole $\frac{1}{2}$ to $\frac{3}{4}$ inch below the insertion of the main lateral nerves; petioles short. This variety only differs from var. *typica* in the markedly cuneate base, and approaches in character var. *insularis*. As seen in cultivation it usually forms imperfect small fruiting heads.

[1] The form with entire lobes is sometimes distinguished as var. *liquidambarifolia*, Spach, *loc. cit.*
[2] *Essences Forestières*, 221 (1903).   [3] Probably var. *undulata*, Aiton, *Hort. Kew*, iii. 364 (1789).

According to Koch,[1] this is a good species, growing as a shrub on the south-eastern slope of the Caucasus, and having a very short stem from which spring many upright branches.   Koch, however, did not find it in flower or fruit; and as no plane has been observed growing wild in the Caucasus, it is probable that what he saw were stunted trees of ordinary *Pl. orientalis*, occurring as escapes from cultivation.

At Grayswood, Haslemere, there are two plants, 4 feet high, with a fastigiate habit, which Mr. Chambers raised eight years ago, from seed sent from Kashmir.

4. Var. *digitata*, Jankó, in Engler, *Bot. Jahrb.* xi. 412 (1890).

Platanus digitata, Gordon, *Garden*, 1872, p. 572.

This is a form of var. *typica*, in which the leaves are smaller than usual, with wider and deeper sinuses, the lobes extending three-fourths the depth of the blade and having large triangular toothed lobules.

Gordon supposed this form to be a native of the Caucasus, and says that it was introduced by Messrs. Loddiges in 1842.   He describes the fruiting heads as only half the size of those of the type; but in the number and structure of the component achenes there is no difference.

5. Var. *acerifolia*, Aiton, *Hort. Kew*, iii. 304 (1799).

Platanus acerifolia, Willdenow, *Sp. Pl.* iv. 474 (1805).   London Plane.   Maple-leaved Plane.

A tree, with a tall upright stem, giving off shorter branches than the typical form.   Leaves (Plate 204, Fig. 1) large, at least 8 inches wide by 7 inches long, with five short, broad, triangular lobes, separated by wide rounded or acute shallow sinuses, which only extend one-third the length of the blade; base truncate or widely cordate, the lamina often descending on the midrib a short distance below the insertion of the two main lateral nerves.   Fruiting heads very variable in size and in number on the peduncle, often badly developed in English trees; achenes similar in structure to those of the typical form, and never resembling those of *P. occidentalis*.

Several forms of the London plane have been distinguished:—
Var. *pyramidalis*.   Pyramidal in habit.
Var. *kelseyana*.   Leaves variegated with yellow.
Var. *Suttneri*.   Leaves creamy-white, more or less splashed or streaked with green, often very large, as much as 12 inches wide by 10 inches long.   This is identical with var. *argenteo-variegata*, which was exhibited at the Royal Horticultural Society in July 1897, by Messrs. Russell of Richmond.   It is one of the handsomest of variegated trees, the variegation usually lasting the whole season.[2]

A form with large leaves has been sent out under the misleading name of var. *californica*.

Though var. *acerifolia* exhibits a wide range of variation in the cutting of the leaf, it always shows very distinct lobes, and cannot be confused with *P. occidentalis*, in which the lobes are indistinctly marked.

[1] *Dendrologie*, II. Part i. p. 470 (1872).   [2] Cf. *Gard. Chron.* xxiv. 190 (1898).

DISTRIBUTION

The oriental plane has been in cultivation from very early times in the Mediterranean region; and the limits of its distribution in the wild state are difficult to determine accurately.

It occurs wild in woods and along torrents in the mountainous regions of Albania and Greece. It grows in chestnut groves and in mountain forests from sea-level to 2500 feet elevation in Macedonia, Thrace, and Bithynia. It is undoubtedly indigenous in the mountains of Crete, Cyprus, and Rhodes; and in Western and Southern Asia Minor up to 5000 feet, not ascending into the zone of the cedars, and also occurs in the Lebanon. M. Gadeceau, in *Rev. Hort.* 1907, p. 207, quotes a letter from a correspondent in Syria to the effect that the plane, known to the Arabs as *Dolbe*, does not grow in the forests, but only in valleys and along the banks of rivers; on Mt. Hermon it reaches 4000 to 5000 feet elevation.

Its occurrence in the wild state elsewhere is very doubtful, as all the specimens which I have seen from other localities were collected near villages. The great plane trees described by travellers as occurring in Persia, Afghanistan, Kashmir, etc., are evidently planted.

Radde,[1] who paid particular attention to the subject, denies its occurrence in the wild state in the Caucasus. Dr. Stapf only saw it cultivated in Persia, where it always grows near villages. In Afghanistan, according to Aitchison,[2] it is certainly not indigenous; but he found it naturalised in one district, where it had been originally planted in a valley and was gradually taking possession of the adjoining hill. According to Brandis,[3] it is cultivated in the North-West Himalaya, particularly in the Kashmir valley, east of the Bias and Sutlej, ascending to 8300 feet in Western Ladak. It grows well at Peshawar; and attains 75 feet in height in Kashmir and Chamba, the largest girth noted by Dr. Stewart at Srinagar being 28 feet. The Nasim Bagh, on the border of the great Kashmir lake, is a large grove planted by Akbar the Great, soon after he had taken Kashmir in 1588. Originally this grove contained 1200 trees, a large proportion of which are still standing. In 1838 Vigne found the average girth to be 13 feet, some of the trees growing near water being as large as 20 feet.

Tchihatcheff states that in Cyprus and Zante it is evergreen; but this statement is not confirmed by other travellers, although occasionally an evergreen variety occurs, as in the case of a tree at Lutraki in Crete, which was mentioned by Admiral Spratt.[4] Mr. Sandwith wrote to Kew in 1884 that this tree had been cut down several years before, but that several vigorous shoots were springing from

---

[1] *Pflanzenverb. Kaukasusländ.* 170, 187, 189 (1899). [2] *Journ. Linn. Soc.* (*Bot.*) xviii. 94 (1881).
[3] *Forest Flora N. W. India*, 434 (1874).
[4] *Travels and Researches in Crete*, ii. 40. Pliny mentioned the existence of an extraordinary evergreen plane tree which grew on the banks of the Lethæus. Tournefort searched for it without success; but its existence was make known to Captain Spratt by Mr. Agnew, an English merchant who owned property at Lutraki, where he showed him two young and branching plane trees growing by a rivulet, which retained their leaves all the winter. Mr. Agnew said that these were suckers from the roots of a very large tree of the same kind, which he had cut down, without knowing its rarity; and Spratt adds that he heard of two others growing near the village of Vourvalete on the banks of the Platanos river in the west of Crete. Spratt speaks, *op. cit.* p. 191, of a grove of large and beautiful plane trees, mixed with elm and oak, and covered with wild vines which climb to their very top.—(H. J. E.)

the stool, and cuttings from these had been grafted by a Cretan farmer on the ordinary plane tree and were preserving the evergreen habit.   (A. H.)

The stellate tomentum, which covers the young leaves of the plane, is gradually cast off; and floating in the air, has been found in some parts of Europe to produce serious bronchial irritation.   This was known to the ancient Greeks,[1] being mentioned by Galen and Dioscorides.   In Alsace-Lorraine, the planting of plane trees is forbidden in the vicinity of schools; and workmen in nurseries on the Continent, where young trees are raised, are often affected.[2]   We have, however, not heard of any complaint of this happening in England.

The young leaves and shoots of var. *acerifolia* are frequently affected by a disease,[3] caused by the fungus known as *Glœosporium nervisequium*, Saccardo, which in early summer attacks the nerves first and soon causes them to wither. Small black spots appear on the dead parts, which are the conidia of the fungus. In England, it is remarkable that the true oriental plane appears to be practically immune from the attacks of this fungus, though its leaves are sometimes blotched between the veins.   Mr. Massee informs us that after a thorough examination of dried and living material, he has failed to find the slightest evidence in support of the statement that *Glœosporium nervisequium* is parasitic on typical *P. orientalis*.   The London plane is almost invariably affected, though less in London than in the country, where almost everywhere some of the leaves and young shoots become brown and wither; but the healthy growth of the tree is scarcely ever seriously interfered with.   Some gardeners believe that this withering is due to cold winds and late frosts; but, though leaves may be injured by climatic conditions, this fungus is undoubtedly the principal cause.   A plane tree, var. *cuneata*, in the Cambridge Botanic Garden, which had the habit of var. *acerifolia*, had the leaves badly attacked in June 1907.   A tree, 30 feet high, of the same variety, at Grayswood, had the young wood seriously injured by a fungus, which Mr. Massee identifies with *Glœosporium*.

The fungus is apparently more severe in its attacks on the Continent; and at Ghent in 1891, all the plane trees lost their leaves.[4]   In the United States, the occidental plane[5] is very liable to be attacked by this fungus, and as a street tree in New England is unsuccessful on that account, though *P. acerifolia* succeeds as well as it does in England.

Klebahn[6] states that *Glœosporium nervisequium* occurs more especially on *P. occidentalis*, less frequently on *P. orientalis*.   He believes that *G. nervisequium* is only a conidial form of a higher fungus, called *Gnomonia Veneta*, Klebahn.

(H. J. E.)

---

[1] Cf. *Gard. Chron.* iii. 370 (1888).   [2] Carrière, *Rev. Hort.* 1890, pp. 370, 435.

[3] Cf. Massee, *Plant Diseases*, 284, f. 76 (1903).   [4] Cf. *Gard. Chron.* x. 491 (1891).

[5] Cf. *Garden and Forest*, 1891, p. 591, 1896, p. 51, and 1897, p. 257.   In an article on Leaf-blight of the Plane Tree, by Murrill in *Journ. N. York Bot. Garden*, viii. 157 (1907), an account is given of an epidemic of the disease occurring this year in New York, believed to have been caused by the late and damp spring.   Murrill observed the oriental plane to be attacked in Italy in 1906; and states that *P. racemosa* is also subject to the disease.

[6] *Jahrb. Wissensch. Bot.* xli. 515 (1905).

History of the Cultivated Planes

The oriental plane was introduced into Italy from Greece about 390 B.C.; and Hehn[1] gives a full account of the classical allusions to the tree.[2] It came into England[3] some time before 1562. In Turner's *Herball*, published in that year, a figure is given, and the author states: "I have sene the leves of that Platanus that groweth in Italy and two very yong trees in England which were called there Playn trees, whose leves in all poyntes were lyke unto the leves of the Italian Playn tree. And it is doubtles that these two trees were either brought out of Italy or of som farr countrie beyond Italy where unto the freres monkes and chanones went a pilgrimage."

The American plane, *P. occidentalis*, was introduced into England by Tradescant,[4] in whose garden two small plants were growing in 1636, when Johnson published his edition of Gerard's *Herball*. It was undoubtedly in cultivation in the eighteenth century, as a specimen from a tree cultivated at Kew in 1781 exists in the British Museum but another specimen from the Chelsea Physic Garden, dated 1789 and labelled *P. occidentalis*, is undoubtedly *P. acerifolia*. The figure given in Evelyn's *Sylva* of *P. occidentalis* really represents *P. acerifolia*. Similarly Loudon's description and figure of the American plane are inaccurate, and in part refer to *P. acerifolia*. The confusion between these two forms is thus shown to have begun early, and has lasted until quite recently; and it is probable that most of the references to the occidental plane in this country and on the continent of Europe refer to *P. acerifolia*.

Platanus acerifolia was first distinguished by Tournefort[5] in 1703. Miller,[6] in 1731, gives an account of three kinds of plane: *P. orientalis vera*, *P. occidentalis*, and *P. aceris folio*; but he was unaware of the real distinctions between the two latter, attributing to *P. occidentalis* the property of being easily propagated by cuttings, whereas it is *P. acerifolia* of which this is true. He asserts that his *P. aceris folio* is only a seminal variety of *P. orientalis*.

Bolle[7] states that Bourgeau found considerable forests of *P. acerifolia* in Lycia. This statement has not been confirmed, and there is no evidence of the occurrence anywhere of this form in the wild state. The difference between it and the wild form of *P. orientalis* (var. *insularis*) is mainly in habit, and taking into account the variability of the leaves on the wild tree, no two of which are alike in the specimens which I have examined, there is little doubt that *P. acerifolia* is a seedling variety of *P. orientalis*, which has been fixed in cultivation. Intermediate forms between it and the ordinary typical variety are not unusual. Bolle states that seedlings of *acerifolia* often exhibit the characters of typical *orientalis*. Further experiments on this point are desirable, as well as a thorough investigation of the range of variation

---

[1] *Kulturpflanzen*, ed. 6, p. 283.

[2] The Romans planted it in their gardens for shade; Ovid calls it *genialis*, and Horace *cælebs* because it did not support the vine.

[3] It was probably introduced into France in Provence about the same time. Cf. *Le Jardin*, 1896, pp. 116, 162.

[4] Cf. Parkinson, *Theat. Bot.* 1427 (1640).
[5] *Coroll.* 41 (1703).

[6] *Gard. Dict.* ed. 1 (1731).
[7] *Gard. Chron.* i. 564 (1876).

of the tree in the wild state in Greece and Asia Minor. At what time *acerifolia* came into cultivation is unknown; but the evidence points to its having become common in the eighteenth century. All the very old trees in the Levant of which we have specimens, are, however, of the typical form.

Considering the difficulty of distinguishing between *acerifolia* and *occidentalis*, in the absence of fruit, a character to which no attention was paid until 1856, when Sir W. J. Hooker[1] cleared up the confusion until then existing between them, it is probable that in the eighteenth century, as at the present time, the commonest plane in cultivation was *P. acerifolia*, and that *P. occidentalis* was very rare. The latter could never have been common, as it is quite unsuitable for our climate It dies at Kew after a few years, and we know of no specimen, older than nursery plants, in England.

This we believe to be the true solution of the difficulty, namely, that *acerifolia* was always the common tree in cultivation in England, that it was perpetually confused with *occidentalis*, and often passed under that name, and that the American plane never reached adult size in this country.

The origin and date of the first cultivation of *acerifolia* must remain in doubt; but I see no grounds for assenting to Schneider's view that it is a possible hybrid between *occidentalis* and *orientalis*. It does not resemble the former in any way in the characters of the fruit; and the similarity in the shape of its leaves to those of the American plane is more apparent than real.   (A. H.)

CULTIVATION

As the true oriental plane is now hardly to be procured in nurseries, where the maple-leaved or London plane has alone been propagated for many years, it is necessary either to sow seeds or make layers from the branch of an old tree. The seeds ripened in this country often (I think I may say usually) fail to germinate, probably on account of the insufficient heat of our autumn, yet I have raised a few seedlings from a tree at Fulham, which, however, soon died when planted out. As the fruiting heads hang on the tree till spring, it seems best not to sow them till then, and I have been successful in raising seedlings from seed gathered at Venice as late as May.

The seeds should be rubbed out with sand, covered very lightly with fine soil, and kept shaded in a greenhouse until they germinate, as they grow very slowly the first year. It is better not to transplant them to the open ground until they are two or three years old, as the young wood does not ripen well; and for these reasons it has been found by nurserymen much better to raise them by layering, or from cuttings, which Boutcher says should be about a foot long and torn asunder at the joints, with a knob of the old wood left on, and buried about eight inches deep. I have struck cuttings from the true oriental plane by this method.

The trees are easy to transplant even when of considerable size, and require a deep rich soil to make them grow well. Though often planted by the side of water

---

[1] *Gard. Chron.* 1856, p. 282.

they do not like cold wet soil or heavy clay; and a seedling which I brought in a bottle from the Temple of Diana at Ephesus in 1874, and planted out at Colesborne, has never been able to make good growth on account of the spring frosts, which cut it back in most seasons; and after thirty-three years' growth is only a stunted bushy tree 20 feet high.

The true oriental plane is always liable to be cut by frost at any age, and the branches and twigs assume a zigzag habit in consequence; but no tree succeeds better in the smoky air of towns than the maple-leaved or London plane, which is by far the largest tree in all the London squares, parks, and gardens, and seems likely to live to a very great age.

As regards hardiness, the plane seems, when of sufficient size, to endure the severest winter frosts, but requires a higher summer temperature than the north and west of England usually affords. In the south-east it is almost everywhere one of the finest exotic trees we have, but should not be planted on poor, stony land or in places exposed to cold winds.

The oriental plane is one of the very few trees that will grow on strongly alkaline soil, and has been successfully planted on the alkaline lands of the San Joaquin valley in California.[1]

REMARKABLE TREES

Var. *typica.*

Perhaps the finest specimen in England of the oriental plane is one which was planted by Bishop Gunning in the palace gardens at Ely, of which he was bishop from 1674 to 1684. This tree seemed to be failing some years ago, probably owing to a succession of dry seasons, and on the advice of Sir W. Thiselton Dyer was liberally top-dressed with good soil; the result was so good that when I saw it in 1903 it was in good health, and measured 104 feet high by 20½ in girth. In 1896 it was found by the shadow to be about 100 feet high and 23 feet in girth at 3 feet from the ground. It forks at about 10 feet, where an immense limb comes off, and the branches almost touch the ground all round. It grows on a low hill 50 to 60 feet above the level of the fens, but is sheltered on the north and east by buildings. On 15th October 1903 (a very wet season), the leaves were still quite green, but there was no fruit. Plate 174 is selected from several photographs taken at different times, as giving the best representation of this noble tree.

In the garden of the Lady Margaret Professor of Divinity at Christ Church, Oxford, there is a very old plane, said to have been introduced by Dr. E. Pococke, who was Professor of Arabic in 1636. It is of no great height and throws out an immense limb close to the ground, where it measures 18 feet 10 inches in girth. When I saw it in September 1907, fruit of the last year was still hanging on it, together with full-grown fruit of the current year, and in some cases six or seven balls were borne on one peduncle.

At Hawsted Old Place, near Bury St. Edmunds, there are three very large

---

[1] Hilyard, *Soils,* 480 (1906).

oriental planes in a grass field near the farm, where an old manor-house formerly stood, and these, according to the Rev. L. Mercer, vicar of Hawsted, are said to have existed in Queen Elizabeth's time, and to be the oldest in England.[1]   They are difficult to measure accurately on account of their broad round tops and the trunks being covered with shoots, but the two largest are 75 to 80 feet high and 17 to 18 feet in girth, and the third is not much smaller.   When I saw them on 24th June 1905, they were still covered with last year's fruit, three to five on a peduncle, and new half-grown fruit was also growing on them.   Their branches spread over a very wide area.

At Corsham Court, Wilts, the seat of General Lord Methuen, there is an oriental plane with very deeply-cut leaves whose branches spread over a larger area than those of any tree I have seen in England (Plate 175).   It is 75 to 80 feet high and 18½ feet in girth.   One of the branches, which is self-layered in several places, extends no less than 27 paces from the main bole, and the total circumference of the branches is 140 paces.   Three of the principal stems grown from the layered branches are 6 feet 3 inches, 4 feet 10 inches, and 4 feet respectively in girth. Lord Methuen believes that this tree was planted soon after his ancestor built the house in 1757, and a cedar recently cut down, probably of the same age, showed about 133 rings and thus tends to confirm his opinion.

Another very large tree of great age is growing close to the banks of the Test, at Mottisfont Abbey, Hants, which has layered its branches in the damp alluvial soil.   When I saw it some years ago it was in good health, and in 1898 measured 29 feet 8 inches at 4½ feet where it forks, the branches having a diameter of 129 feet.[2]   At Bisterne Park near Ringwood, in the same county, there is a very fine old tree measuring, in 1906, 100 feet by 18 feet.

At Weston Hall, Staffordshire, the seat of the Earl of Bradford, there is a very fine plane which I have not seen myself, but which was measured[3] in 1875 as 80 feet by 18½ feet with a bole 11 feet high, and twelve large limbs from 4 to 7 feet in girth.

At Blickling Hall, Norfolk, there are two very old trees growing close together, the largest of which was, in 1907, 11 feet in girth and almost 60 feet high, with remarkably spreading branches, covering a space of 140 feet in the largest diameter, and 127 paces in circumference.   The lowermost branches lying upon the ground had taken root, and growing erect for a time had again bent down and taken root a second time.

At Chiswick House, London, there is another example of a large oriental plane with very spreading pendulous branches, many of which are lying upon the ground and one has taken root.   This tree was, in 1903, 13½ feet in girth, 74 feet in height, and the diameter of the spread was 100 feet.

At Greycourt, Ham, in Surrey, the residence of Colonel Biddulph, there is a tree which measured, in 1906, 16 feet 11 inches in girth and 72 feet in height, and had a short bole of 7 feet dividing into three great limbs, the diameter of the spread of branches being 94 feet.

[1] Cf. *Woods and Forests*, 1884, p. 153.   [2] Cf. *Gard. Chron.* xxiii. 24, figs. 9, 10 (1898).
[3] Geo. Berry, in *Garden*, xx. 370 (1881).

The well-known tree in Kew Gardens, near the Palace, is about 80 feet high and measures 14 feet 8 inches in girth.

In Scotland, the oriental plane is rare; and seems to be much injured by late spring frosts. The Rev. D. Landsborough[1] speaks of one planted in 1864 in Kay Park, Kilmarnock, which was killed by the frost of 15th April 1903; others, however, in a sheltered position had not suffered. A large oriental plane at Kelso, which was a favourite tree of Sir Walter Scott's, was destroyed by the great frost of 1814. This was probably the tree referred to by Walker[2] in 1812, who states that the oriental plane grows at Mount Stewart, Bute, like a willow, and nowhere else so good in Scotland except at Kelso.

In Ireland the oriental plane is rather a rare tree. The finest specimen we know of is at Carton, the seat of the Duke of Leinster, which was, in 1904, 82 feet in height by 11 feet 9 inches in girth. Mr. W. E. Gumbleton has sent us specimens from a very fine tree growing in his grounds at Belgrove, near Queenstown in County Cork. Another good tree is reported to be growing at Curragh Chase, in Co. Limerick. There are smaller trees at Clonmannin, in Co. Wicklow.

The oriental plane attains an enormous size and great longevity in the eastern parts of Europe, Asia Minor, and Persia.

One of the most remarkable was a tree growing in the village of Vostiza, on the Gulf of Lepanto, in Greece, which measured,[3] in 1842, 37 feet 4 inches in girth at 5 feet from the ground, and was estimated to be 130 to 140 feet in height. This tree is supposed to be the one referred to by Pausanias, who wrote in the second century A.D.; yet in 1842 the trunk appeared to be perfectly sound, though many of the larger branches have succumbed to age and storm. Sir F. Elliot, British Minister at Athens, was good enough to make enquiry about this tree from Mr. Wood, British Consul at Patras, who informed him that when he last saw the tree, only a few feet of hollow stump remained. There are two remarkable oriental planes at Cannosa, near Ragusa in Dalmatia, which measure at breast height 32 feet 1 inch and 30 feet in girth. They are about 120 feet high.[4]

The famous plane at Bujukdere on the Bosporus is not a single trunk, but is formed of nine stems fused together. According to Ch. Martins,[5] in September 1856, the height was 200 feet—evidently an exaggeration—with a spread of branches 373 feet in circumference. One trunk girthed 18 feet; two trunks united together for some distance girthed 36 feet, the remaining six trunks being in an ellipse of 76 feet. One of the stems was hollow and afforded stable room for two horses. This tree is typical *orientalis*.[6] This tree is also sometimes called the "Seven Brothers" or the plane tree of Godfrey de Bouillon, as tradition states that he and his crusaders encamped in its vicinity in 1096.[7] Sir N. O'Conor informed me in 1903 that it has suffered much within the last few years owing to some excavations made close to its roots, and is evidently declining rapidly. A sketch of it made for the late Mr. C. Ellis, is now in Lady Emlyn's possession.

[1] *Kilmarnock Glenfield Ramblers' Soc. Annals*, 1901-1904, p. 33. [2] *Hebrides*, ii. 199 (1812).
[3] D. H. in *Woods and Forests*, i. 174 (1884). [4] Beck v. Mannagetta, *Veget. Verhält. Illyrisch. Länd.* 185 (1901).
[5] *Du Spitzberg au Sahara*, 474 (1866). [6] Bentham and Viscount Downe, in *Gard. Chron.* 1856, p. 118.
[7] Mouillefert, *Essences Forestières*, 215 (1903); and *Garden and Forest*, ii. 349 (1889).

The tree of the Janissaries, the ancient plane, which stands in the Court of the Janissaries in the Old Seraglio at Constantinople, was 39 feet in girth at 3 feet from the ground in 1890; but the trunk was hollow, the branches and foliage, however, being sound and vigorous.[1]

In the *British Medical Journal* of 21st June 1902, there is an excellent account, with illustrations, of a plane tree in the island of Cos, which from its appearance must be one of the oldest trees in the Mediterranean, if not so old as its somewhat mythical history alleges. Local tradition says that under this tree Hippocrates, the celebrated Greek physician, taught the art of healing no less than 2300 years ago. The tree grows near the landing-stage, between an ancient castle and a mosque, close to a drinking-fountain. Mr. von Holbach, who measured it, gives the girth of its hollow trunk as 18 metres; but all the upper part has decayed away, and the lower part of the tree now consists of immense branches which are supported on antique marble columns, over the tops of which their great weight has caused them to grow. Dr. Clapton, of 41 Eltham Road, Lee, procured a section of one of the branches of this tree, and has presented a photograph of it to the Hunterian Museum.

Bonvalot,[2] on his way from Samarcand to Amu, states that he halted at Sarijui, near the residence of the chief, under a plane tree, which was about 37 feet in diameter at 6 feet above the ground. In his book, a picture of the tree is given, and a great limb comes off low down, which evidently was included in the above measurement. The tree appears to be about 50 feet in girth at the base below where the limb comes off. Another enormous tree,[3] 49 feet in girth, stands in the grounds of the mosque of Tajrish, a village in the Elburz Mountains, north of Teheran, in Persia.

Var. *acerifolia*.

The variety *acerifolia* seems to have generally replaced the cut-leaved form at some period above a hundred years ago, but we cannot find any certain evidence of this, because it was generally confused with the western plane.

All the planes that we have seen in the squares, parks, and gardens about London of less age than about 100 years are *acerifolia*, and the finest specimen that I know of is the one in the Ranelagh Gardens, which measured, in 1903, 105 feet high, with a girth of 20 feet 4 inches.

The planes in Berkeley Square are worth notice on account of their uniform burry trunks swelling at the base. They all appear to have been propagated from the same stool and to have retained this peculiarity throughout. They were planted by Mr. Edward Bouverie in 1789 and are probably the oldest plane trees in London.[4] According to Mr. R. Birkbeck the two largest, in 1906, girthed at 5 feet, 13 feet 10 inches and 13 feet 4 inches, and were about 85 feet high. Tradition says that this area was a burial-ground during the Plague of London in 1665.

On the banks of the river Rother at Woolbeding Rectory, Sussex, in the garden

---

[1] *Garden and Forest*, iv. 85, fig. 19 (1891), where a full account and good picture of this remarkable tree are given.
[2] *Through the Heart of Asia*, i. 207, fig. on p. 209. Sarijui is a village 96 miles S.S.E. of Samarcand.
[3] Figured in *Woods and Forests*, i. 375 (1884).   [4] Hare, *Walks in London*, ii. 74 (1894).

of Archdeacon Elwes, are two of the largest planes in England, one of them, which is partly hollow at the base, but has been filled up with brickwork, being about 105 feet high by no less than 25 feet in girth. The other, a better-shaped and very vigorous tree, is 100 feet by 10 feet, with a bole 12 feet long. In Colonel Lascelles' grounds close by, there is another splendid plane, 110 feet high by 11 feet in girth, with a bole 30 feet long. At Cowdray, in the same neighbourhood, there are some fine old planes at the bottom of the park, two which I measured being about 80 feet high and over 15 feet in girth.

There is a very fine tree at Rickmansworth which, when measured by Henry in 1904, was 105 feet by 16 feet 3 inches. Another at the same place was 103 feet by 15 feet. Sir Hugh Beevor tells us of a fine tree at Shotesham, Norfolk, the seat of R. Fellowes, Esq., which, in 1904, was 100 feet by 17 feet 8 inches.

I measured a very tall and handsome tree, which appears to be growing fast, at Albury, in 1905, when it was at least 105 feet high by 11 feet 3 inches (Plate 178).

A tree which appears from its leaves to be more or less intermediate between the oriental and London planes is growing at Boconnoc in Cornwall, and is decaying, the climate being probably too damp for it, as I have seen no very large plane trees in the far west. In 1905 it was about 85 feet by 11 feet 2 inches.

At Hampton Court, Herefordshire, there is a fine tree on the lawn, which was measured in 1881 by Hogg, when it was 80 feet by only 8 feet 6 inches, and when I saw it in 1905 had increased to 95 feet by 14 feet 6 inches. I cannot help suspecting a mistake in the earlier measurement of girth, as an increase of 6 feet in twenty-four years seems extraordinary.

At Eastwell Park, Kent, there is a tree in a shrubbery drawn up by surrounding trees to a height of 105 feet, though the trunk is only 8 feet 6 inches in girth.

At Fawley Court, Oxfordshire, there is a row of four large trees about 100 feet high, and from 16 to 18 feet in girth, which are probably of no great age. In this damp alluvial soil they seem likely to become as large as any in England.

At Ashleigh College, on the east side of Mortlake, close to the Thames, a group of five trees form a conspicuous feature in the landscape. One of these is 110 feet in height and 10 feet in girth.

At Pains Hill, Cobham, Surrey, there are six or seven very large trees, one of which measured, in 1904, 100 feet high by 15 feet in girth.

At Chipleigh, near Minehead, a very fine plane, said to be an occidental plane, but undoubtedly a London plane, which was planted in 1760, is recorded by Mr. E. C. Batten[1] as being, in 1888, 105 feet high by 15½ in girth, with a spread of 120 feet.

At Heron Court, Ringwood, the seat of the Earl of Malmesbury, there is a London plane, which measures 90 feet by 11 feet 5 inches, and has a stone at the base with the date 1707 cut on it. A woodpecker's hole in the trunk, which is clean for about 35 feet, shows that this tree has begun to decay.

At Ribston Park,[2] Yorkshire, there is a tree which appears to be as old as any that I have seen. Tradition says that it was brought over by the Knights Templars,

[1] *Trans. Eng. Arb. Soc.* ii. p. 221.   [2] *Gard. Chron.* xxx. 34 (1901).

who had a preceptory here. It measured, in 1906, 72 feet high by 13 feet 9 inches in girth, but has the habit of *typica* rather than of *acerifolia*.

In Wales the finest plane that I have seen is a tall tree growing near the icehouse at Dynevor Castle. In 1906 it measured about 100 feet by 10½ feet with a clean trunk about 40 feet long.

In Scotland we have seen no plane remarkable for size, but there is one growing in the grounds at Benmore, Argyllshire, which has a curious resemblance in its foliage to the occidental plane.

In Ireland, the largest London plane, seen by Henry, is growing at Lismore Castle, and measures 12 feet 10 inches in girth, with an estimated height of about 90 feet. Mr. R. D. O'Brien informs us that a tree at Cooper Hill, near Limerick, is 10 feet 5 inches in girth, with a spread of 74 feet in diameter.

TIMBER

The wood of the plane is so little known in the timber trade of this country that it is not even mentioned by "Acorn," except as a name in use for sycamore, which is commonly called plane in Scotland; and in a recent letter in the *Timber Trade Journal*, what is known as lace-wood in the trade is spoken of as wood of the sycamore, imported from America, though it is really that of *Platanus occidentalis*. The ignorance which prevails among English timber merchants and builders about many of our useful woods is remarkable, and has led to many lawsuits, but there is no doubt that the wood of the oriental plane is one of much greater value than is supposed, both for ornamental work and for coach-building.

Mr. George Berry of Longleat[1] says that the timber of the plane tree is used almost exclusively by coach-builders and pianoforte-makers. No wood takes the paint and stands so well for the sides of large waggonettes as this. In the case of pianos, it was used exclusively for bridges, the toughness and hardness enabling the pins to be most securely held. He considers that plane timber exported from America is of very inferior value as compared with that of English growth.

Dr. Day sent me from the Lebanon a large board which shows a very beautiful and varied figure produced by the medullary rays, and I have seen in Prof. Sargent's house at Brookline, near Boston, very handsome panelling made of the wood of the western species. This wood is converted into veneer or three-ply, and sold as lace-wood, for covering the walls of rooms, and would make very pretty furniture if properly cut and seasoned. Gamble says that the wood is not valued in Kashmir, except to make boxes, trays, pencases, and similar articles, which are lacquered or painted. I have seen very ornamental boxes made from this wood in Russia.

(H. J. E.)

[1] *Garden*, xxii. 83 (1882).

PLATANUS OCCIDENTALIS, Western Plane, Buttonwood

Platanus occidentalis, Linnæus, *Sp. Pl.* 999 (1753); Loudon, *Arb. et. Frut. Brit.* iv. 2043 (1838)
    (in part); Sargent, *Silva N. America*, vii. 102, tt. 326, 327 (1895), and *Trees N. Amer.* 344
    (1905).
Platanus lobata, Moench, *Meth.* 358 (1794).
Platanus hybridus, Brotero, *Fl. Lusit.* ii. 487 (1804).
Platanus vulgaris, ε. *angulosa*, Spach, *Ann. Sci. Nat.* xv. 291 (1841).

A tree, attaining in America a height of 170 feet and a girth of 35 feet. Bark furrowed near the base of large trunks into broad rounded scaly ridges, higher on the tree separating in large thin scales, exposing irregular surfaces of pale-yellow or whitish inner bark. Branchlets and buds indistinguishable from those of *P. orientalis*.

Leaves (Plate 204, Fig. 2) large, about 8 inches broad by 7 inches long, broadly ovate, obsoletely three- to five-lobed, the lobes much less marked than in *P. acerifolia*, not reaching to one-third the length of the blade, separated by very broad and very shallow sinuses, entire or with long acuminate teeth; base widely cordate or truncate, with often a slight cuneate part decurrent on the petiole.

Fruiting heads solitary [1] at the end of a glabrous stalk (3 to 6 inches in length), about an inch in diameter, not bristly, but smooth, the persistent styles being minute or obsolete. Achene truncate or rounded at the apex, which has a central depression bearing the short remains of the style, glabrous on the body, surrounded by a ring of long stiff hairs arising from the base only.

This species differs from all forms of *P. orientalis* in the solitary fruiting head, composed of glabrous achenes, with only a basal ring of long hairs. The foliage is nearest to *P. acerifolia*, but differs in the obsolete ill-defined lobes.   (A. H.)

DISTRIBUTION

According to Sargent the western plane grows on the borders of streams and lakes in rich alluvial soil, and is widely distributed throughout the eastern half of the United States, crossing into Canada, where it is confined to the northern shores of Lake Ontario. In the United States it extends from South-Eastern New Hampshire and Northern Vermont southward to Northern Florida, Central Alabama, Mississippi, and Texas, as far as the valley of the Devil's River; and westward to Eastern Nebraska and Kansas. It attains its largest size in the bottom-lands of streams in the basins of the lower Ohio and Mississippi rivers.

The buttonwood, or sycamore, as it is often called in the United States, is the most massive, if not the tallest, deciduous tree of North America. The younger Michaux [2] states that his father measured, on an island in the Muskingum valley, a "palm-tree or *Platanus occidentalis*," 40 feet 4 inches in girth at 5 feet. The same

---

[1] Sargent in a letter to Kew, dated 1884, states that he had received from three different collectors specimens showing two heads on the peduncle; but this is rare and abnormal.
[2] *Travels in the Alleghanies*, 86 (1805).

tree had been measured twenty years earlier by Washington, when it was nearly the same size. Michaux measured one, 36 miles from Marietta on the road from Wheeling, on the Ohio, 47 feet in girth at 4 feet, which kept the same size for 15 to 20 feet, and then forked into several branches. This tree was hollow.

Ridgway records[1] a tree in Gibson County, Indiana, 160 feet high, 30 feet in girth at the smallest part of the trunk, with a spread of 134 feet by 112 feet; another in Wabash County 168 feet high, 25 feet in girth, and 68 feet to the first branch; another $83\frac{1}{2}$ feet to the first branch, but only 9 feet in girth. He found the prostrate trunk of a tree near Mount Carmel, in Illinois, which was much larger than any of those above mentioned. The decayed base measured 60 feet in circumference; and at 20 feet from the ground, where the tree divided into three large limbs, it was still about 62 feet round. Each of the three limbs was about 70 feet long by 5 feet in diameter, so that the total cubic contents by quarter-girth measurement must have been over 8000 feet. None of these were quite so tall, but much larger in girth than, the largest Tulip tree on record, and I know of no broad-leaved tree in the northern hemisphere which equals these dimensions. The largest which I actually saw myself, shown me by Dr. Schneck near Mount Carmel, measured 150 feet by 25 feet, and was standing in a cornfield. In New England it does not attain anything approaching these dimensions, the largest mentioned by Emerson, near Lancaster, Mass., being 18 feet in girth at 6 feet, and holding its size for 20 feet, and with a broad head of great height.

CULTIVATION

This tree is unsuited to our climate, and though seedlings are frequently raised at Kew, they never live more than a few years, and suffer severely from frosts. One tree raised from Michigan seed attained a height of about 12 feet, but became badly attacked by disease, and was removed about a year ago. So far as we know, there is not a single tree of this species of any size now growing in Britain.

Thomas Rivers, in an interesting article,[2] states that in 1820 there were in his nursery stools of *P. occidentalis*, which had been planted by his grandfather in 1780. These stools gave shoots with enormous, almost circular slightly-lobed leaves; but the young shoots always died down. From 1830 to 1840 he imported seeds of the American tree, which gave plants like these stools, but never lived for any length of time. He quotes Sir W. J. Hooker's statement: "We often raise young plants of *P. occidentalis* from American seed; but the annual shoots are killed every winter."[3] Rivers believed that the American plane had never existed in England so as to form large specimens, and that those mentioned as being large trees by Miller in 1759 and by Loudon in 1838 were not the true *occidentalis*.

The tree is equally rare on the Continent. M. Gadeceau, who wrote two papers[4] on the differences between the occidental and oriental planes, knew of

---

[1] *Proc. U.S. Nat. Museum*, 1882, p. 288.   [2] *Gard. Chron.* 1856, p. 86, and 1860, p. 47.

[3] But this is not always the case, as I planted out two seedlings of *P. occidentalis*, raised at Kew in the autumn of 1906, and they have remained healthy throughout the cold wet summer of 1907.

[4] *Bull. Soc. Sc. Nat. Nantes*, iv. 105 (1894), and *Rev. Hort.*, 1907, p. 205.

the existence of only one true American plane in France, which is growing in the Jardin des Plantes at Angers; but I saw in the rich collection of M. Allard at La Maulevrie near Angers, a tree about 30 feet high which, though it bore no fruit, seemed quite healthy. M. L. Henry[1] says that another tree exists in the Botanic Garden at Nantes. There are three trees of this species, about 50 feet high, growing in the garden of Messrs. Simon Louis at Plantières-lès-Metz.[2] Schneider knows of no trees in Germany or Austria. (H. J. E.)

[1] *Le Jardin*, 1903, p. 212.   [2] Schelle, *in litt.*

ACER

Acer, Linnæus, *Sp. Pl.* 1054 (1753); Bentham et Hooker, *Gen. Pl.* i. 409 (1862); Pax, in Engler, *Pflanzenreich*, iv. 163, *Aceraceæ*, 6 (1902); Schneider, *Laubholzkunde*, ii. 192 (1907). *Negundo*, Ludwig, *Gen. Pl.* 308 (1760); Bentham et Hooker, *loc. cit.*

TREES or shrubs, belonging to the natural order Aceraceæ, which is often considered to be a division of Sapindaceæ. Leaves usually deciduous, rarely evergreen, opposite, without stipules—simple, in which case they are undivided or palmately lobed—or compound with three to five leaflets. Buds covered by several scales arranged in decussate pairs, or protected by two valvate scales, sessile or occasionally stalked. Twigs with epidermis persisting for more than one year and remaining green in the second year; or becoming corky on the surface and changing colour in the first season. Inflorescence terminal on two- to four-leaved branchlets, or arising out of lateral buds without leaves, in racemes, corymbs, or fascicles. Flowers appearing at the same time as the leaves or earlier; regular, diœcious, or with male and perfect flowers on the same tree, or with male flowers on one tree and perfect flowers on another tree. Parts of the flowers in fours or fives or multiples of those numbers. Calyx with four, five, to twelve sepals, usually free, occasionally connate. Petals equal in number to the sepals, absent in some species. Disc secreting honey usually present, absent in a few species, annular, lobed, or reduced to small teeth. Stamens four to ten, usually eight, inserted either outside the disc, inside it, or upon it. Ovary, two-lobed, two-celled, each cell containing two ovules. Styles or stigmas two, free or connate at the base. Fruit of two samaræ, attached by their bases, with long and diverging wings. Seeds one or two in each samara, without albumen; cotyledons appearing above the ground on germination.

About 110 species of maple are known, occurring usually in mountainous regions; in Europe, south of lat. 62°, in Algeria, Asia Minor, the Caucasus, Persia, Turkestan, the Himalayas, China, Manchuria, Japan, Formosa, the Philippines, Java, Sumatra, Celebes, and in North America from Southern Canada and Oregon to Mexico and Guatemala. A large number have been introduced into cultivation, fifty-seven species being enumerated in the Kew Hand-List; but many of these are shrubs or small trees, the detailed treatment of which does not come within the scope of our work.

The genus is divided by natural characters into thirteen sections by **Pax**, whose monograph and that of Schneider should be consulted by cultivators of the rarer species. The following synopsis, in which the species are arranged artificially according to the shape of the leaves, will help to distinguish most of the species in cultivation.

Synopsis of the Maples in Cultivation

I. Leaves simple, not lobed.

* *Leaves entire in margin.*

1. *Acer oblongum*, Wallich. Himalayas, China.

Leaves coriaceous, 3 to 6 inches long, narrowly elliptical, long acuminate, glabrous and glaucous beneath; nerves, eight to twelve pairs, not reaching the margin; stalks 2 to 3 inches long.

A tree, about 50 feet high. The Himalayan form was introduced in 1824, and has been long in cultivation in the temperate house at Kew; it is doubtfully hardy. The Chinese form, introduced by Wilson in 1901, is growing rapidly out-of-doors at Coombe Wood; but is slightly tender. The young leaves of the latter are bright-red in spring, and are slightly toothed with distinct serrations.

** *Leaves serrate.*

2. *Acer carpinifolium*, Siebold et Zuccarini. Japan.

Leaves about 4 inches long, plicate, and resembling those of the Japanese hornbeam (*Carpinus japonica*), long acuminate, pubescent beneath, sharply bi-serrate; nerves, twenty pairs, extending to the margin; stalks $\frac{1}{4}$ to $\frac{1}{2}$ inch long.

A tree, attaining 50 feet in height. Introduced by Maries in 1881. Trees at Coombe Wood are about 15 feet high, and are remarkably distinct in foliage from the other species of maple.

3. *Acer distylum*,[1] Siebold et Zuccarini. Japan.

Leaves 5 inches long, 4 inches broad, ovate, long acuminate, finely serrate, pubescent on both surfaces when young, later glabrescent; nerves, eight to ten pairs, looping and not reaching the margin; petioles pubescent, about $1\frac{1}{2}$ inch long.

A tree, the height of which is not stated, introduced by Maries in 1881. At Coombe Wood, a specimen is about 25 feet high, and produces fruit, borne in erect racemes.

4. *Acer tataricum*, Linnæus. South-Eastern Europe, Southern Russia, Asia Minor, Caucasus.

Leaves (Plate 207, Fig. 33) 3 inches long, 2 to $2\frac{1}{2}$ inches wide, ovate, rounded or slightly cordate at the base, rounded or shortly acuminate at the apex, unequally bi-serrate, green and scattered pubescent beneath. The leaves, usually without lobes, show occasionally slight and irregular lobes.

Introduced in 1759. A shrub or small tree,[2] coming early into leaf.

---

[1] Figured in *Gard. Chron.* xv. 499, f. 93 (1881).

[2] At Arley Castle a good specimen is 29 feet high and 1 foot 9 inches in girth.

5. *Acer Fargesii*, Rehder.[1]   Central China.

Leaves 3 to 4 inches long, ¾ to 1 inch broad, coriaceous, lanceolate, cuneate or rounded at the base, caudate or long acuminate at the apex, serrate on young plants, entire on old trees; nerves indistinct, about ten pairs, looping and not reaching the margin; green and glabrous on both sides, except for minute axil-tufts beneath, developed occasionally on leaves of old trees.

A small tree, introduced by Wilson in 1901; spring foliage and fruit bright crimson. Young plants at Coombe Wood are rather tender, and have only attained 1½ foot high in six years.

6. *Acer Davidi*, Franchet.[2]   Central China.

Leaves 3 to 4 inches long, 1½ inch broad, ovate, cordate at the base, long acuminate, slightly lobulate and crenately bi-serrate; nerves prominent, about twelve pairs, looping before reaching the margin; upper surface dark-green, shining, glabrous; lower surface pale, glabrous, except for reddish-brown axil-tufts.[3]

A tree, 50 feet in height, introduced by Maries in 1879. Young plants raised from seed sent by Wilson in 1902 are 10 to 12 feet high and perfectly hardy at Coombe Wood. Bark green, striped with white.

6A. *Acer cratægifolium*.   See No. 10.

Leaves of this species, without lobes, are distinguishable from *A. Davidi* by the absence of the conspicuous axil-tufts on the lower surface.

II. Leaves simple, three-lobed.

* Leaves entire in margin.

7. *Acer monspessulanum*, Linnæus.·   Southern Europe, North Africa, Asia Minor, Caucasus.

Leaves (Plate 207, Fig. 31) 1¼ inch long, 2¼ inches broad, coriaceous, greyish beneath; lobes ovate, obtuse. Petiole without latex. (See description, p. 665.)

** Leaves minutely crenulate.

8. *Acer creticum*, Linnæus.   Greece, Crete, Lycia.

Leaves (Plate 207, Fig. 32) ¾ to 1½ inch long, coriaceous, variously three-lobed, or with the lobes obsolete on some of the leaves, short-stalked, cuneate, or rounded at the base; margin non-ciliate; bright-green and glabrous beneath; petiole without latex. Fruit small, glabrous; wings parallel and not diverging.

A small tree, scarcely exceeding 30 feet in height, introduced in 1752. Old trees exist at Syon (mentioned by Loudon), Barton, and White Knights.

*** Leaves irregularly toothed.

9. *Acer trinerve*, Dippel.   A juvenile form of the Chinese *Acer trifidum*, Hooker et Arnott, cultivated in Japan.

Leaves 2½ inches long, 2 inches broad, variable in lobing; lobes acuminate;

---

[1] In Sargent, *Trees and Shrubs*, i. 180 (1905). Cf. *Acer lævigatum*, Wallich, var. *Fargesii*, Rehder, in J. H. Veitch, *Journ. Roy. Hort. Soc.* xxix. 353, fig. 91 (1904).

[2] Cf. J. H. Veitch, *op. cit.* 348, figs. 86, 90.

[3] These axil-tufts are partly covered by a small membrane at the junction of the lateral nerve and midrib.

three-nerved at the rounded base ; under surface glaucous, glabrous, without axil-tufts.   Petiole without latex.   Young branchlets slightly pubescent.

A young tree is growing vigorously at Kew.

10. *Acer cratægifolium*, Siebold et Zuccarini.   Japan.

Leaves 3 inches long, coriaceous, ovate, cordate at the base, acuminate at the apex, usually with two short lateral lobes, occasionally undivided, dark-green and shining above, pale and glabrous beneath.   Petiole without latex.   Bark striate.

A small tree, attaining about 20 feet in height, pyramidal in habit.   Introduced by Maries in 1881.   Var. *Veitchii* is variegated with irregular blotches of pink and white.

11. *Acer ginnala*, Maximowicz.   Amurland, Southern Mongolia, China, Japan.

Leaves 3 inches long, 2½ inches wide ; terminal lobe elongated ; under surface bright-green, scattered pubescent or glabrescent, without axil-tufts.   Petiole without latex.

A small tree, with leaves assuming in autumn a brilliant red tint.   Var. *Semenovii*, from Turkestan, has smaller leaves, occasionally five-lobed.

**** *Leaves regularly bi-serrate.*

12. *Acer pennsylvanicum*, Linnæus.   North America.

Leaves (Plate 206, Fig. 13) 7 inches long, 6 inches wide ; lobes arising from the upper part of the leaf, triangular, acuminate ; under surface with scattered minute pubescence ; petiole without latex.   Young branchlets not glaucous.

Introduced in 1759, and remarkable for its striated bark, green branchlets, and fruit in drooping long racemes.   It rarely attains a height of 40 feet.

13. *Acer rufinerve*, Siebold et Zuccarini.   Japan.

Leaves 3½ inches long, 3 inches wide, resembling in shape those of *A. pennsylvanicum*, but smaller, and with reddish pubescence along the sides of the primary nerves, forming axil-tufts.   Racemes erect and reddish-tomentose.   Young branchlets glaucous.

This is the representative in Japan of *A. pennsylvanicum*, which it resembles in bark, habit, and size.   The type was introduced by Maries in 1881, and there are thriving small specimens at Kew.   A variety, with the leaves white on the margin, was exhibited by Standish in 1869 at the Horticultural Society, and was figured in the *Botanical Magazine*, t. 5793.

14. *Acer capillipes*, Maximowicz.   Japan.

This species strongly resembles the last, but differs in the glabrous leaves, and non-glaucous young branchlets ; racemes glabrous, pedicels long.

A small tree or shrub, introduced by Sargent in 1892, and represented at Kew by a thriving young specimen.   The autumn tint is purplish-brown, suffused with yellow along the nerves.

15. *Acer spicatum*, Lamarck.   North America.

Leaves about 3½ inches long and wide, broadly ovate, acuminate, ciliate in margin, glabrous above, pubescent beneath.   Petioles and young branchlets pubescent.

A bushy tree, occasionally 30 feet high. According to Loudon,[1] introduced in 1750, and common in his day in ornamental plantations, one tree at Croome, Worcestershire, being reported as 40 feet high. This tree, which was probably incorrectly named, no longer exists. This maple[2] is extremely rare at the present day, and small specimens at Kew have a miserable appearance.

III. Leaves five-lobed, the basal lobes very small, obscure or obsolete in some of the leaves.

* Leaves quite glabrous beneath.

16. *Acer coriaceum*, Tausch. Hybrid between *A. Pseudoplatanus* and *A. monspessulanum*.

Leaves $2\frac{1}{2}$ inches long, 3 inches wide, coriaceous, pale beneath, often deeply cordate, slightly crenulate in margin; lobes broadly ovate, acute or obtuse.

A small tree, rarely seen except in botanical gardens.

17. *Acer glabrum*, Torrey. Western North America.

Leaves 3 inches long and wide, membranous, dark-green and shining above, pale beneath, bi-serrate; lobes acuminate; sinuses very acute at their base; petiole without latex.

A small tree, young specimens of which are thriving at Kew. In the wild state the leaves are extraordinarily variable, being often tri-partite or tri-foliolate. It is readily distinguished by its perfectly glabrous thin leaves.

** Leaves glabrous, except for slight pubescence on the nerves at the base.

18. *Acer rotundilobum*, von Schwerin.[3] Hybrid between *A. Opalus*, var. *obtusatum*, and *A. monspessulanum*.

Leaves $3\frac{1}{2}$ inches long and broad, pale beneath, slightly crenate in margin; lobes short, broadly ovate, cuspidate; sinuses rounded at their base; petiole without latex.

This species, which has leaves thinner in texture and much more glabrous than *Acer Opalus*, is represented at Kew by a fast-growing young specimen, obtained from Simon Louis of Metz.

19. *Acer hybridum*, Spach. Hybrid between *A. Pseudoplatanus* and *A. Opalus*.

Leaves $3\frac{1}{2}$ inches long, 4 inches broad, pale beneath, irregularly and slightly serrate; lobes broadly ovate, shortly acuminate; sinuses acute at their base; petiole with latex.

There are small trees in the collection at Kew.

*** Leaves pubescent beneath.

20. *Acer Opalus*, Miller, var. *obtusatum*. Southern Europe, Caucasus, North Africa.

Leaves (Plate 206, Fig. 16) $3\frac{1}{2}$ inches long, 4 inches wide; under surface pale with the pubescence densest on the nerves; lobes broadly ovate or

---

[1] *Arb. et Frut. Brit.* i. 407 (1838).

[2] Seedlings which I raised in 1905, from seed collected by me near Ottawa in 1904, have grown vigorously at Colesborne, and are now 3 to 4 feet high.—(H. J. E.)

[3] In *Mitt. Deut. Dend. Gesell.* 1894, p. 76, von Schwerin considers the parents to be *A. Opalus*, var. *obtusatum*, and *A. Pseudoplatanus*.

rounded, short, obtuse or acute ; margin with small irregular teeth ; petiole without latex. In var. *neapolitanum*, the leaves (Plate 206, Fig. 15) are more obscurely lobed, with very dense long pubescence beneath. (See description, p. 663.)

21. *Acer grandidentatum*, Nuttall. Montana southwards to New Mexico.

Leaves (Plate 205, Fig. 4) 3 inches long, 4 inches broad, with three large oblong lobes, separated by sinuses extending half-way to the base of the blade ; margin with a few large obtuse lobules, otherwise entire ; under surface covered with pale dense pubescence ; basal lobes represented by the lowest pair of the marginal lobules.

A small tree, rarely forty feet in height, representing the sugar maple in the West. There is a small specimen thriving in the Kew Collection.

22. *Acer rubrum*, Linnæus. North America.

Leaves (Plate 207, Fig. 27) averaging 3 inches long and broad, variable in shape ; under surface silvery white with scattered pubescence ; lobes usually triangular, acute or acuminate, sharply toothed or bi-serrate in margin ; sinuses acute at the base, variable in depth ; base of the leaf truncate or rounded, rarely cordate. (See description, p. 671.)

23. *Acer tetramerum*,[1] Pax. Central China.

Leaves 3 inches long, 2 inches broad, ovate, cordate at the base, indistinctly five-lobed ; basal lobes obscure or obsolete ; lateral lobes short, triangular, acute and sharply serrate ; terminal lobe with two or three pairs of serrated teeth, and prolonged into a long narrow acuminate apex ; margin ciliate ; upper surface dark-green, scattered pubescent ; lower surface pale, covered with white pubescence, densest in the axils ; petiole without latex.

A small tree, introduced by Wilson in 1901. Young plants at Coombe Wood are perfectly hardy and free in growth, having already attained 16 feet in height.

IV. Leaves five-lobed ; basal lobes well-developed ; white or pale beneath ; petiole without latex.

Leaves not serrate.

24. *Acer Opalus*, Miller. Southern Europe.

Leaves (Plate 206, Fig. 14) 2½ inches long, 3 inches wide ; lobes short, acute, irregularly toothed ; under surface with scattered pubescence, denser on the nerves and forming axil-tufts. (See description, p. 663.)

25. *Acer saccharum*, Marshall. North America.

Leaves (Plate 206, Fig. 12) 5 inches long, 6 inches wide ; lobes triangular, acuminate, with one or two pairs of sinuate teeth ; lower surface with axil-tufts of pubescence, elsewhere glabrous or more or less pubescent. (See description, p. 677.)

26. *Acer hyrcanum*, Fischer et Meyer. South-Eastern Europe, Crimea, Asia Minor, Caucasus.

[1] Cf. J. H. Veitch, *Journ. Roy. Hort. Soc.* xxix. 353, fig. 97 (1904).

Leaves (Plate 207, Fig. 22) 2½ inches long, 3 inches wide; lobes oblong, acute or acuminate, with one to four small teeth; sinuses reaching half the length of the blade, usually rounded at the base; lower surface glabrous except for pubescence along the nerves and forming axil-tufts.

A small tree, representing *A. Opalus* in the Orient and Balkan Peninsula. There are good specimens in the collection at Kew.

27. *Acer Heldreichii*, Orphanides.   Balkan Peninsula, Greece.

Leaves (Plate 206, Fig. 17)[1] 4 inches long, 5 inches broad, deeply five-lobed; the middle sinuses narrow, acute at the base, and reaching nearly to the base of the blade; lobes acuminate, with two or three pairs of triangular teeth; under surface with brown pubescence along the primary nerves, forming axil-tufts, glabrous elsewhere.

There is a small tree of this species in the Maple Collection at Kew.

28. *Acer Trautvetteri*, Medwedjeff.   Caucasus.

Leaves (Plate 206, Fig. 19) averaging 5½ inches long and 6 inches broad, deeply five-lobed; the middle sinuses acute at the base, reaching two-thirds the length of the blade; lobes long acuminate, with four or five pairs of irregular teeth; under surface glabrous except for conspicuous tufts of reddish-brown pubescence in the axils, at the base, and at the junctions of the primary and secondary nerves.   (See description, p. 669.)

** *Leaves distinctly serrate.*

29. *Acer Pseudoplatanus*, Linnæus.   Europe, Asia Minor, Caucasus, Northern Persia.

Leaves (Plate 206, Fig. 20) averaging 5 inches long and 6 inches wide; lobes acuminate, coarsely and irregularly serrate and lobulate; sinuses acute at the base, and extending half-way the length of the blade; under surface pubescent along the primary nerves.   (See description, p. 641.)

30. *Acer insigne*, Boissier et Buhse.   Caucasus, Northern Persia.

Leaves (Plate 206, Fig. 18) resembling those of the sycamore, but usually longer than broad, averaging 7 inches long and 6 inches wide; under surface with pubescence dense along the nerves, forming axil-tufts, and spreading over the leaf between the nerves.   (See description, p. 667.)

31. *Acer Volxemi*, Masters.   Central Caucasus.

Leaves considerably larger than in the preceding species, attaining 10 to 12 inches in width and length, glaucous beneath, with pubescence confined to the sides of the nerves.   (See description, p. 668.)

32. *Acer dasycarpum*, Ehrhart.   North America.

Leaves (Plate 207, Fig. 28) about 5 inches long and wide; sinuses rounded at the base, and concave on the sides, extending half-way the length of the blade; lobes long acuminate, with serrated triangular teeth or lobules; under surface silvery white, scattered pubescent, without axil-tufts.   (See description, p. 674.)

---

[1] In the mountains of Greece, the leaves are smaller than is the case in trees growing in Southern Servia, Montenegro, Herzegovina, and Bulgaria, and average only 2 to 3 inches in diameter.

V. Leaves simple, five-lobed; basal lobes well developed; green beneath; margin entire or with a few teeth and without cilia; petiole containing latex.

* *Margin entire.*

33. *Acer pictum*, Thunberg.   Asia Minor to Japan.

Leaves (Plate 205, Fig. 9) about 4 inches long and 4½ inches broad; lobes acuminate, bristle-pointed; basal lobes pointing outwards; glabrous beneath, except for pubescent tufts in the basal axils.   Young branchlets green and not glaucous, turning grey in the second year in the type, remaining green in var. *colchicum.*   (See description, p. 660.)

34. *Acer Lobelii*, Tenore.   Italy.

Leaves (Plate 205, Fig. 8) 4 inches long, 4½ inches wide; lobes acuminate, ending in long sharp points; basal lobes directed forwards; glabrous beneath, except for pubescent tufts in the axils of the primary and secondary nerves and at the base; young branchlets glaucous, remaining green in the second year.   (See description, p. 659.)

35. *Acer truncatum*, Bunge.   Northern China.

Leaves (Plate 205, Fig. 6) about 2½ inches long and 3 inches wide, truncate or widely cordate at the base; lobes[1] acuminate, bristle-pointed; basal lobes directed outwards; glabrous beneath, except for a slight trace of pubescence at the base.   Young branchlets not glaucous, becoming brown in their first winter.

A small tree, attaining 25 feet in height.   Introduced some years ago by seeds received from Dr. Bretschneider, and thriving at Kew.

36. *Acer Dieckii*,[2] Pax.   Hybrid between *A. platanoides* and *A. pictum*, var. *colchicum.*

Leaves (Plate 207, Fig. 30) 3 inches long, 4 inches broad; lobes five, shortly acuminate, not bristle-pointed; brown pubescent beneath at the base and in the axils of the primary and secondary nerves.   Young branchlets not glaucous, becoming brown in their first winter.

** *Margin toothed.*

37. *Acer platanoides*, Linnæus.   Europe, Asia Minor, Caucasus.

Leaves (Plate 206, Fig. 11) 5 inches long, 7 inches wide; lobes acuminate, bristle-pointed; sinuses wide, rounded and open; margin with a few sinuate pointed teeth; glabrous beneath, except for pubescence at the base and in the axils of the primary and secondary nerves.   Young branchlets not glaucous, becoming brown in their first winter.   (See description, p. 656.)

38. *Acer neglectum*, Lange.[3]   Hybrid between *A. campestre* and *A. pictum*, var. *colchicum.*

Leaves (Plate 205, Fig. 7) 4 inches long, 5 inches wide; lobes acuminate, not bristle-pointed, the upper three with one or two short teeth; glabrous beneath, except for pubescence along the nerves, densest at the base.   Young branchlets not glaucous, pubescent, becoming brown in their first winter.

---

[1] The terminal lobe in leaves of young trees has often one or two sharp teeth.

[2] Sometimes known in cultivation as *Acer platanoides*, var. *integrilobum*, Zabel.   There is a small tree at Kew, the bark of which is striped with white lines.

[3] There are small trees in the Botanic Gardens at Kew and Edinburgh, the bark of which is striped with white lines.

VI. Leaves simple, five-lobed; basal lobes well developed; green beneath; margin serrate; petiole without latex.

39. *Acer Oliverianum*,[1] Pax.   Central China.

Leaves 3½ inches long, 4 inches broad; lobes long acuminate, finely and simply serrate; glabrous beneath, except for pubescent tufts at the base and in the axils of the primary and secondary nerves.

A small tree, 20 feet in height.   Introduced by Wilson in 1901.   Plants at Coombe Wood are thriving, and are about 12 feet high.

40. *Acer argutum*, Maximowicz.   Japan.

Leaves about 3 inches long and broad; lobes triangular, acuminate, sharply bi-serrate; lower surface with scattered white pubescence, dense on the nerves and veinlets.   Young branches densely pubescent.

A small tree, introduced by Maries in 1881.   There are small specimens in the Kew Collection, and a good-sized one at Westonbirt.

40A. *Acer palmatum*, Thunberg.   (See No. 46.)

VII. Leaves simple, five-lobed; basal lobes well developed; green beneath; margin with a few teeth or lobules, ciliate.

* Petiole containing latex.

41. *Acer campestre*, Linnæus.   Europe, Caucasus, Northern Persia.

Leaves (Plate 207, Figs. 23, 24, 25) 2½ inches long, 3 inches broad; margin irregularly and obtusely dentate; dark-green, and pubescent on the nerves above; light-green beneath with scattered pubescence, densest on the nerves, and tufted in the axils.   (See description, p. 651.)

42. *Acer macrophyllum*, Pursh.   Alaska to California.

Leaves (Plate 205, Fig. 3) about 9 inches long and broad; margin with large triangular lobules or teeth; upper surface dark-green, shining, scattered pubescent; lower surface light-green, glabrescent between the nerves, with tufts of white pubescence in the axils.   (See description, p. 681.)

43. *Acer Miyabei*, Maximowicz.   Yezo.

Leaves 5 inches long, 6 inches broad; lobes with long acuminate obtuse-tipped apex, and with one or two pairs of obtuse lobules; sinuses narrow, acute at the base; both surfaces pubescent, densest on the nerves; petiole and young branchlets pubescent.

A tree, attaining 40 feet in height.   Introduced by Sargent, who obtained seeds from Miyabe in 1892.   There are young trees at Kew and Coombe Wood; and one at Grayswood, near Haslemere, was about 18 feet high in 1906.

** Petiole without latex.

44. *Acer diabolicum*,[2] Blume.   Japan.

Leaves (Plate 207, Fig. 26) 6 inches long, 6½ inches wide; lobes short,

---

[1] *Acer sp.* in J. H. Veitch, *Journ. Roy. Hort. Soc.* xxix. 354, fig. 98 (1904).

[2] Specimens of reputed *Acer Francheti*, Pax, from Coombe Wood, introduced by Wilson in 1901, are indistinguishable from this species.   Cf. *Journ. Roy. Hort. Soc.* 353, fig. 88 (1904).

ovate, with obtusely tipped acuminate apex, and one or two pairs of coarse teeth; upper surface dark-green, shining, with scattered pubescence densest on the nerves; lower surface pubescent, dense on the nerves, and forming axil-tufts; petioles and young branchlets pubescent.

A tree, 30 feet high, remarkable for the stinging hairs on the fruit-carpels. Introduced by Maries in 1881, and about 20 feet high at Coombe Wood.

45. *Acer villosum*, Wallich. North-Western Himalayas.

Leaves about 8 inches long and wide; lobes broadly ovate, caudate-acuminate, with a few crenate teeth; pubescent on the primary nerves above; lower surface, petioles, and young branchlets densely pubescent.

A large tree in its native home. A specimen, the only one seen in cultivation, at Grayswood, Haslemere, remains shrubby.

VIII. Leaves simple; lobes more than five, sharply bi-serrate; petiole without latex.

* Petioles glabrous.

46. *Acer palmatum*, Thunberg. Japan, Central China.

Leaves about 3 inches in length and breadth; lobes usually seven, occasionally five, long acuminate; sinuses extending half the length of the blade or to near the base; glabrous on both surfaces, except for minute axil-tufts of pubescence beneath. Young branchlets glabrous.

A tree, rarely attaining 50 feet in height. The type was introduced[1] in 1820, and in cultivation is a small tree, occasionally 25 feet high, with numerous small branches and extremely dense foliage. A very large number of horticultural varieties[2] have been produced in Japan, which are highly valued on account of the varied shape and colour of their leaves, and are commonly cultivated in Europe.

47. *Acer circinatum*, Pursh. British Columbia to California.

Leaves (Plate 205, Fig. 5) about 4 inches long and broad; lobes seven to nine, acute; sinuses reaching about one-third the length of the blade; scattered pubescent on both surfaces at first, ultimately glabrous except for minute traces of pubescence at the base on both sides. Young branchlets glabrous.

A shrub or small tree, rarely 40 feet high. Introduced in 1826, perfectly hardy, and producing fruit freely.

** Petioles, with dense white long pubescence in spring, more or less persistent till autumn.

48. *Acer japonicum*, Thunberg. Japan.

Leaves about 4 inches long and broad; lobes usually nine, acuminate; sinuses reaching one-third the length of the blade; both surfaces scattered pubescent, with a tuft at the junction of the blade and petiole above, and axil-tufts beneath. Young branches glabrous.

A small tree, attaining about 20 feet in height. Several varieties are

[1] Loudon, *Encycl. Trees*, 90. The largest specimen which we have seen is at Waterer's Nursery, Knaphill, Woking.
[2] Cf. J. H. Veitch, *Journ. Roy. Hort. Soc.* xxix. 338 (1904).

mentioned by Mr. J. H. Veitch, one of which, growing well at Mount Usher, Wicklow, Ireland, has deeply-cut leaves, figured on Plate 207, Fig. 29.

49. *Acer Sieboldianum*, Miquel. Japan.

Leaves about 2 inches in length and breadth; lobes usually nine, acuminate; sinuses reaching nearly half the length of the blade; glabrous above with a tuft of hairs at the base; pubescent beneath on the main nerves. Young branchlets covered with dense white pubescence.

A small tree or shrub, very rare in cultivation.

IX. Leaves compound; leaflets three or five.

A. *Leaflets entire in margin.*

50. *Acer Henryi*,[1] Pax. Central China.

Leaflets three, narrowly elliptical, acuminate; under surface green, pubescent on the midrib and lateral nerves, forming axil - tufts. Young branchlets pubescent.

A small tree, attaining about 30 feet in height. Introduced by Wilson in 1900. Young plants at Coombe Wood are now about 10 feet high, and are hardy and thriving.

B. *Leaflets serrate or toothed.*

* *Leaflets green beneath.*

51. *Acer Negundo*, Linnæus. North America.

Leaflets (Plate 205, Fig. 2) usually five, occasionally three; coarsely and irregularly toothed and serrate, sometimes three-lobed; under surface with scattered pubescence, dense on the primary and secondary nerves; petioles glabrous. Young branchlets glabrous, green or glaucous. (See description, p. 684.)

52. *Acer Negundo*, var. *californicum*, Wesmael. California.

Leaflets (Plate 205, Fig. 1) three or five, resembling those of the last species, but with coarser teeth and serrations; lower surface covered with a dense white pubescence. Petioles and young branchlets pubescent. (See description, p. 684.)

53. *Acer cissifolium*, Koch. Japan.

Leaflets three, obovate or oblong, with a long acuminate or cuspidate apex; margin ciliate with serrations ending in fine points; under surface glabrous, except for axil-tufts and slight pubescence along the midrib. Young branchlets pubescent.

A small tree, perfectly hardy, and fruiting freely in this country. A specimen of this species about 30 feet high, which Prof. Sargent says is as large as he saw it in Japan, is growing at Westonbirt.

---

[1] *Acer sutchuense*, Franchet, also from Central China, differs in the glabrous branchlets and serrate leaflets, which are pale beneath, and has not yet been introduced, the plants referred to this species in *Journ. Roy. Hort. Soc.* xxix. 353, figs. 93, 96, being *A. Henryi*, Pax. *Acer mandschuricum*, Maximowicz, a native of Manchuria, which differs little from *Acer sutchuense* in foliage, is said to be in cultivation in Germany (*Mitt. Deutsche Dendrol. Ges.* 1906, p. 30).

** *Leaflets pale beneath.*

54. *Acer nikoense*, Maximowicz.   Japan, Central China.

Leaflets three; terminal one about 4 inches long; lateral leaflets slightly smaller and unequal-sided; elliptical, acute; margin crenate, ciliate; under surface villous on the midrib and nerves, scattered pubescent between the nerves.   Petioles stout, and like the young branchlets, densely woolly.

A tree, attaining 50 feet in height, with smooth, dark, slightly furrowed bark; leaves turning brilliant scarlet in autumn.   Introduced by Maries in 1881.   A tree at Coombe Wood is about 30 feet high.

55. *Acer griseum*, Pax.   Central China.

Leaflets three; terminal leaflet about 2½ inches long; lateral leaflets smaller and unequal-sided; coarsely toothed and ciliate in margin; woolly pubescent on the midrib and nerves beneath.   Petioles slender, and, like the young branchlets, pilose.

A tree, attaining 40 feet in height, with bark peeling off like a birch. Introduced by Wilson in 1901.   Young plants at Coombe Wood are about 3 feet high.   (A. H.)

ACER PSEUDOPLATANUS, Sycamore

Acer Pseudoplatanus, Linnæus, *Sp. Pl.* 1054 (1753); Loudon, *Arb. et Frut. Brit.* i. 414, 448, (1838); Willkomm, *Forstliche Flora*, 749 (1887); Mathieu, *Flore Forestière*, 37 (1897).

A large tree, attaining about 100 feet in height and 20 feet in girth,   Bark[1] smooth and greyish on young trees, fissuring and scaling off in large strips on old trunks.   Leaves (Plate 206, Fig. 20) 4 to 8 inches in length and width, cordate at the base; lobes five, ovate, acuminate, coarsely and irregularly serrate, lateral lobes larger than the basal ones; sinuses extending about half-way to the midrib, and very acute at the base; upper surface dark-green, shining, glabrous; lower surface paler green and glaucous or sometimes reddish, pubescent along the principal nerves; petiole without latex.   The leaves usually turn brownish in autumn, and are often disfigured by black blotches, caused by the fungus known as *Rhytisma acerinum*, Fr., which, however, does little or no harm to the vitality of the tree.[2]

Flowers in long pendent racemes, composed of umbellate cymes of three flowers each, the central flower in the cyme usually perfect, the two lateral flowers staminate, with longer stamens and abortive ovaries; pedicels short; sepals five, deciduous, greenish-yellow; petals five, greenish-yellow, imbricate, inserted at the margin of a fleshy hypogynous disc.   Stamens eight, inserted on the disc; filaments subulate; pubescent below, ovary tomentose.   Fruit: keys divergent at a varying

[1] In the Edinburgh Botanic Garden there are trees about a foot in diameter, which have remarkably white bark, resembling that of a birch.   The largest sycamore in the garden measured, in 1906, 78 feet in height, and 13 feet 7 inches in girth.   [2] Cf. *Board of Agriculture, Leaflet* No. 183.

angle, generally directed forwards, about 1½ inch long; wings green, narrow below, scimitar-shaped. Seed obovate, without albumen; cotyledons narrow, long, either spirally coiled or plicate.

Seedling.[1]—Cotyledons normally two, rarely three, carried above ground on germination, 1½ to 2 inches long, oblong, sessile, obtuse, entire in margin, obscurely three-nerved, glabrous, pale-green. Caulicle, ½ to 1 inch long, glabrous, ending in a tapering, flexuose, primary root, which gives off a few lateral fibres. Young stem terete, glabrous. First pair of leaves, ovate, cordate at the base, palmately five-nerved, acuminate, irregularly serrate. Second and third pairs similarly cordate and five-nerved, distinctly three-lobed; terminal lobe long, triangular-ovate; lateral lobes short, broad, with two indistinct lobules or teeth. Succeeding pairs resemble those of the adult plant.

Seedlings with three cotyledons, observed by Sir W. Thiselton Dyer,[2] bore leaves in whorls of threes in their first and second years; but in the third and following years they reverted to the ordinary type with opposite leaves.

IDENTIFICATION

In summer the sycamore is readily distinguishable by the shape of the leaves, and can only be confused with *A. insigne*, from which it differs markedly in the buds, as explained under the latter species.

In winter the twigs are shining, glabrous. Buds sessile, ovoid; terminal buds larger than the lateral buds, which arise from the twigs at an angle of 45°; scales, six to eight visible externally, in opposite decussate pairs, green with dark edges, glabrous or slightly pubescent near the tip, ciliate in margin. Leaf-scars not joining around the twig, crescentic or V-shaped, three-dotted. The bud-scales are homologous with leaf-bases, and fall off when the bud opens, showing at this stage a minute three-lobed projection at the tip, which corresponds to a leaf-blade.

VARIETIES

The sycamore is remarkably constant in foliage in the wild state,[3] only one well-marked geographical form being known:—

Var. *villosa*, Parlatore, *Fl. Ital.* v. 404 (1872). Leaves coriaceous; base widely cordate, margin coarsely toothed, lower surface pubescent throughout in early spring. Fruits usually tomentose, with very broad wings. This variety occurs in the mountains of Sicily, Calabria, and Dalmatia.

In the common wild form the leaves are scarcely coriaceous, and are only pubescent along the nerves beneath, while the fruit is glabrous. From the common form, numerous varieties have arisen in cultivation, as many as fifty being

[1] Cf. Lubbock, *Seedlings*, i. 360, f. 252 (1892).
[2] *Ann. of Botany*, xvi. 553, plate xxiv. (1902). In plate xxv. an abnormal seedling is figured, in which three cotyledons are followed by two leaves, one of which is bi-partite. Cf. Loudon, *op. cit.* p. 415.
[3] On a tree growing in Tullymore Park, Co. Down, Ireland, many of the branchlets bore the leaves alternately and not in pairs. Specimens of this were sent to the Kew Herbarium in 1871.

enumerated by Pax and von Schwerin;[1] but many of these show only trifling variations, often inconstant in character; and only the most important varieties will be mentioned here :—

1. Var. *pyramidale*, Nicholson.[2] Pyramidal in habit with ordinary foliage; originated in a row of trees, close to the gate of Roger, M'Clelland and Co.'s Nursery, Newry.

2. Var. *erythrocarpum*, Carrière. Fruits red in colour and abundant, producing a fine effect in autumn. This variety is said to have originated in M. Ferrand's nursery at Cognac; but according to von Schwerin is of common occurrence in the wild state in the Bavarian Alps.

3. Var. *palmatifidum*,[3] Duhamel. Leaves deeply cut, being five-partite.

4. Var. *ternatum*, von Schwerin. Leaves tri-partite to the base, so that they are ternate or nearly so.

5. Var. *vitifolium*, Tausch. Leaves large, deeply cordate at the base, with three broad rounded lobes, the basal lobes being very small or obsolete.

6. Var. *clausum*, von Schwerin. Leaves deeply cordate, with the sinus closed by the overlapping of the basal lobes.

7. Var. *crispum*, von Schwerin. Leaves wrinkled in margin. Forms also occur in which the leaves are rolled either backwards or inwards from the margin.

8. Var. *cruciatum*, von Schwerin. Leaves three-lobed, the lateral lobes being exactly at right angles to the median lobe.

9. Var. *jaspideum*, von Schwerin. Bark yellowish; leaves of the ordinary form.

10. Var. *euchlorum*, Nicholson.[2] Leaves deep-green above, normal in shape. This variety originated in Späth's nursery at Berlin.

11. Var. *splendens*, von Schwerin. Leaves red at the time of opening, afterwards becoming green. This includes var. *Rafinesquianum*, Nicholson, in which the young leaves are blood-red; var. *cupreum*, Behnsch, in which they are copper-red at the time of opening; and var. *metallicum*, von Schwerin, with the leaves at first yellowish, afterwards copper-coloured, and ultimately green.

12. Var. *albo-variegatum*, Loudon. Stripe-leaved sycamore. Leaves splashed and marked with white. This variety is mentioned in the *London Catalogue of Trees*, published in 1730, which states that it comes true from seed. It often grows to a large size, a tree at Highmore Hall, Oxfordshire, the seat of G. T. Inman, Esq., being 72 feet high by 9 feet in girth.

13. Var. *corstorphinense*, von Schwerin.[4] Corstorphine or Golden Sycamore. Leaves often three-lobed, coming out yellow in spring, and in some places appearing about a fortnight earlier than those of the ordinary form. The colour is a fine golden one, and usually lasts till summer is advanced.

Loudon named this variety *flava-variegata*, variegated with yellow, which is

---

[1] *Gartenflora*, xlii. 258 (1893). Cf. also Nicholson, *Gard. Chron.* xv. 300 (1881).
[2] *Gard. Chron.* xv. 300 (1881).
[3] This is var. *longifolium*, Loudon, *Trees and Shrubs*, 86 (1842).
[4] *Gartenflora*, xlii. 263 (1893).

scarcely a happy designation, and states that the original tree grew in the grounds of Sir T. Dick Lauder in the parish of Corstorphine, near Edinburgh.

The Corstorphine sycamore is well illustrated and described by Sargent,[1] who quotes from a book, published locally by Mr. G. Upton Selway, called *A Midlothian Village*.   The tree has a romantic history, as being the only survivor of an avenue which formerly led to an old manor-house belonging since 1376 to the Forrester family.   James Baillie, second Lord Forrester, who took an active part against the Commonwealth, and became involved in difficulties on account of a heavy fine laid on him by Cromwell, is said to have quarrelled with his sister-in-law on August 26, 1679, and to have been murdered by her at the foot of this tree.

Mr. R. Galloway, Secretary of the Royal Scottish Arboricultural Society, informs us that this tree now stands in a garden attached to one of the houses in the village of Corstorphine, and measured in 1905, 61 feet in height and 11 feet in girth.   A tree of this variety, which in 1904 measured 62 feet high by 5 feet 10 inches in girth, grows at Kilmarnock, and is believed by the Rev. Dr. Landsborough to be 112 years old.   He says that this variety does not grow so fast or attain such a large size as the common sycamore, as owing to its early leafing, the golden sycamore is liable to suffer from spring frosts.

14. Var. *Worlei*, Rosenthal.   Another form of the golden sycamore, with reddish leaf-stalks and bright-yellow leaves, which are orange-coloured at the time of opening.

15. Var. *aucubæfolium*, Nicholson.   Leaves marked with yellow spots, similar in appearance to those on the leaves of the common *Aucuba japonica*.   Originated as an accidental seedling in Little and Ballantyne's nursery at Carlisle.

16. Var. *Leopoldi*, Lemaire, *Illust. Horticole*, xi. t. 411 (1864).   Leaves deep pink at the time of unfolding, afterwards variegated with pink and purple. This originated in the seed-bed in the nursery of M. Vervaene, at Ledeberg-les-Gand, by whom it was sold to Van Geert.

17. Var. *Webbianum*, Nicholson.   Leaves with silvery streaks.   This originated in the nursery of C. Lee and Son at Isleworth.

18. Var. *purpureum*, Loudon.   Purple-leaved sycamore.   Leaves purple beneath.   The petiole and wings of the fruit are often also bright-red.   This variety originated in 1828 in Sanders' nursery in Jersey.

Various sub-varieties are known as var. *atropurpureum*, Späth, under surface of the leaves very dark purple ; var. *Nizeti*, von Schwerin, leaves purple beneath, spotted with yellow above ; var. *Handjeryi*, Späth, vinous-purple beneath, upper surface with yellow minute spots ; var. *purpureo-variegatum*, Nicholson, with rose-coloured or white stripes on a purple ground.   The latter originated in Van Volxem's nursery at Perck, and is considered by Nicholson to be identical with var. *variegatum*, Carrière, *Rev. Hort.* 1877, p. 334, which originated as a branch sport from the ordinary purple-leaved variety in the Bois de Boulogne nurseries at Longchamps in 1874.

19. Var. *flavo-variegatum*, Hayne.[2]   Leaves splashed and marked with yellow.

[1] *Garden and Forest*, vi. 202, f. 32 (1893).   [2] *Dendrol. Flora*, 212 (1822).

DISTRIBUTION

The exact limits of the distribution of the sycamore are difficult to define, as the tree has been extensively planted for centuries outside of its original home, which in Europe may be roughly described as the great central chain of the Pyrenees, Alps, and Carpathians, with the mountains and hilly districts radiating from them in all directions. It is truly wild in the Pyrenees, and reaches its western limit in the Iberian Peninsula in the Cantabrian Mountains, being absent from the greater part of Spain and all Portugal.[1] It occurs in all the mountainous and hilly districts of France except in the north-west; in the Alps generally; in the mountains of Germany, as far north as the Harz Mountains; in the Carpathians, Apennines, mountains of Sicily, Dalmatia, Bosnia, Servia, and in the mountains of Thessaly and Epirus in Greece.[2] In Russia it occurs in the provinces along the Black Sea, extending inland along the banks of the great rivers, and in the mountains of the Crimea. It is widely spread in the mountains of Asia Minor, Armenia, and the Caucasus,[3] where it grows at all altitudes from sea-level to 4000 feet. Its extreme easterly point is near Astrabad, south-east of the Caspian Sea, about lat. 37°.

The tree is not a native of the British Isles, North-West France, Belgium, Holland, the North German plain, Denmark, Scandinavia, or the greater part of Russia. In these countries the sycamore, however, flourishes, and is extensively cultivated, reaching its northerly limit as a planted tree, according to Schübeler, in Norway and Sweden about lat. 64°.

It is usually met with, in the wild state, as an isolated tree or in small groups, being only known to form pure woods, and those of no great extent, in the Thuringian forests. It does not occur naturally on light soils, on heavy clay soils, or on wet ground; and apparently, in order to compete with other trees, must grow on a soil rich in mineral constituents, such as often occurs in valleys or ravines, or in pockets here and there in forests, where the soil is generally poor. It forms a part of the great beech and silver-fir forests; but reaches higher than either of these species on the mountains, where it dwindles to a sub-alpine shrub near the timber line. In the Bavarian forests it grows in the zone between 1000 and 4400 feet altitude, and a peculiar form occurs, with twisted curved branches, which Dr. Christ[4] has not observed elsewhere. In Switzerland,[4] it ascends to 5300 feet, and is most plentiful at Sernfthal above Elm. In France it ascends to about 5000 feet, and is most generally met with in the beech forests, its abundance and flourishing condition being considered a sure index of a fertile soil.

The sycamore has not been found in the fossil state in the British Isles. Clement Reid[5] hazards the suggestion that it was perhaps introduced by the

[1] Willkomm, *Pflanzenverbreitung auf der Iberischen Halbinsel*, 94 (1896).
[2] Halácsy, *Consp. Fl. Grææ*, i. 285 (1900).
[3] Radde, *Pflanzenverbreitung in den Kaukasusländern*, 175 (1899).   [4] Christ, *Flore de la Suisse*, 278, 279 (1907).
[5] *Origin British Flora*, 16 (1899).

Romans.   Ray states, *Synopsis*, 230, published in 1690, that the sycamore was then planted in cemeteries and about the houses of the nobility, and that it was nowhere wild in England.   It would appear from this that it was by no means a common tree in Britain in the seventeenth century.   (A. H.)

CULTIVATION

The sycamore, or plane as it is commonly called in Scotland, is a tree which thrives in almost any dry soil, and seems to reach its greatest size or perfection in the colder hilly parts of England and Scotland, where nearly all the finest specimens we know of are to be found.   In the Cotswold hills it is the only tree, except the wych-elm and the beech, which attains a maximum of size, and even here there are none quite equal to some trees in Scotland.

It is absolutely unaffected by the severest frosts [1] at any season, and is rarely attacked by insects or fungoid diseases, ripens seed profusely almost every year, and reproduces itself almost everywhere with such ease that in my own district I believe it might overrun the country if allowed to do so.   It grows rapidly when young, and, though not usually planted as a forest tree, is well suited to produce timber in windy situations, where more valuable trees will only languish. Its foliage in spring and summer is very handsome, but assumes a dirty and ragged aspect in autumn, especially in smoky districts, and therefore it is not suitable for town planting.   It does not grow so well, or live so long on sand, gravel, or on heavy clay as on limestone.   Its branching habit makes careful pruning or close crowding necessary if clean tall stems are desired, and as its timber is most valuable in the form of clean boles of considerable girth, it must be looked on as of somewhat uncertain economic value as a forest tree.   But I have found the sycamore a very useful tree for filling up blanks in thin woods, where, when once established, it grows on dry soil at least twice as fast as the ash, and four or five times as fast as the oak.

No tree can be raised from seed more cheaply and easily than the sycamore, and grafting or budding is only resorted to in the case of varieties. The seed falls in the late autumn and winter, and grows in abundance in gravel paths, so that when only a few are wanted, self-sown seedlings can usually be obtained.   The seed germinates very early, often in February, though if kept dry it should not be sown before March.   It is very liable to be smothered by grass in the first year, and is so easy to transplant that it will be found better to move self-sown seedlings at a year old to the nursery.

In the spring of 1900 I found great quantities of young plants recently germinated on the top of a bare hill pasture, where I wished to renew a clump of trees forming a conspicuous landmark, and had a fence put round them, in order to protect the seedlings from rabbits and cattle; but in the summer I found that every one had been suppressed by grass.   In the following year I sowed sycamore seeds with many other tree seeds in lines in cultivated soil, where I

---

[1] According to Lord Leicester it is on the coast of Norfolk the hardiest tree, except *Quercus Ilex*, which bears better the force of the gales from the sea.—(A. H.)

wished to establish a plantation by sowing, and they were the only species of which the plants showed up well in the lines the first summer. As they were much too thick in the rows, about 10,000 were transplanted the following winter, when 4 to 8 inches high, and these grew so fast in the nursery that in two years more I had trees 4 to 6 feet high, whilst those which were left where they were sown, after four years' growth had made very little progress, few exceeding 12 to 18 inches in height, and many remaining so stunted that they could hardly be recognised among the grass.

Though rabbits will not eat it so readily as beech or ash, yet where they are found, the sycamore is not safe from their attacks until it is a foot or more in diameter ; after which I have not seen them touch it ; and in a park, deer, however hungry, do not bark this tree, though they will peel the branches when cut. It shoots freely from the stool when treated as coppice wood, and on dry soil produces a much greater bulk of poles than ash or lime will do, but in this form is not so valuable as ash or oak, because the poles are neither strong nor durable, and are not used for hurdle-making.

REMARKABLE TREES

Among the many large sycamores which I have measured, it is hard to say which is the finest, but in England I think the palm must be awarded to a tree near the Marquess of Ripon's house at Studley Royal, Yorkshire. This tree is about 104 feet high, by $17\frac{1}{2}$ feet in girth. It has a very large burr close to the ground, where it is $29\frac{1}{2}$ feet round, and a clear bole of about 30 feet.

An almost equally fine tree, growing in front of the Earl of Darnley's house at Cobham Hall, Kent, was figured by Strutt, plate xxx. He gave its height as 94 feet, its girth at the ground 27 feet, and its cubic contents 450 feet. When I measured it in 1905, I found that it was about 105 feet by 17 feet 9 inches at 5 feet up, and still quite healthy. At Penshurst, in the same county, there is a fine tree 104 feet by 13 feet 10 inches.

The tree figured in Plate 179 grows on my own lawn, constantly in sight as I write, and though not quite so large as some others, is still a beautifully-shaped tree, 100 feet by 15 feet. Its top, I grieve to say, has been dying back for some years. At Lypiatt Park, near Stroud, there is a fine tree close below the house, of which Sir John Dorington has sent me a photograph. It measures about 90 feet by 18 feet, dividing near the base into three main stems.

At Essendon Place, Herts, Mr. H. Clinton Baker measured a tree in 1906 as 94 feet high by 9 feet 9 inches in girth. At Fawley Court, near Henley, close to the Thames, in a dense plantation, a tree, estimated by Henry in 1907 to be 100 feet or more in height, was 12 feet 8 inches in girth. At Shiplake House, also near Henley, there is a widespreading tree, 85 feet high and 10 feet 3 inches in girth.

At Lowther Castle, Westmoreland, the seat of the Earl of Lonsdale, there is an immense tree no less than 19 feet 9 inches in girth, but not so tall as, some I have mentioned; and at this fine place I also remarked a large sycamore, with the bark scaling off in a very unusual way, so that it resembled the bark of an old hickory. I am indebted to Capt. Parkin for a good photograph of this curious tree.

At Mitford Castle, near Morpeth, Northumberland, the seat of Edward Mitford, Esq., Mr. W. H. Mason informs me that there is a very fine tree, about 100 feet in height, with a girth of 18½ feet, and a spread of branches 102 yards in circumference.

In Wales there are many fine trees, the largest that I know being at Dynevor Park, where the Hon. Walter Rice has measured one 81 feet by 15 feet 3 inches in girth. At Gwydyr Castle, Carnarvonshire, a tree, dividing into two stems at ten feet from the ground, is 86 feet high by 14 feet 8 inches in girth.

In Scotland nearly every large place has fine sycamores, some of great age; but I have seen none to surpass in size, shape, and perfection the one which I figure (Plate 180) in front of Newbattle Abbey. This tree in 1904 was about 95 feet high by 16 feet 6 inches in girth at 5 feet, below which its roots spread out very widely. There are other large sycamores here, said by Loudon to have been planted before the Reformation, and one, which I could not identify, before 1530. Judging, however, from a beautiful photograph taken by Col. Thynne, the Newbattle tree is exceeded in height and equalled in girth by one (Plate 181) at Drumlanrig Castle, the seat of the Duke of Buccleuch, of which he gives the height as 105 feet and the girth 19 feet 6 inches. Henry, however, who measured the same tree in December 1904, made it 101 feet by 18 feet, with a bole 22 feet high dividing into two large stems.

At Castle Menzies, Perthshire, there are two magnificent trees, the largest of which is figured by Hunter.[1] I found it in April 1904 to be 92 feet high by 19 feet 2 inches in girth. Probably this is the same of which Hunter says that it contains upwards of 1000 feet of timber. I have a good photograph of it, but prefer to figure (Plate 182) another sycamore, also at Castle Menzies, which, though not so tall, is 20 feet 4 inches in girth, and is remarkable on account of its short trunk, covered with great burrs and excrescences.

The tallest tree of this species which we know in any country is one at Blair Drummond, Perthshire, the seat of H. S. Home Drummond, Esq., which is drawn up in a dense wood behind the house. It was measured by Henry in 1904 as no less than 108 feet in height, though only 10 feet in girth. In open ground on the same property there is another tree[2] of an entirely different character, which is 19½ feet in girth with a short bole of 9 feet, dividing into two great limbs and forming a very widespreading crown. Its height is only 85 feet. Large variegated sycamores, one 70 feet high by 13 feet 10 inches in girth, also are growing at Blair Drummond.

At Kippenross, near Dunblane, the seat of Captain Stirling, there are the

[1] *Woods, Forests, and Estates of Perthshire*, 397 (1883).
[2] In Loudon, *Gard. Mag.* 1840, p. 505, there is a list of measurements of trees at Blair Drummond; and a sycamore, 80 feet high and 16 feet in girth in 1841, is probably this tree.

remains of an immense sycamore which was struck by lightning in 1860. It measured in 1821 as follows :—

	Feet.	Inches.
Height	100	0
Girth at smallest part of trunk . . .	19	6
„ where branches separate . . .	27	4
„ at ground	42	7
Extreme width of branches	114	0
Cubic contents	875	0

Captain Stirling informs me that its age, which can be determined from an entry in the records of the estate, is about 440 years. Hunter[1] says that this tree was known as the big tree of Kippenross in the time of Charles II., and in 1806 was described by Mr. Ramsay of Ochtertyre, as " now much the greatest in this country." He adds that in 1740 the late John Stirling of Keir was told by a woman over eighty, that though all the other trees had grown much in her recollection, she knew of no change in the great tree which many people came to see as a curiosity.

At Keir, Perthshire, there is a remarkable sycamore stool, with eleven stems, 85 feet in height and averaging in girth about 6 feet.

Sir Herbert Maxwell sends me the measurement of a sycamore behind the Birnam Hotel at Dunkeld, which Hunter figures on the frontispiece of *Woods and Forests of Perthshire*, and says[2] that it was supposed to be 1000 years old ; this of course is without foundation, and the girth, 19 feet 8 inches, as taken by Sir H. Maxwell, is precisely the same as Hunter gave 25 years before.

In the parish of Cramond, county of Edinburgh, a tree of this species is stated in the *Old and Remarkable Trees of Scotland*, p. 198, to have attained 130 feet in height, but I must doubt the accuracy of this statement, of which I can obtain no confirmation.

Many large trees in the south-west of Scotland are recorded by Renwick,[3] of which one at Erskine House, Renfrewshire, is 75 feet high and 19 feet 4 inches in girth, but this has only a short bole of 7 feet. Another at Logansraes, in the same county, is 80 feet high and 18 feet 3 inches in girth, with a spread of 95 feet. Strutt figures, on plate 3 of *Sylva Scotica*, a sycamore at Bishopton, in Renfrewshire, which was about 60 feet by 20 feet and contained 720 feet of timber.[4]

At Inveraray Castle, close to the big Scots pine, there is a fine sycamore about 100 feet high, with a bole of about 35 feet, but its girth in 1906 was only 12 feet 2 inches.

At Invergarry Castle there is an avenue of sycamores, said to date from 1689.

In Ireland, the sycamore is commonly planted and grows with great vigour, but Henry has seen no specimens rivalling those of Scotland in size. At

---

[1] *Woods, Forests, and Estates of Perthshire*, 286 (1883).   [2] *Ibid.* 73.

[3] *Measurements of Notable Trees*, Glasgow, 1901.

[4] Sir C. Renshaw informs me that this tree still survives, and that, according to local tradition, John Knox formerly preached under it, but that it is not so fine a tree as one at Erskine House, the property of W. A. Baird, Esq.

Kilmacurragh, Co. Wicklow, there is a fine tree, 15 feet in girth with an estimated height of 90 feet. This is supposed to be the tree referred to by Hayes,[1] as being in 1794 the largest then living in Wicklow; but if this is the case the tree must have remained stationary in growth for many years. At Powerscourt there is a fine widespreading tree 80 feet high by 14 feet in girth. At Carton, a sycamore, remarkable for its small leaves, which are only half the ordinary size, measured, in 1903, 87 feet in height and 11 feet in girth. At Woodstock, Co. Kilkenny, a tree was, in 1901, 73 feet high by 11½ feet in girth; and, according to the records kept here was 7 feet 10 inches in 1825, 8 feet 2 inches in 1830, and 8 feet 11 inches in 1846. At Cushendun, Co. Antrim, in a situation completely exposed to the blasts from the sea, in the garden of Miss M'Neil, a sycamore is 60 feet high by 13 feet in girth.

On the Continent the sycamore is not so often planted as in England, but in Switzerland and the Austrian Alps it attains a great size. Two are figured in the *Baum-Album der Schweiz*,[2] of which one, formerly growing at Truns in the Oberland at an elevation of 853 metres, close to the old chapel of St Anna, is interesting on account of its great age. Under this tree the Grey League, one of the three bodies which, when confederated in 1525, formed the canton of Grisons, were sworn in 1424; and though the last remnant of the veteran was torn up by a storm in 1870, it shows that the sycamore may attain an age of about 600 years. A figure of it, taken from a painting in the possession of M. Descurtins of Coire, is given in the work from which I quote. A young tree raised from its seed was planted in 1870 on the spot, and was in 1896 already over 30 feet high and 4 feet 4 inches in girth.

By the kindness of M. Coaz, Director of the Swiss Government Forests, I am able to reproduce a beautiful photograph (Plate 183) of an even finer tree, now standing on the land of the commune of Kerns, in Melchthal, canton of Unterwalden, at an elevation of 1350 metres, in deep loamy soil, on a formation described as *calcaire schrattique*. This immense tree measures 12.20 metres in circumference above the point where its trunk expands, and at 5 feet from the ground 8.85 metres, equal to about 29 feet, thus exceeding any tree of which we have a record in this country. At 12 feet from the ground a branch about 9 feet in girth is given off. The height is not stated, but the branches spread to a diameter of about 25 yards, and though the trunk is hollow and covered in places with a moss (*Leucodon sciaroides*), the tree still bears fruit. Its aspect reminds me strongly of many sycamores which grow on the Alps of the Vorarlberg in Austria, and especially of one, from the cover of whose trunk I shot my last chamois, a cunning old buck, which for four seasons I had hunted in vain.

TIMBER

The wood of the sycamore is of a white colour, close grain, and moderately hard; and when of large size is one of the most valuable woods we have, as it has been found the most suitable for making the large rollers, technically called

---

[1] *Practical Essay on Planting*, 121 (1794).   [2] Published by Schmid, Francke & Co., Bern, 1896.

"bowls," which are used in cotton-dying and washing machines. For this purpose it must be cut early in the winter, in order to preserve the purity of its colour, and removed as soon as possible, for if left standing till the sap begins to rise, which it does early in spring, or left lying exposed to the weather, it is soon depreciated in value. Butts of moderate age, free from branches or knots and over 18 or 20 inches quarter-girth, are worth from 3s. 6d. to 5s. per foot, or even more when near their best market, which is in Lancashire. The measuring of this timber presents a difficulty when, as often happens, the logs are not round or quite straight, as in conversion they have to be turned down to a true cylinder, and in trees grown in the open, as is usually the case, large buttresses and swellings often occur, for which allowance must be made.[1]

Smaller and rougher trees are worth much less than large clean ones, and are converted into planks and smaller rollers, which are used by manufacturers of dairy utensils and mangles, brush-makers, toy-makers, and turners, for bobbins and many small articles. From 1s. to 2s. per foot is a fair price for such timber, but the price varies much, according to the locality. A certain quantity of sycamore is cut into veneers, and when the wood has a wavy grain, like that of the so-called fiddle-backed maple, it is very ornamental, and may be used with good effect for the interiors of cabins, railway carriages, and furniture. What is known in the furniture trade as "hare wood" is, I believe, nothing more than fine wavy sycamore, which by age or staining has taken a pinkish-brown colour.

(H. J. E.)

ACER CAMPESTRE, Common Maple

Acer campestre, Linnæus, *Sp. Pl.* 1055 (1753); Loudon, *Arb. et Frut. Brit.* i. 428 (1838); Willkomm, *Forstliche Flora*, 764 (1887); Mathieu, *Flore Forestière*, 42 (1897).

A tree, rarely attaining 70 feet in height, usually smaller. Bark, corky on young trees, ultimately becoming fissured and scaly. Young branchlets usually pubescent, in some forms glabrous, and not remaining green throughout the first year. Leaves variable in size, averaging $2\frac{1}{2}$ inches long and 3 inches broad, cordate at the base, five-lobed, the two basal lobes occasionally obsolete; lobes shortly acuminate; margin plainly ciliate, usually with a few coarse obtuse teeth; upper surface dark-green, pubescent on the nerves; lower surface light-green, with scattered pubescence, dense on the nerves and tufted on the axils. Petiole with milky sap. Plate 207, Figs. 24 and 25, taken from adult trees growing in England, show considerable variation in the shape of the leaves and the amount of pubescence on the branchlets. Fig. 23 represents the foliage of a coppice shoot in a French forest.

Flowers, in corymbs, at first erect, afterwards pendent, opening with or soon after the leaves, green in colour, with pubescent pedicels and sepals; lateral flowers

---

[1] William Low, Esq., of Monifieth, Scotland, informs me that in his neighbourhood there is a large consumption of sycamore for making rollers used in the jute and flax-spinning industry. These are from 7 to 9 inches in diameter, and $1\frac{3}{4}$ inch thick. They cost about 30s. per gross, and are preferred when made of hard and slowly grown Scotch timber, which is considered to be less liable to crack in drying, when cut in transverse sections.

staminate, others mixed; ovary in perfect flowers glabrous.   Fruit[1] with horizontally spreading wings, glabrous or pubescent.

In summer it is distinguishable from all the other small-leaved maples by the leaves being ciliate in margin, with petioles containing milky sap.   In winter the corky ridges on older twigs are characteristic; young branchlets are glabrous or pubescent towards the tip, with crescentic three-dotted leaf-scars, which join at their ends around the twigs.   Terminal buds sessile, about ⅛ inch long, with scales pubescent at the apex and fringed with white cilia.   Lateral buds almost appressed to the twigs.

VARIETIES

Two varieties occur in the wild state both in England and on the Continent; var. *hebecarpum*, in which the carpels of the fruit are pubescent, and var. *leiocarpum*, with glabrous fruit.   The leaves show considerable variation in shape and in the amount of pubescence on the lower surface; and six sub-varieties are distinguished by Schneider,[2] as follows :—

1. *subtrilobum*.   Leaves three-lobed; fruit pubescent.
2. *lobatum*.   Leaves with five obtuse, toothed lobes; fruit pubescent.
3. *acutilobum*.   Leaves with five acute, almost entire lobes; fruit pubescent.
4. *pseudomarsicum*.   Leaves three-lobed; fruit glabrous.
5. *normale*.   Leaves with five obtuse, toothed lobes; fruit glabrous.   Var. *collinum* is a form of this, with the leaves glabrous beneath.
6. *austriaca*.   Leaves with five acute, almost entire lobes; fruit glabrous.

In var. *pulverulentum*, as cultivated at Kew, the leaves are spotted with white. This appears to be a very slow-growing tree.   In var. *variegatum* the leaves are white in margin.

A hybrid[2] between this species and *A. monspessulanum* has been found wild in Herzegovina, and has been named *A. Bornmülleri*, Borb.   *A. neglectum*, Lange, described above in the Synopsis, p. 637, is a hybrid between *A. campestre* and *A. pictum*, var. *colchicum*.

DISTRIBUTION

Acer campestre is spread generally throughout Europe, with the exception of the greater part of Scandinavia, Finland, Northern Russia, and the south of the Iberian Peninsula; and extends into Western Asia, where it is found in the Caucasus and in the province of Astrabad in Persia, where it reaches its most easterly and southernmost point.

In Norway, according to Schübeler, it is not indigenous; but it lives as far north as Trondhjem and grows as tall as 25 feet in the south.   Its northern limit as a wild tree, beginning in the province of Scania in South Sweden, crosses into the province of West Prussia in Germany, where it grows at Thorn, and extends through Poland and Central Russia to Vladimir, where it reaches its northernmost

---

[1] A series of abnormal fruits, each with three to eight keys, instead of two, the normal number, is exhibited in the Kew Museum.   Cf. Sir W. Thiselton-Dyer in *Ann. of Bot.* xvi. 556 (1902).

[2] Schneider, *Laubholzkunde,* ii. 230, 231 (1907).

and most easterly point in Russia. Its easterly limit extends from here through Voronej and Kharkof to the Crimea, where it grows in the mountains. It is also met with in the region of the steppes, growing on the banks of streams. The southern limit in Europe is not exactly known; but the tree occurs in the mountains of Turkey, Dalmatia, Italy, and Sicily, and in the Pyrenees and the mountainous parts of the northern provinces of Spain and Portugal.

Inside these limits, its distribution over the continent of Europe is not at all uniform, and is very scattered, as it is totally absent from many districts where the climate or the conditions of the soil are unfavourable. It is rather a tree of the plains, valleys, and hills, than of the mountains; and is especially met with in the broad-leaved forests, often growing as underwood in coppice with standards, and on the edges of woods, on the banks of streams, and in hedges. It ascends in Southern Bavaria to 2500 feet elevation. In France, it is scattered through coppice woods on the plains and low hills; but is rather rare in the Mediterranean region, and is not a native of Corsica. It has been found in Algeria in one or two restricted localities. It grows throughout the Caucasus[1] at elevations ranging from sea-level to 6000 feet.

Acer campestre is abundant as a wild tree in Southern England, and is recorded by Watson[2] from most of the counties of England and Wales, as far north as Durham. It is clearly native, according to Baker,[3] in the denes of the magnesian limestone of Durham, but is doubtfully so north of the Tyne, though it may be indigenous in the woods of the steep banks of the Wansbeck about Morpeth and Mitford, where there are trees about 30 or 40 feet high; but in the Cheviot Hills it seems to have been introduced. Most of the English county records[4] mention it as common in woods, hedges, and on the banks of streams; and in North Yorkshire[5] it ascends to 300 feet and in West Yorkshire[6] to 600 feet.

It is probably not indigenous in Scotland, though Woodforde[7] records it in woods at Queensferry, near Edinburgh, and Gardiner[8] says it grows in a wood at Mains of Hallerton in Forfarshire. In Ireland,[9] though it grows in hedges and woods in many places, it is in all cases planted or derived from plantations.

It has been found in the fossil state[10] in neolithic deposits at Crossness in Essex, and in preglacial deposits at Pakefield in Suffolk. (A. H.)

CULTIVATION

The maple is common in hedgerows in many parts of England, but can hardly be considered as a forest tree, though it forms a considerable part of the under-wood in some woods in the Cotswold Hills, and attains considerable size even on

[1] Radde, *Pflanzenverb. Kaukasusländ.* 175 (1899).   [2] *Topographical Botany,* 104 (1873).
[3] Baker and Tate, *New Flora of Northumberland and Durham,* 141 (1868).
[4] Jones and Kingston, *Flora Devoniensis,* 69 (1829); Ley, *Flora of Herefordshire,* 63 (1889); Bromfield, *Flora Vectensis,* 95 (1856); Hind, *Flora of Suffolk,* 93 (1889); Druce, *Flora of Oxfordshire,* 65 (1886); Leighton, *Flora of Shropshire,* 163 (1841).
[5] Baker, *North Yorkshire,* 276 (1906).   [6] Lees, *Flora of W. Yorkshire,* 187 (1888).
[7] *Catalogue,* 23 (1824).   [8] *Flora of Forfarshire,* 39 (1848).
[9] *Cybele Hibernica,* 482 (1898).   [10] Reid, *Origin British Flora,* 113 (1899).

poor soil. Though the autumn colour of the leaves is pretty, yet as an ornamental tree it is inferior to the Norway, American, and Japanese maples. It seeds itself freely in hedges, growing slowly and living to a great age. Lees[1] mentions a hollow tree at Powick which he estimated to be near 600 years old, and says that one almost as ancient stood near Hanley Castle.

It likes a dry, somewhat stony soil, and is most commonly seen in open sunny places in the form of a large shrub of very irregular growth. It is often pollarded, and in such cases becomes hollow, usually with a burry stem and spreading roots. So far as I have observed, the seeds germinate in the first year if sown when ripe, and are easy to raise. It bears pruning very well and is therefore suitable for hedges, which in France are often made of this tree.

According to Mouillefert,[2] it suffered much at Grignon in the severe winter of 1879-1880, when the thermometer fell to − 27° Cent.; and is much less hardy than the sycamore, which sustained this low temperature without being injured in the least; but I have never seen it damaged by frost in this country.

Remarkable Trees

There are many good-sized trees of this species in England, of which one at Cobham Hall, Kent, is the tallest we have seen. This is a twin tree in a wood near the house, with two tall straight stems from the same root, which are 6 feet 4 inches and 6 feet respectively in girth, and about 75 feet high. Another tree, in the deer park here, is 70 feet high, with a trunk girthing, at three feet from the ground, 8 feet 11 inches, and dividing at 4 feet up into four stems. At Chilham Castle, Kent, the seat of C. S. Hardy, Esq., there is a splendid tree 55 feet high by 13 feet 8 inches in girth which covers an area 86 paces round, but the bole is only 7 feet high.

One of the best-shaped large trees that I have seen, grows in the park at the Mote near Maidstone: it measures 60 feet by 10 feet 3 inches. I lately discovered a magnificent tree past its prime at Langley Park, Norfolk, which, though only 45 feet high, girths 9 feet 5 inches and has branches which spread to a width of 24 paces.

At Hursley Park, Hants, the property of Sir G. Cooper, Bart., there is a tree which Sir Joseph Hooker told me was the finest that he had ever seen. Mr. J. Clayton, who was forester-there, tells me that it has a short bole 9 feet 6 inches in girth, with ten large spreading limbs, and contains about 111 feet of measurable timber. The tree in Boldre Churchyard, figured by Strutt, and said by him to be the largest in England, was, however, only 45 feet by 7 feet 6 inches, but I cannot learn that it still exists.

In Cassiobury Park there is a very well shaped tree on the golf ground, which measured, in 1907, 60 feet by 9 feet 6 inches (Plate 184), being little less than the one at the Mote. At Moor Park, Herts, Sir Hugh Beevor found in 1902 a quite sound tree, which was 10 feet 3 inches in girth at 4 feet from the ground, and no less than 76 feet in height.

---

[1] *Botany of Worcester*, p. xxxviii (1867).   [2] *Essences Forestières*, 208 (1903).

At Casewick, Lincolnshire, the seat of Lord Kesteven, a maple measured, in 1907, 53 feet high by 9 feet 1 inch in girth, with a bole of 9 feet. At Arley Castle, near Bewdley, a slender tree measured, in 1906, 66 feet high by 4 feet 8 inches in girth. At Colesborne, just below the church, there is one about 60 feet high by 9 feet in girth ; and this is the largest that I know in Gloucestershire.

None of the trees recorded by Loudon approach those mentioned above in size, and though the tree is so hardy in the south of England, it usually does not attain a considerable size in the north. I have myself seen no specimens in Scotland worth recording ; and it is not mentioned either in the *Old and Remarkable Trees of Scotland* or in Hunter's *Woods and Forests of Perthshire*. Loudon records a tree at Hopetoun House, near Edinburgh, 46 feet high, and another at Blairlogie in Stirlingshire, said to have been 302 years old and no less than 55 feet high by 4 feet in diameter. The Hon. Vicary Gibbs, however, measured in 1905 a pollarded tree at Armadale Castle, in the Isle of Skye, 42 feet high by $7\frac{1}{2}$ feet in girth at $2\frac{1}{2}$ feet from the ground. Mr. Renwick also reports a large one at Ardgowan, Renfrewshire, which was, in 1904, 12 feet 2 inches in girth at 1 foot up, and another at Auchentorlie, Dumbartonshire, which was, in 1907, 41 feet high by 9 feet 5 inches in girth at 3 feet from the ground. Henry measured in 1905 a tree in the Edinburgh Botanic Garden, 54 feet high by 6 feet 2 inches in girth, dividing into two stems at 8 feet from the ground.

The common maple is often planted in Ireland ; but Henry has seen no trees of great size. It thrives well at Castlewellan, where there is a young tree about 30 feet in height.

TIMBER

Though the wood is one of the best of its class, on account of its fine grain, close texture and hardness, and though it sometimes shows a most beautiful figure, which when polished is highly ornamental, this wood, formerly much sought after for turning, inlaying, and cabinetmaking, is now hardly known in commerce, and is not mentioned by most recent writers. Stevenson,[1] however, says that waved or mottled specimens when cut into veneer are little, if anything, inferior to American bird's-eye maple.

The so-called mazer bowls which in ancient times were carried by every pilgrim to drink from, just as they now are by the Tibetans,[2] were turned from the roots and burrs of the common maple, and when mounted in silver the few remaining specimens of these bowls are very highly valued by collectors. The colour of the wood is normally white, but in old trees it turns to a pinkish or brown colour, and so far as my experience goes it is a wood which shrinks and warps very little. For parquet flooring it would be admirable, and might be very well used for table legs. In *English Timber and its Economical Conversion* it is said to be subject to the attacks of worms, but I do not know whether this statement is based on experience. (H. J. E.)

---

[1] *Trees of Commerce*, p. 112.　　　　　　　　[2] Cf. p. 662, note 3.

ACER PLATANOIDES, NORWAY MAPLE

Acer platanoides, Linnæus, *Sp. Pl.* 1055 (1755); Loudon, *Arb. et Frut. Brit.* i. 408 (1838); Willkomm, *Forstliche Flora*, 757 (1887); Mathieu, *Flore Forestière*, 41 (1897).

A tree, attaining occasionally 90 feet in height, but usually smaller. Bark smooth on young trees, but ultimately becoming rough and fissured longitudinally. Young branchlets glabrous, not remaining green throughout the first year. Leaves (Plate 206, Fig. 11) averaging 5 inches long by 7 inches wide, five-lobed; lobes oblong with an acuminate bristle-pointed apex; sinuses wide, rounded and open, not reaching the middle of the leaf; base cordate; margin non-ciliate and with a few large sinuate pointed teeth; both surfaces shining, green, and glabrous, except for tufts of pubescence in the axils of the primary and secondary nerves beneath; petiole with milky sap.

Flowers, opening early before the leaves expand, in erect corymbs, yellowish-green; the earliest mostly staminate, those opening later perfect; stamens 8, as long as the sepals; pedicels, calyx, corolla, filaments, and ovary glabrous. Fruit pendulous, on long stalks, glabrous; keys about $1\frac{3}{4}$ inch long; wings widely divergent.

In summer the Norway maple is readily distinguishable by the leaves shining on both surfaces, with long pointed lobes and teeth, and by the milky sap in the petioles. In winter the twigs are shining, glabrous, with very narrow three-dotted leaf-scars, the opposite pairs of which are joined at the ends around the stem. Terminal buds $\frac{1}{3}$ inch long, sessile; scales shining, either green at the base and reddish-brown above, or reddish-brown throughout, glabrous, ciliate. Lateral buds appressed to the stem.

VARIETIES

A large number of varieties have appeared in cultivation, of which the most noteworthy are :—

1. Var. *laciniatum*, Aiton,[1] Eagle's Claw or Hawk's-foot Maple. Said by Loudon to have originated in the seed-bed. Leaves (Plate 205, Fig. 10) about half the size of the type, cuneate at the base; lobes acutely, deeply, and irregularly cut; margin rolled up. This variety usually attains to no great size, but Sir Hugh Beevor tells us of a tree at Gelderstone Hall near Beccles, Suffolk, 50 feet high by 2 feet 8 inches in girth; and Renwick in 1907 measured one at Auchendrane, Ayrshire, 48 feet by 3 feet 2 inches.

2. Var. *dissectum*, Jacquin fil. (var. *palmatum*, Koch[2]) (*A. Lorbergi*, Van Houtte). Leaves (Plate 206, Fig. 21) deeply cut to near the base, which is cordate; lobes five, ending in long sharp points, the three upper lobes again divided into three lobules; margin with a few sharp-pointed teeth. First introduced from Belgium in 1845 by Knight of Chelsea, it grows to be a fair-sized tree, and is worth cultivating on account of its elegantly cut foliage.

[1] *Hort. Kew*, iii. 435 (1789).

[2] *Dendrologie*, i. 530 (1869).

3. Var. *heterophyllum*, Nicholson.[1] Leaves asymmetrical and irregularly cut.

4. Var. *Schwedleri*, Koch.[2] A tree, vigorous in growth, with large leaves, bright-red when young, changing to dark-green. A valuable ornamental tree.

5. Var. *Reitenbachii*, Nicholson.[1] Leaves pale-red when unfolding, turning a dark blood-red in late summer. This variety originated in Reitenbach's nursery at Plicken in Prussia, and was sent out by Van Houtte. It comes fairly true from seed, and grows at Colesborne as fast as the type.

6. Var. *rubrum*, Herder.[3] Leaves green when young, but towards autumn assuming a fine red colour. This variety is cultivated in the Imperial Park at St. Petersburg, and in gardens and parks in Germany.

7. Var. *albo-variegatum*. Leaves irregularly blotched with white. In a form[4] sent out by Messrs. Drummond of Stirling, the leaves are green in the centre, with a very deep edging of white. In var. *maculatum* the leaves are spotted with white, and are very pretty.

8. Var. *aureo-marginatum*.[5] Leaves with a yellow margin.

9. Var. *cucullatum*. Leaves irregularly and shortly lobed, crimpled, dark-green.

10. Var. *columnare*. Of erect columnar habit, with small leaves. Raised[1] by Simon Louis at Metz in 1855 and first sent out in 1879.

11. Var. *globosum*. With a dense, compact, rounded head of foliage.

12. *Acer Dieckii*, Pax, a supposed hybrid between this species and *A. pictum*, var. *colchicum*, has been described above in the Synopsis, No. 36, p. 637.

DISTRIBUTION

The Norway maple has a wide distribution, as it inhabits the greater part of Europe, and extends eastwards into the Caucasus and North Persia.

In Norway, where it is called *Lon* (in Swedish *Lönn*), it is indigenous, according to Schübeler, from the extreme south to Romsdal on the west coast and Elvedal in Osterdal, but has been planted and exists as a shrub as far north as Tromsö, lat. 69° 40′; and in Sweden it is met with growing wild as far north as lat. 63° 10′ on the east coast. Elwes has never seen any large trees in Norway; but Schübeler figures one (his fig. 129) at Triset in Laurdal, which was 60 feet high by 9 feet 8 inches in girth, and had very much the appearance of a sycamore. He mentions a tree at Lid at 1000 feet elevation which was 120 years old; and says that the largest known to him were at Drobak on the Christiania fjord, 60 feet high by 14 feet in girth, and at Mollendorf near Drammen, a tree which was called the great Lon, and was 70 feet high by 11 feet in girth. In former times the peasants used to tap this maple, and make a sort of beer with the sap; and Loudon states that in Germany the sap has been found to contain more saccharine matter than that of the sugar maple.

---

[1] *Gard. Chron.* xv. 564 (1881).

[2] *Dendrologie*, i. 530 (1869). Koch could throw no light on the name of this variety, the origin of which is unknown.

[3] *Gartenflora*, 163, t. 545 (1867). [4] *Gard. Chron.* xxxiv. 24 (1903).

[5] Var. *aureo-variegatum*, a form in which the leaves are mottled green and white, is described in *Gard. Chron.* xiv. 241 (1880).

It grows wild in Finland as far north as 62° lat. on the west coast; and the northern limit of the species as a tree extends from Lake Ladoga, through Central Russia, where it is, with birch and aspen, the commonest deciduous tree, to Orenburg, where it reaches, but does not cross the Ural, in lat. 54°. To the north of this limit it is often met with as a shrub in the pine forests, and has been found even on the west coast of Onega Bay. Southward it grows in the Crimea, and is common in the Caucasus, at 2000 to 6000 feet altitude, extending into Armenia and North Persia.

It is common throughout Germany, especially in the north, where it grows mainly in the plains, valleys, and low hills; and extends southwards through Austria and the Balkan States to Epirus. It occurs in Northern Italy. In Switzerland[1] the tree is only met with in the lower part of the beech forests, ascending to about 3300 feet. In France it is spread throughout the forests of the low hills and mountains, as far south as the Cevennes, Auvergne, and the Central Pyrenees, where it crosses into Northern Spain.

The Norway maple is not a native of the British Isles, where it has not yet been found in the fossil state.

It always grows in company with other trees, never forming pure forest, and generally solitary or in small groups. It is often associated with the sycamore, and like it thrives best on soils rich in soluble mineral matter. It does not ascend in mountainous regions as high as the sycamore; but succeeds better than that species in wet situations. It bears a great degree of cold and is entirely unaffected by late frosts.

Mistletoe[2] is occasionally found on the Norway maple.   (A. H.)

CULTIVATION

Though the tree has been known in England for a very long period—Loudon says it was introduced in 1683—and is one of the hardiest trees in cultivation, it is seldom seen of any size, and is not nearly so commonly planted as it deserves to be. It will grow on the driest and poorest soil, and on my own land there are many trees over 50 feet high by 6 to 7 feet in girth which have not been planted more than fifty years.

It ripens seed abundantly almost every year, and is as easy to raise as the sycamore, and though it does not attain nearly the size and age of that tree, is far more ornamental both in spring, when covered with pale-yellow flowers, before the leaves come out, and in autumn when they turn a brilliant red or yellow colour.

REMARKABLE TREES

The largest tree of this maple that we know of in England is not far from the stables at Cassiobury Park, the seat of the Earl of Essex, and measures over 90 feet, perhaps as much as 95 feet high, by 13 feet 9 inches in girth. But it is not

[1] Christ, *Flore de la Suisse*, 181 (1907).   [2] Cf. *Gard. Chron.* xxxix. 238 (1906).

so handsome or well-shaped a tree as the one figured (Plate 186), from a photograph for which I am indebted to Mr. W. M. Christy, who tells me that this tree, growing at Watergate, near Emsworth, Hants, is no less than 14 feet 4 inches in girth at 4½ feet from the ground, the branches spreading over an area about 102 to 105 paces round.

Another very fine tree grows in a wood at Park Place near Henley-on-Thames, by the side of the drive from Templecombe, on loamy soil overlying chalk, and this in 1905 measured 80 feet by 9 feet 9 inches, with a bole about 12 feet long (Plate 187).

At Syon there is a healthy vigorous tree, 68 feet by 7 feet 3 inches, with a bole 10 feet long. At Colesborne there are several of about this size, and some younger trees, one of which was photographed when covered with hoar-frost in winter, and shows the habit of branching which this tree sometimes assumes (Plate 188).

There are many handsome specimens in the park and woods at Highclere, which produce a beautiful effect when in flower. The largest that I measured is about 80 feet by 7 feet 2 inches, with a trunk 40 feet high.

At Arley Castle there is a tree, which in 1906 measured 70 feet high by 6½ feet in girth, dividing into two stems at 6 feet from the ground; and self-sown seedlings are growing near it. At Croome Court, Worcestershire, a tree in 1905 was 50 feet high by 4 feet 8 inches in girth. Sir Hugh Beevor measured in 1906 a tree at Newnham Paddox, Worcestershire, which was 63 feet high by 12½ feet in girth at two feet from the ground, dividing above into three stems, girthing each about 6 to 7 feet; and another at the Cranleigh Cottage Hospital, Surrey, 70 feet high by 8½ feet in girth.

At Pampisford Hall, Cambridgeshire, there is a group of four trees, the largest of which was in 1906 65 feet by 4½ feet; but the specimen at Kew, which in Loudon's time was 76 feet high at 70 years old, no longer exists.

In Scotland the largest record we have is of a tree at Bowhill, near Selkirk, one of the Duke of Buccleuch's properties, which Col. H. Thynne tells me is no less than 84 feet high by 7 feet in girth, having been drawn up by surrounding trees. At Smeaton-Hepburn, East Lothian, Henry saw in 1905 two more, one 58 feet by 6 feet, the other 50 feet by 7 feet 2 inches. Another in the Edinburgh Botanic Gardens measured in the same year 48 feet by 4 feet 11 inches.

In Ireland, Henry has seen no Norway Maples of considerable size; but Loudon mentions one at Charleville which was 78 feet high.

The timber of this species is very similar to that of the Sycamore and may be used for the same purposes.   (H. J. E.)

ACER LOBELII

Acer Lobelii, Tenore, *Cat. Hort. Neap.* append. 2, p. 69 (1819), and *Fl. Nap.* v. 291 (1835).
Acer platanoides, Linnæus, var. *Lobelii*, Loudon, *Arb. et Frut. Brit.* i. 409 (1838).

A tree attaining about 50 feet in height, forming a narrow pyramid with ascending branches. Bark striped longitudinally. Young branchlets glaucous, glabrous, remaining green throughout the first and second years. Leaves (Plate 205, Fig. 8) about 5 inches long by 5½ inches wide; lobes five or three, with their apices pointing away from the base of the leaf and ending in long sharp points ; margin repand, non-ciliate ; upper surface dark green, shining, glabrous ; lower surface light green, dull, glabrous except for pubescent tufts in the axils of the nerves ; petiole with milky sap.

Flowers smaller than those of *A. platanoides*, in corymbs. Fruit, glabrous, with horizontally spreading wings, each key about an inch long.

A variety with deeply-cut leaves is described by Tenore ; but it is not apparently in cultivation.

This species is readily distinguished by the pyramidal habit, striped bark, and glaucous shoots. It grows in woods in the mountains around the Bay of Naples, and according to Spach [1] is also found in the mountains of Calabria.   (A. H.)

Acer Lobelii, which Loudon treats as a variety of the Norway maple, is so distinct in its habit of growth and in its bark that it is well worth cultivation. The largest we have seen in England is at Grayswood, a handsome tree with erect branches about 40 feet high. It is quite hardy as far north as Yorkshire, where Sir Charles Strickland has planted a good many which are now from 25 to 40 feet high and growing vigorously. They all have an erect, fastigiate habit. There are two good specimens in Kew Gardens. A large tree was reported [2] to be growing in 1839 at Croome Court, near Worcester ; but when this place was visited by us in 1905, it could not be found.

At Verrières, near Paris, in M. P. de Vilmorin's grounds, this tree has attained 55 feet in height by 5 feet in girth.   (H. J. E.)

ACER PICTUM

Acer pictum, Thunberg, *Fl. Jap.* 162 (1784) ; Shirasawa, *Icon. Ess. Forest. Japon*, text 105, t. 65. ff 1-12 (1900) ; Brandis, *Forest Flora N.W. India*, 112 (1874), and *Indian Trees*, 183, 705 (1906).

Acer lætum, C. A. Meyer, *Verz. Kaukas. Pflanz.* 206 (1831).

Acer cultratum, Wallich, *Pl. As. Rar.* ii. 4 (1831).

Acer colchicum, Booth, in Loudon, *Gard. Mag.* 1840, p. 632.

Acer Mono, Maximowicz, *Bull. Acad. St. Péters.* xv. 126 (1857), and *Prim. Fl. Amur.* 68 (1859).

Acer Mayri, von Schwerin, *Mitt. Deut. Dendr. Ges.* 1901, p. 58 ; Mayr, *Fremdländ. Wald- u. Parkbaüme*, 460, f. 161 (1906).

A tree attaining 60 feet in height ; bark smooth, usually striped with white lines or bands. Young branchlets green, glabrous, not glaucous except in one variety, remaining green and smooth in the second year in some varieties, becoming grey or brown with irregular fissures in others. Leaves (Plate 205, Fig. 9), averaging 4 inches long and 4½ inches broad, cordate at the base, entire in margin ; lobes long, cuspidate or caudate-acuminate, bristle-pointed, five or seven in number

---

[1] *Ann. Soc. Nat.* 2 sér. ii. 168 (1834).   Cf. also Tenore, *Essai Géog. Roy. Naples*, 81 (1827).

[2] Loudon, *Gard. Mag.* 1840, p. 44.

in the latter case, with the two basal lobes very small and directed downwards; membranous; shining, green, and glabrous on both surfaces, except for pubescent tufts in the basal axils beneath; petiole containing latex.

Flowers in corymbs, appearing with the leaves, yellow or greenish-yellow, variable as regards the pubescence of the calyx and the relative length of the petals and sepals. Fruit, variable as regards the length and divergence of the wings.

Acer pictum is the representative in Asia of *A. platanoides*, and is very uniform in foliage, though it extends over a wide area. Owing, however, to the remarkable variation in the characters of the fruit, it has been usually divided into two species, which cannot be maintained,[1] as there are numerous connecting links; and the different forms are best treated as geographical varieties.

> ** Branchlets remaining smooth and greenish in the second year. Wings of the fruit two to three times as long as the carpels.*

1. Var. *colchicum* (*A. lætum*, C. A. Meyer). Asia Minor, Caucasus.

Leaves, five- to seven-lobed, light green in colour, thin in texture. Fruit-wings usually spreading at a wide angle. In the ordinary form of this variety, the leaves are green on opening. This was introduced[2] in 1838 by Messrs. Booth of Hamburg, plants being in cultivation in the London Horticultural Society's garden in 1840.

In var. *colchicum rubrum*, introduced[3] in 1846, the young leaves and young branchlets are deep red in colour.

2. Var. *cultratum* (*A. cultratum*, Wallich). Persia, Himalayas, Central China.

Leaves thicker in texture than the last, usually five-lobed, more truncate at the base. Fruit usually with horizontally spreading wings.

This was introduced[4] from China in 1901 by E. H. Wilson, and is in cultivation at Coombe Wood, where there are plants now 10 feet in height.

3. Var. *tricaudatum*, Rehder.[5] Central China.

Leaves, three- to four-lobed; basal lobes small or obsolete. This is a peculiar form, with leaves smaller than in the type, scarcely exceeding two inches long by three inches wide; and was introduced[6] by Wilson in 1901. Young plants at Coombe Wood are already 14 feet high and are growing very vigorously.

4. Var. *tomentosulum*, Rehder.[5] Central China. A rare form with the young leaves covered beneath with dense whitish tomentum.

> *** Branchlets[7] becoming grey or brown and fissured in the second year. Wings of the fruit about 1½ times as long as the carpels.*

5. Var. *eu-pictum* (*A. pictum*, Thunberg). Japan.

Leaves darker green and thicker in texture than in var. *colchicum*; lobes

[1] Cf. Rehder, in Sargent, *Trees and Shrubs*, i. 178 (1905).   [2] Cf. Loudon, *Gard. Mag.* 1840, p. 632.
[3] Nicholson, *Gard. Chron.* xvi. 375 (1881).   [4] *Journ. Roy. Hort. Soc.* xxix. 354, f. 101 (1904).
[5] In Sargent, *Trees and Shrubs*, i. 178 (1905).   [6] *Journ. Roy. Hort. Soc.* xxix. 354, ff. 100, 102 (1904).
[7] Rehder, *loc. cit.*, points out that this character is inconstant, as he has found in several Japanese specimens the bark of the branchlets similar in colour to that of var. *colchicum*.

occasionally short and broad.   Fruit-wings erect and parallel or diverging only at an acute angle.   Introduced by Maries in 1881.

6. Var. *Mono* (*A. Mono*, Maximowicz).   Japan, Saghalien, Amurland, Manchuria, Northern and Central China.

Differs from the last in the wings of the fruit diverging at about a right angle.   Introduced[1] from Central China by Wilson in 1901.   Plants at Coombe Wood are about 6 feet high.

*** *Young branchlets glaucous.*

7. Var. *Mayri* (*A. Mayri*, von Schwerin).   Yezo.   This differs from var. *eu-pictum* in the young branchlets being glaucous.   Mayr, who discovered the tree in 1886, says that the bark is almost white in colour and hard and smooth.   Apparently not yet introduced.

Acer pictum is widely distributed, occurring from Asia Minor through the Caucasus and the Himalayas to China, Manchuria, and Japan.   In Asia Minor it is met with in the mountains near Trebizond, where, as in the Caucasus, it grows in mixed forests and beech woods, ascending from sea-level to 5600 feet. It has been collected in Armenia and in the Elburz Mountains of Northern Persia. According to Gamble,[2] it is the commonest maple in the Western Himalaya, but extends throughout the middle and outer ranges from the Indus to Assam, where it grows as a moderate-sized tree with thin grey bark at elevations ranging from 4000 to 9000 feet.   The wood is used in India for construction, ploughs, bedsteads, and carrying-poles; and the Tibetan drinking cups are turned from the knotty excrescences which are often found on this tree.[3]   Further east the tree is spread throughout the mountains of Western China in the provinces of Yunnan, Szechwan, and Hupeh; and it is found northward in the province of Chihli and throughout Manchuria.   It also grows in the island of Saghalien, and is the most common and largest species of maple in Japan, where, according to Sargent,[4] it is one of the most abundant trees in Hokkaido, occasionally attaining a height of 50 feet and a girth of 5 feet.   Elwes, however, saw none as large as this.   The tree is beautiful in May, when the flowers are just opening, as the large lengthened inner scales of the winter buds are then bright orange-yellow, and very showy.   The autumnal colour of the leaves is described as yellow and red.

This species is usually seen in England as a small tree in botanic gardens and public parks, var. *colchicum rubrum* appearing to be the commonest variety in cultivation, the form from Japan being very rare.   The finest trees we have seen are two at Tortworth, one of which is 49 feet high, and 5½ feet in girth, with a very spreading top 45 to 50 paces round, and many suckers from the roots with reddish leaves.   Another in the park, is grafted on *A. platanoides*, and has very

---

[1] *Journ. Roy. Hort. Soc.* xxix. 348, ff. 87, 89 (1904.)   [2] *Indian Timbers*, 202 (1902).

[3] Hooker in *Himalayan Journals*, i. 132, 133, says that some of these cups are supposed to be antidotes against poison, and fetch a very high price.   The knotty excrescences are produced on the roots of oaks, maples, and other mountain forest trees in the Himalaya by a parasitical plant known as Balanophora.

[4] *Forest Flora of Japan*, 29 (1894).

handsome bark with purple and green streaks, smoother than that of the stock. At Park Place, Henley, there is a tree 51 feet high, and 4½ feet in girth at 4 feet from the ground, dividing above this into several stems. The bark is smooth and grey, and close to the trunk are several suckers about 4 feet high.   (A. H.)

ACER OPALUS, ITALIAN MAPLE

Acer Opalus, Miller, *Dict.* ed. 8, No. 8 (1768); Aiton, *Hort. Kew.* iii. 436 (1789); Loudon, *Arb. et Frut. Brit.* i. 420 (1838).

Acer italum, Lauth, *De Acere*, 32 (1781); Willkomm, *Forstliche Flora*, 762 (1887).

Acer opulifolium, Villar, *Hist. Pl. Dauph.* i. 333 (1786); Loudon, *Arb. et Frut. Brit.* i. 421 (1838); Mathieu, *Flore Forestière*, 40 (1897).

Acer rotundifolium, Lamarck, *Encycl.* iii. 382 (1789).

A tree attaining about 50 feet in height, often met with in the wild state as a mere shrub. Bark smooth and grey on young trees, fissured and darker in colour on old trees. Young branchlets glabrous, becoming dark red in their first autumn. Leaves (Plate 206, Fig. 14), variable in size and shape, usually about 2½ inches long by 3 inches wide, cordate at the base, five-lobed; lobes short, ovate-triangular, acute at the apex, irregularly toothed; sinuses shallow, usually rounded at the base; upper surface dark green, shining; lower surface dull, pale, with scattered pubescence, denser on the nerves and forming axil-tufts, in some forms glabrescent; petiole without milky sap.

Flowers appearing very early, before the leaves, in sessile corymbs, yellow; pedicels long, glabrous or pubescent. Fruit, ripening in autumn, brown, glabrous; keys about an inch long; wings more or less divergent, only slightly narrowed at the base.

In winter the twigs are shining, glabrous. Buds conical, obtuse at the apex; outer scales about twelve, pubescent and ciliate. Lateral buds shortly stalked, arising from the twigs at an acute angle. Leaf-scars very slender, crescentic, three-dotted, and fringed on their upper margins with white hairs; opposite pairs of leaf-scars often joined around the stem.

VARIETIES

This species is very variable as regards the foliage. *A. hispanicum*, Pourret, which grows in Spain, and *Acer Martini*, Jordan, a rare tree in Savoy and Basses-Alpes in France, are connecting links between *A. Opalus* and *A. hyrcanum*.

1. Var. *obtusatum*.

Acer obtusatum, Kitaibel, in Willdenow, *Sp. Pl.* iv. 984 (1805); Loudon, *Arb. et Frut. Brit.* i. 420 (1838); Willkomm, *Forstliche Flora*, 763 (1887).

Leaves (Plate 206, Fig. 16) larger, 4 inches or more in width, more rounded in outline, more coriaceous, more densely pubescent beneath; lobes short, broad, slightly and crenately toothed; basal lobes very short.

This is kept up as a distinct species by Pax and Schneider; but intermediate forms are common, and there are no distinctive characters in the flowers or fruit.

2. Var. *neapolitanum*.

Acer neapolitanum, Tenore, *Fl. Napol.* ii. 372 (1820).

This variety differs from the last in the lobes being still shorter, with the basal lobes often obsolete; and the lower surface of the leaves (Plate 206, Fig. 15) is covered with dense whitish tomentum.

DISTRIBUTION

This species is widely distributed in Southern Europe, extends eastwards into the Caucasus, and also occurs in Algeria and Morocco. The typical form is found in the mountain forests of the south and south-east of France, ascending as high as the silver fir, and is recorded[1] from the Jura, Burgundy, Lyonnais, Dauphiné, Savoy, Alpes Maritimes, Provence, Aveyron, Pyrenees, and Corsica. It rarely attains a height of more than 30 feet, and is often only a bushy small tree. It also grows in South-western Switzerland, extending along the Jura as far north as Neuchâtel, and is also found in the Apennines of Northern and Central Italy.

Var. *neapolitanum* is found in wooded regions in the mountains around Naples, ascending as high as the beech, and attains, according to Tenore,[2] large dimensions in the Basilicata and Calabria.

Var. *obtusatum* is widely spread through Italy, as far south as Calabria and Sicily, and is common in the Balkan peninsula, extending from Croatia, through Istria, Dalmatia, Bosnia, Servia, and Herzegovina to Roumelia, reaching its most southerly point in the Pindus range. In the Caucasus, according to Radde,[3] it is confined to the province of Talysch, where it grows at elevations between 1500 and 5000 feet. It is also found in Algeria and Morocco.   (A. H.)

CULTIVATION

This species was introduced, according to Loudon, from Corsica in 1752, and though little known in general cultivation and rarely found in nurseries, has apparently a first-rate constitution, and is perfectly hardy. It ripens fruit in this country; and I have raised plants from seed sent me in 1901 by the Earl of Ducie, which grew the first season as fast as a sycamore, and are now about 10 feet high. It is one of the first of the maples to come into flower, early in March at Kew; and when in full flower the tree has a most handsome appearance. Its leaves colour nicely in autumn; and this maple is well worth a place in pleasure grounds, where it is not particular about soil, if this is well drained.

[1] Rouy et Foucaud, *Flore de la France*, iv. 150 (1897).   [2] *Essai Géog. Roy. Naples*, 81 (1827).
[3] *Pflanzenverb. Kaukasusländ*, 184 (1889).

REMARKABLE TREES

The only trees which we know to exceed 50 feet in height are two at Arley Castle, which Mr. Woodward measured in 1905—one 57 feet by 5 feet 1 inch at 5 feet from the ground, and the other 56 feet by 7 feet near the base, where it divides into two stems. The one which I figure (Plate 189) is growing in Sir Hugh Beevor's park at Hargham, Norfolk, and when I saw it in 1905 was 45 feet by $7\frac{1}{2}$ feet with a bole of about 7 feet. There is a tree in Kew Gardens, not far from the Director's office, which is 45 feet high, by 5 feet 10 inches in girth at two feet from the ground, dividing at four feet up into four or five stems. Lord Ducie's tree at Tortworth, not more than forty or fifty years old, is 40 feet by 6 feet· There is a younger one almost as large at Grayswood.

In Scotland the only large one we know is at Smeaton-Hepburn, East Lothian, measured by Henry in 1905 as 45 feet high by 6 feet 3 inches in girth.

At Glasnevin, Dublin, a tree measures 46 feet by $7\frac{1}{2}$ feet; and another at Glenstal, near Limerick, was 47 feet by 4 feet in 1905.

TIMBER

The wood is said by Mouillefert[1] to be like that of the sycamore, but pinkish or pale red in colour, closer in the grain, heavier, and more lustrous, and is esteemed in France by turners, cabinetmakers, and wheelwrights. (H. J. E.)

ACER MONSPESSULANUM, MONTPELLIER MAPLE

Acer monspessulanum, Linnæus, *Sp. Pl.* 1056 (1753); Loudon, *Arb. et Frut. Brit.* i. 427 (1838); Willkomm, *Forstliche Flora*, 769 (1887); Mathieu, *Flore Forestière*, 43 (1897).
Acer trifolium, Duhamel, *Traité des Arbres*, i. t. 10 (1755).
Acer trilobatum, Lamarck, *Encycl.* ii. 382 (1786).
Acer trilobum, Moench, *Meth.* 56 (1794).
Acer rectangulum, Dulac, *Fl. Haut. Pyr.* 242 (1867).

A small tree, in the wild state rarely attaining 40 feet, and often only a shrub. Bark smooth on young trees, ultimately fissuring. Young branchlets glabrous, green, becoming dark brown in the first autumn. Leaves (Plate 207, Fig. 31) coriaceous, small, averaging $1\frac{1}{4}$ inch long and $2\frac{1}{4}$ inches broad, cordate at the base, three lobed; lobes ovate, obtuse; sinuses wide, acute at the base; margin non-ciliate, usually entire, rarely toothed; upper surface dark green, shining, glabrous; lower surface pale or greyish, with tufts of pubescence in the axils at the base, else-where glabrous; petiole without milky sap.

Flowers, appearing before or with the leaves, in small corymbs, at first erect, afterwards pendulous, yellowish-green; pedicels long. Fruit, ripening in

[1] *Essences Forestières*, 206 (1903).

autumn; keys, ¾ inch long, erect, convergent; carpels glabrous; wings brownish or reddish.

The Montpellier maple in the wild state varies in the amount of pubescence on the leaf, the apex of which may be sharp or rounded; and the margin, usually entire, is occasionally toothed.   The keys of the fruit are occasionally so convergent as to cross each other in their upper part (var. *rumelicum*, Grisb.).

A hybrid between this species and *A. Opalus*, known as *A. Peronai*, von Schwerin,[1] has been found in the Apennines at Vallombrosa.   Another hybrid, *A. rotundilobum*, von Schwerin,[2] occurs between this species and *A. Opalus*, var. *obtusatum*, and is mentioned in the synopsis, p. 634.

The Montpellier maple is widely spread throughout Southern Europe, from Portugal to Turkey.   It occurs also in the mountainous regions of Algeria and Morocco, and extends eastwards through Asia Minor to the Caucasus,[3] where it grows at elevations between 3000 and 5000 feet, and to Turkestan.   In France, it is common in the south in dry, rocky situations; and ascends on the west as far north as Poitiers and Niort, and on the east to Gap, Lyons, Grenoble, and Chambery; and, according to Christ,[4] grows at two spots near Bugey, in the southern Jura.   It is found in Germany in the mountains of Rhineland, as far north as Coblenz and in the valleys of the Moselle and Nahe rivers; and also grows at Würzburg in Bavaria.   In Switzerland, it is wild near Geneva, in the Jura, at Fort de l'Ecluse.[5]   In Austria, its northern limit extends from the Southern Tyrol, through Carinthia, Carniola, Istria, and Croatia to Banat; and the tree is spread southward through the Balkan peninsula to Greece.

This species has been found[6] in the fossil state in England, in interglacial deposits at Stone, Hants, and Selsey, Sussex.

It was introduced into England, according to Loudon, in 1739, and in the southern counties thrives very well, ripening its fruit perfectly, and attaining a larger size than any wild trees recorded in Southern Europe.

There are two fine trees in Ricksmansworth Park, Herts, growing in a good loamy soil overlying chalk, which, in 1904, measured 50 feet in height by 8 feet 3 inches (Plate 190) and 45 feet by 8 feet 1 inch.   There are two good specimens in the grounds behind the Herbarium at Kew, which were covered with flowers on April 1, 1907.   These are about 45 feet high, and girth respectively 5 feet 10 inches and 5 feet.   There are also trees of a considerable size at Oxford, Fulham Palace, and Bicton.   At Ewelme Rectory, Wallingford, there is a wide-spreading tree, which the Rev. Canon Cruttwell informs us is 36 feet high, and 9 feet in girth near the base.   At Arley Castle, near Bewdley, there is a tree 35 feet high, by 6 feet 9 inches in girth at one foot from the ground, above which it divides into two limbs.

In the Edinburgh Botanic Garden a tree measures 34 feet by 4 feet 2 inches.

In the Jardin des Plantes at Paris there is a fine specimen, which I measured

---

[1] *Mitt. D. D. Gesell.* 1901, p. 59.

[2] *Ibid.* 1894, p. 50.

[3] Radde, *Pflanzenverb. Kaukasusländ.* 184 (1899).

[4] Christ, *Flore de la Suisse*, Suppl. 64, 65 (1907).

[5] Christ, *op. cit.* 466.

[6] Reid, *Origin Brit. Flora*, 113 (1899).

in 1904 as 60 feet by 8 feet. At Grignon,[1] a tree resisted without injury the low temperature of − 23° cent. in the winter of 1870-1871 ; and in 1879-1880, when the thermometer fell to − 26° cent., it only lost a few of its branches. (A. H.)

ACER INSIGNE

Acer insigne, Boissier et Buhse, *Aufzähl. Transkaukas.*, 46 (1860) ; Boissier, *Flora Orientalis*, i. 947 (1867) ; Masters, *Gard. Chron.* x. 189, f. 24 (1891).

A large tree ; bark of young stems dark grey, smooth, and marked with longitudinal whitish lines. Young branchlets glabrous, green, becoming dark reddish in autumn. Leaves (Plate 206, Fig. 18) resembling those of *A. Pseudoplatanus* in form and size, scarcely exceeding 6 inches wide and 7 inches long, but usually with somewhat shorter lobes, acute or acuminate ; serrations and teeth more rounded than in the sycamore ; under surface pale in colour, scarcely glaucous, with loose white or brown pubescence, dense along the sides of the primary and secondary nerves, forming tufts in the axils, and scattered over the surface between the nerves.

Flowers, very distinct from those of *A. Pseudoplatanus*, in erect many-flowered terminal corymbs, appearing with the leaves, small, greenish ; bracteoles minute, about $\frac{1}{50}$ inch ; filaments glabrous. Fruit, ripening in autumn ; keys $1\frac{1}{2}$ to 2 inches long ; carpels brown, pubescent on the upper side ; wings broad, divergent at an angle of 45°.

In the absence of flowers or fruit, this species is best distinguishable from the sycamore by the buds, which are long and sharp-pointed, with eight to ten external scales, ciliate in margin and with a tuft of pubescence at the tip ; lateral buds plainly stalked, arising from the twigs at an acute angle ; opposite leaf-scars not joined round the twig, which is glabrous throughout.

Boissier considered that there were two forms of the species growing wild :— var. *velutina*, with leaves velvety pubescent beneath ; and var. *glabrescens*, with glabrous leaves. Bornmüller,[2] who studied the tree, while collecting in Persia, is of opinion that these varieties are unstable, as the amount of pubescence is variable ; and all the specimens in the Kew herbarium are more or less pubescent.

(A. H.)

Acer insigne was discovered by Buhse in the eastern Caucasus, in the mountains of Talysch, where, according to Radde,[3] it is common in the forests from sea-level to 2000 feet altitude, and grows to a large size, developing a wide crown of foliage on good soil, and thriving best in moist situations. It is also recorded by Radde from the valley of the Alasan river in the central Caucasus. It grows

---

[1] Mouillefert, *Essences Forestières*, 214 (1903). Cf. also *Actes Premier Congrès Internat. Bot.* 385 (1900), where it is stated that this species sustained at Paris without injury the severe winters of 1879-1880 and 1890-1891.

[2] *Bull. Herb. Boissier*, v. 643 (1905). Bornmüller recognises three varieties, based on the shape of the leaf :—*typica*, Bornmüller, lobes of the leaf acute or acuminate ; *obtusiloba*, Freyn et Sint., *Bull. Herb. Boissier*, iii. 843 (1902), lobes obtuse ; and *longiloba*, Bornmüller, lobes three, elongated, acuminate.

[3] *Pflanzenverb. Kaukasusländ.* 184 (1899).

wild in woods in the provinces of Ghilan and Astrabad in Northern Persia, and is often planted along roads; and in Teheran is cultivated in gardens.

This species was introduced by van Volxem, at the same time as *A. Volxemi*. Dr. Masters received in 1877 from van Volxem three plants, one each of *A. insigne*, *A. Volxemi*, and *A. Trautvetteri*, which he planted in his garden at 9 Mount Avenue, Ealing. His specimen of *A. insigne* first flowered in 1889 and 1890, and is now a very fine tree, 43 feet high and 3 feet 7 inches in girth, as measured by Henry in October 1907. It has a good straight bole, with slender ascending branches, and produces fruit regularly. Two seedlings near it are about three feet high. There are two small trees of *A. insigne* in the collection at Kew, which are less striking in appearance than *A. Volxemi*; but are apparently growing as fast.

At Colesborne, plants raised from seed, received from Lagodechi in the central Caucasus, under the name of *A. Trautvetteri*, and sown in December 1902, are now about 5 feet high.　　　　　　　　　　　　　　　　　　　　　　　　　　　(H. J. E.)

ACER VOLXEMI

Acer Volxemi, Masters, *Gard. Chron.* x. p. 18, figs. 1, 2, and p. 188 (1891).
Acer Van Volxemi, Masters, *Gard. Chron.* vii. 72, fig. 10 (1877).
Acer insigne, Boissier et Buhse, var. *Van Volxemi*, Pax, in Engler, *Jahrb.* xvi. 395 (1892).

This tree is referred by Schneider and von Schwerin to *A. insigne*, var. *glabrescens*, Boissier; but it differs considerably from *A. insigne*, and is either a distinct species or is possibly a hybrid between *A. insigne* and *A. Trautvetteri*, resembling the former more in the shape of the leaves and the latter in the structure of the flowers.

In this species, the leaves are extremely large, often 10 inches wide and 9 inches long, resembling those of the sycamore on a large scale, pale and glaucous beneath, with white pubescence along the sides of the primary and secondary nerves, forming axil-tufts, but not scattered over the surface. The buds resemble those of *A. insigne*; but the twigs differ in being pubescent at the nodes and on the upper edges of the leaf-scars. The flowers resemble those of *A. Trautvetteri*, having long bracts and bracteoles.

This tree is little known in the wild state, the only account[1] being that of van Volxem, who collected seed of it some years before 1877, in the valley[2] of a tributary of the Kura, above the military station of Lagodechi, on the southern slope of the central Caucasian chain. According to van Volxem, " it is a very large tree, very distinct from *A. Pseudoplatanus* in its larger size and its paler green colour, by which it is recognisable hundreds of yards away. The winged fruits are also smaller. It grows intermixed with *A. Pseudoplatanus* in the same forests, but with no intermediate forms, hence it is not a local form, nor would a mere variety remain distinct

---

[1] *Gard. Chron.* vii. 72 (1877). Van Volxem's specimens from the Caucasus, from which Dr. Masters drew up his description, cannot now be found; and the species has apparently not been collected by any one except van Volxem.

[2] Van Volxem names this tributary the Yora (or Jora); but this river lies much to the southward of Lagodechi, the Alasan river intervening.

in the wild state. It does not grow at so great a height on the Caucasus by at least 1000 or 1500 feet, so far as I was able to observe. The form of the tree is more columnar. The light green colour of the leaves makes the difference between the two conspicuous and remarkable. The colour of the bark and the shape of the buds are different." (A. H.)

Introduced from seeds collected by van Volxem in the locality mentioned above, a tree, sent to Dr. Masters in 1877 and planted in his garden at Ealing, grew very rapidly, producing splendid large foliage, silvery white on the lower surface. This tree is still living, but has been headed down, as there was no room for it to develope, and is now only about 15 feet high and 1 foot 10 inches in girth. It has not borne fruit. Dr. Masters told me that all efforts to propagate it by grafts, cuttings or layers failed, though tried by some of the leading nurserymen; and he considered this tree, which he watched from 1877, to be the fastest-growing and the noblest of the maples.

A. Volxemi flowered for the first time in 1891, in Belgium; and its distinctness from the sycamore was then clearly established. A small tree at Kew, now about 20 feet in height, has flowered several times and produced fruit. There are also healthy young trees at Frensham Hall, and in the garden of Mr. Chambers at Grayswood, both of which places are near Haslemere.

M. E. Louis, of Simon Louis Frères, the well-known nurserymen at Metz, informed me in a letter dated October 1902, that he cultivates the true *A. Volxemi*, which is sometimes erroneously called *A. Trautvetteri*.

In November 1902, through the kindness of the Grand Duke Nicholas Mikhailovitch of Russia, I obtained a quantity of fresh seed of this species, as well as of *A. insigne*, from Lagodechi, the original locality; and have raised a number of plants from them. These grow rapidly, but have not as yet ripened their autumnal growths well, and in consequence are rather bushy. The tree may, however, be considered perfectly hardy, and is well worth growing on account of its rapid growth and splendid foliage.

Acer insigne, var. *Wolfi*, von Schwerin,[1] raised from seeds sent from the Caucasus by Herr Wolf of St. Petersburg, is apparently, from the description, a variety of *A. Volxemi*, distinguished by the very large leaves, perfectly glabrous and deep purple in colour beneath. (H. J. E.)

ACER TRAUTVETTERI

Acer Trautvetteri, Medwedjeff, ex Trautvetter, *Act. Hort. Petrop.* vii. 428 (1880); Wolf, *Gartenflora*, xl. 263, figs. 58-61 (1891).

Acer insigne, Nicholson, *Gard. Chron.* xvi. 75, f. 14 (1881), and J. D. Hooker, *Bot. Mag.* 6697 (1883). (Not Boissier and Buhse.)

A tree attaining, in the Caucasus, 50 feet in height and 6 feet in girth; bark grey, smooth. Young branchlets glabrous, green, becoming dark red in the first

[1] *Mitt. Deut. Dendr. Ges.* 1905, p. 210.

autumn or winter.   Leaves (Plate 206, Fig. 19), about 6 inches long and 8 inches wide, cordate at the base, deeply five-lobed; lobes, oblong or ovate, acuminate at the apex, each with three or four small teeth on the margin, which is neither serrate nor ciliate; sinuses, reaching two-thirds the length of the leaf, acute at the base; upper surface dark green, shining, glabrous; lower surface light green, glabrous, except for conspicuous tufts of reddish-brown pubescence in the axils of the primary and secondary nerves; petioles without milky sap.

Flowers, appearing with the leaves, in erect, long-stalked corymbs; bracts and bracteoles conspicuous, $\frac{1}{8}$ inch long; filaments glabrous, ovary pubescent.   Fruit, ripening in autumn; keys $1\frac{3}{4}$ inch long, narrowly divergent; carpels scurfy pubescent when young, glabrous when mature; wings broad.

The leaves are variable as regards the depth of the sinuses, being described by Medwedjeff as either five-lobed or five-partite; and the description above applies to the deeply-cut form, which is in cultivation.

The foliage resembles in size and shape some forms of the sycamore, but can be readily distinguished by the margin being simply dentate and not serrate.   The buds are also different.   In winter the terminal buds are ovoid, obtuse, with six outer scales, the lower pair of which are shining, dark red and glabrous, with the middle and upper parts ciliate; lateral buds distinctly stalked, arising at an acute angle; twigs polished, dark red, glabrous.

This species was discovered by Radde[1] in 1864 in the Caucasus at an elevation of 6000 feet, and was at first identified by Trautvetter with *A. platanoides*, which it resembles in no respect.   It is allied to *A. insigne*, and has a more westerly distribution than that species, growing on both sides of the main chain of the Caucasus, but not extending into Talysch or Persia.   It is a tree of high elevations, growing at 6000 to 8000 feet altitude in company with birch and subalpine shrubs, or mixed with *Abies Nordmanniana* on the edges of alpine meadows, and flowers in May.   It ascends in many places to the timber line, and at lower levels is replaced in the forests by the Norway maple.[2]   According to Wolf the tree attains 50 feet in height and 6 feet in girth; but Radde[1] gives the measurement of a tree, probably of this species, which was 120 years old and 62 cubic feet in volume.

This species was raised in van Volxem's nursery from seeds collected in 1866 by Balansa in Lazistan, and for a long time was confused with *A. insigne*,[3] being described and figured under that name in the *Botanical Magazine*.   Van Volxem informed Sir J. Hooker that it was the hardiest of the eighty species and varieties of maple cultivated by him, having withstood without injury the disastrous winters of 1879-80 and 1880-81; and being a late grower, it had never even been nipped by spring frosts.   At Kew, where there are two healthy trees, it is one of the latest maples to come into leaf.   The tree sent by van Volxem to Dr. Masters flowered

[1] *Pflanzenverb. Kaukasusländ*, 108, 175, 225, 310 (1899).

[2] *Ibid*. 245.   Radde speaks of *A. platanoides* and *A. Trautvetteri* growing together in impassable thickets, which are beaten down by the heavy snow.

[3] *A. Trautvetteri* has also been confused with *A. Volxemi*, as in *Gard. Chron.* x. 188, note, and 189 (1891).   Rehder, in *Cycl. Am. Hort.* 15 (1890), agrees with me that the tree, figured in *Bot. Mag.* 6697, is the true *A. Trautvetteri*, though it differs from wild specimens preserved in the Kew herbarium, in having the leaves more deeply cut.

at Ealing on May 23, 1882, but never throve, and has been dead for some years. This species is very rare in cultivation. I saw a young tree, about 20 feet high, in 1906, at Grignon in France, where it exceeds the sycamore in rate of growth.

(A. H.)

ACER RUBRUM, RED MAPLE

Acer rubrum, Linnæus, *Sp. Pl.* 1055 (1753); Loudon, *Arb. et Frut. Brit.* i. 424 (1838); Sargent,
　　Silva N. Amer. ii. 107, tt. 94, 95 (1892), and *Trees N. Amer.* 639 (1905).
Acer coccineum, Michaux f., *Hist. Arb. Am.* ii. 203 (1810).
Acer sanguineum, Spach. *Ann. Sc. Nat.* ser. 2, ii. 176 (1834).

A tree attaining in America 120 feet in height and 15 feet in girth; with ascending branches. Bark of young stems smooth and light grey, becoming on old trunks darker, ridged, and separating on the surface into plate-like scales. Young branchlets green or red, slightly pubescent towards the tip. Leaves (Plate 207, Fig. 27) very variable in size, averaging 3 inches long and broad; either five-lobed, with two very small basal lobes, or three-lobed, the middle lobe the longest; lobes short, triangular, acute or acuminate at the apex; sinuses very shallow, acute at the base; base of the leaf truncate, slightly cordate or shortly cuneate; margin non-ciliate, irregularly toothed, or doubly serrate; upper surface dark green, glabrous; lower surface silvery white, scattered pubescent, without axil-tufts; petioles without milky sap. The leaves turn scarlet or orange in autumn.

Flowers appearing early in spring before the leaves, in few-flowered, umbel-like clusters encircling the branchlets of the previous year; diœcious or monœcious; reddish; pedicels long; petals present; ovary glabrous. Fruit hanging on drooping stalks, ripening in June, and germinating as soon as it falls upon the ground; keys glabrous, about an inch long, at first convergent, afterwards divergent, brown or reddish in colour.

The red maple can only be confused with the silver maple, from which it differs in the ascending branches and in the shape of the leaves, which are usually only three-lobed, always have very shallow acute sinuses, and are less cordate (often truncate) at the base than in *A. dasycarpum*. In winter the twigs are glabrous, reddish; leaf scars very narrow, three-dotted, opposite pairs not united around the stem. Buds small, shortly stalked, reddish; external scales, six to eight, fringed with whitish cilia; lateral buds arising from the twigs at an angle of 45°.

VARIETIES

In addition to the typical form, above described, Sargent admits two well-marked varieties, occurring wild in America.

1. Var. *Drummondii*, Sargent. (*Acer Drummondii*, Hooker and Arnott, *Journ. Bot.* i. 200 (1834).) Leaves three-lobed, with short broad lobes, and covered on

the under surface, like the young branchlets and petioles, with hoary tomentum. Flowers and fruit bright scarlet. The variety is found in deep river swamps of Southern Arkansas, Eastern Texas and Western Louisiana.

2. Var. *tridens*, Wood.[1] Leaves three-lobed at the apex, rounded or cuneate at the base, thick and firm in texture, serrate except towards the base with remote incurved glandular teeth. Flowers sometimes yellow; fruit usually much smaller and rarely also yellow. This variety occurs in the coast region from Southern New Jersey to Southern Florida, and along the Gulf Coast to Eastern Texas.

A large number of varieties, based on trivial characters, are given by Pax and von Schwerin as occurring in cultivation, the most noteworthy of which is var. *sanguineum*, with the leaves deeper green above, bluish-white beneath, and turning a brilliant red in autumn. In cultivated trees in England there are marked differences in the size and shape of the leaves and in the amount of pubescence on their under surface; but these differences are not worth naming. Columnar and bushy rounded forms are known; and a pendulous form is also mentioned in the Kew Hand List, which seems, however, to be a form of *A. dasycarpum*.

(A. H.)

DISTRIBUTION

In America this tree is one of the commonest and most widely distributed, extending from about lat. 49° N. in Quebec and Ontario, south to Florida and west to Wisconsin, Iowa, and the Trinity river in Texas; abundant in the Mississippi valley, and attaining its largest size on the lower Ohio and in the Wabash valley, where Ridgway measured a tree 108 feet by 15 feet, with a clean bole 60 feet long, and says that larger trees could be found. In New England it grows abundantly in swamps and low ground, and is usually a tree of no great size, so far as I have seen. Emerson records no large trees, while Michaux says that he nowhere saw it larger than in the swamps of New Jersey and Pennsylvania, where it is often 70 feet high and 3 to 4 feet in diameter. It is the earliest tree in flower and was nearly over in the middle of May, near Boston, when the leaves were partly developed. Emerson says that they vary remarkably in size and shape, being sometimes broad and five-lobed, sometimes long and narrow, and are liable to become of a scarlet, crimson, or orange colour at all seasons, sometimes at midsummer, long before other trees have changed colour. He thinks that the frost has little to do with the autumnal coloration of leaves, and that the greater intensity of the light and transparency of the air is the reason why the leaves of trees usually turn so much more brilliant in colour in America than in Europe.

A fastigiate tree of this species is illustrated in *Garden and Forest*, vii. 65 (1894), where it is erroneously[2] called a sugar maple. It grows in the grounds of Mrs. Leavitt at Flushing, New York, and is 80 feet high. Sargent adds that only two other American trees, the tulip tree and the Robinia, are known to have produced forms with fastigiate branches.

---

[1] Cf. Rehder, in *Rhodora*, ix. 116 (1907).
[2] Mr. W. A. Stiles, in a letter to Kew, dated March 28, 1894, corrects this error.

CULTIVATION

This is the commonest in cultivation of all the American maples, except *A. Negundo*, and was the first to be introduced, having been cultivated by Tradescant as long ago as 1656. Miller says that a tree produced seed in his time from which plants were raised in the Bishop's garden at Fulham; and, according to Loudon, one of these in 1793 was 40 feet by 4 feet 3 inches, but was dead before 1809. It was often confused with the silver maple, and even Loudon says that they are only varieties of one species, though he treats them under separate names. No one, however, who has seen them in their native country could doubt their distinction, which was first established by Linnæus.

The red maple is perfectly hardy everywhere in Great Britain, but requires considerable summer heat and a good soil to bring it to any size. On dry sandy soils it is a stunted tree of no beauty. Its seed, like that of the silver maple, ripens early and must be sown at once, but Loudon says that in his time it was propagated by layers, which, coupled with insufficiency of moisture in the soil, may account for the rarity of fine specimens.

I have raised seedlings this year from seeds sent me from Arley by Mr. R. Woodward in July, when he found them germinating freely below the parent tree; and Mr. Knowles, gardener to H.R.H. the Duke of Connaught, tells me that he has found self-sown seedlings at Bagshot Park.

REMARKABLE TREES

At no place in England, so far as we know, are there so many fine red maples as at Bagshot Park, the seat of H.R.H. the Duke of Connaught. When I visited this place on May 22, 1907, the fruit was so abundant on trees in an open wood that it gave them quite a red appearance. The largest of these that I measured was 82 feet by 9½ feet, with a bole about 20 feet long (Plate 177). There is another on the bank of the lake at Claremont which measures about 75 feet by 9½ feet.

At Whitton, near Hounslow, there is a large tree, probably 150 years old, near the group of *Taxodium distichum*, in ground which has moisture beneath, and in 1904 it measured 80 feet by 8 feet 5 inches, but as this tree is not mentioned by Loudon, it may not be so old as we think, though decay has already commenced (Plate 192). At Walcot there is a tree which in March 1904 was in flower, and measured 68 feet by 6½ feet. At Arley Castle there is a fine tree with mistletoe growing on it, which produced seed freely in 1907, and measures about 60 feet by 7½ feet.

In a wood south of Virginia Water in Windsor Park, Henry measured, in 1906, a tree 80 feet by 6 feet 2 inches; and at South Lodge, Enfield Chase, there is a tree which was 50 feet by 6 feet 7 inches in 1904.

A variety under the name of *globosum*, which I saw growing in an ornamental plantation in Silk Wood, Westonbirt, Gloucestershire, was conspicuous among all other trees for its brilliant scarlet leaves and upright habit, in October 1907.

In Scotland it grows well as far north as Brahan Castle, Ross-shire, the seat of Col. Stewart Mackenzie of Seaforth, where in 1907 I measured a tree about 50 feet by 12 feet 2 inches; and at Gordon Castle an old tree at the west end of the holly bank, was in the same year 55 feet by 6 feet 2 inches. There is also a good-sized tree close to the lodge at Moncreiffe House, near Perth, which I believe to be a red maple.

In Ireland Loudon mentions one at Woodstock, which at 60 years old was 50 feet high, but Henry could not find it now living.   (H. J. E.)

ACER DASYCARPUM, SILVER MAPLE

Acer dasycarpum, Ehrhart, *Beit.* iv. 24 (1789).
Acer saccharinum,[1] Linnæus, *Sp. Pl.* 1055 (1753); Sargent, *Silva N. Amer.* ii. 103, t. 93 (1892), and *Trees N. Amer.* 638 (1905).
Acer eriocarpum, Michaux, *Fl. Bor. Am.* ii. 203 (1803); Loudon, *Arb. et Frut. Brit.* i. 423 (1838).

A tree attaining in America 120 feet in height and 12 feet in girth, the stem usually dividing at a short distance from the ground; ultimate branches pendulous, long, and slender. Bark of young trees smooth and grey, on old trunks dark in colour, ridged, and separating on the surface into thin loose scales. Young branchlets glabrous, green, becoming shining brown in the first autumn. Leaves (Plate 207, Fig. 28) about 5 inches long, 6 inches wide, usually cordate at the base, five-lobed; basal lobes well developed; lobes long acuminate at the apex, with serrated triangular teeth or lobules; sinuses rounded at the base and concave on the sides, extending halfway or more to the base of the blade; upper surface glabrous, shining green; lower surface silvery white, scattered pubescent, without axil-tufts; petiole without milky sap. The leaves turn yellow in autumn.

Flowers appearing before the leaves, earlier even than those of *A. rubrum*, greenish yellow, diœcious or monœcious, in crowded fascicles on the branchlets of the previous year; pedicels very short, petals absent, ovary pubescent. Fruit on slender drooping stalks, ripening in America in May or June, earlier than that of *A. rubrum*, and germinating as soon as it falls; keys woolly when young, ultimately glabrous, widely divergent, pale brown, $\frac{3}{4}$ to $1\frac{1}{2}$ inch long.

The silver maple can only be confused with the red maple, and the marks of distinction are given under the latter species. In winter, the twigs are indistinguishable from those of *A. rubrum*.

---

[1] This name, which was first given to the silver maple by Linnæus, was subsequently transferred to the sugar maple by Wangenheim and remained in universal use for the latter species during many years. In 1889, Sargent, in *Garden and Forest*, ii. 364, re-established it as the name for the silver maple, and he has been followed in this by most American botanists and foresters. The usage, however, of *A. saccharinum* for the silver maple, and of *A. saccharum* for the sugar maple, is confusing; and we have adopted *A. dasycarpum* for the former, as being a name long in use, and one which has never been applied to any other species.

VARIETIES

The most remarkable are :—var. *laciniatum*, leaves deeply divided into narrow lobes; and var. *tripartitum*, in which the division of the leaves is carried to the midrib. Various intermediate forms, as regards the shape of the leaf, have also received names, which are not worth recognition. Variegated forms are also known in cultivation. (A. H.)

DISTRIBUTION

The silver maple extends from New Brunswick through Southern Ontario to Eastern Dakota, Nebraska, Kansas, and Indian Territory on the west, and southward to Florida; but is rare near the Atlantic coast and on the higher Alleghany Mountains. Sargent gives an excellent article [1] on this species, with an illustration of a tree growing in the open near Boston, and says that it is an inhabitant of low sandy river banks, and grows to its greatest size on the tributaries of the lower Ohio, where it sometimes attains 120 feet in height and 9 to 12 feet in girth. Ridgway measured one in the lower Wabash Valley, which was 118 feet by 14 feet. Michaux says that near Pittsburg, trees of 12 to 15 feet in girth were common on the bank of the river, sometimes alone and sometimes mixed with the willow. Emerson states that in Massachusetts he measured a tree [2] 12½ feet in girth in a meadow near Northampton, and that another near Lancaster was 16 feet 8 inches round at 6 feet from the ground.

In Canada, where I saw it on the sandy banks of the Gatineau River, near Ottawa, close to its northern limit, it was no larger than in England, but the colour of the leaves was more beautiful than it ever becomes in cultivated trees with us, as is usual in the case of deciduous trees in America. In the open situations which it usually frequents, it is a wide-spreading tree; and Michaux says that it forms a more spacious head than any other tree that he knew.

The fruit, if not destroyed by spring frost, which often happens, ripens in a few weeks after the time of flowering; and if it falls on moist open ground, germinates at once, and sometimes produces plants nearly a foot high before winter. Sargent suggests that this rapid ripening, which is peculiar to the red and silver maples, is a provision of nature for their preservation in situations where the seed, if it ripened in autumn, like other maples, would be water-logged by floods and lose its vitality.

Sargent considers it a valuable tree for ornamental planting, only in deep moist meadow land, or by the banks of streams, where it can spread its long and graceful branches and show its brilliant foliage. This is quite borne out by the specimens which I have seen in England. It is one of the favourite trees for planting in many of the northern cities of the United States.

[1] *Garden and Forest*, iv. 133 (1891).

[2] Fifty-two years later, in 1890, we learn from a note in *Garden and Forest*, iii. 36, that this tree was 17 feet 4 inches in girth at the same height, having made an annual increase in girth of more than an inch. Though the trunk was partly hollow and some of the branches were gone, the tree was still growing vigorously and might live for many years more.

CULTIVATION

This tree was introduced by Sir Charles Wager in 1725, and is still known in some catalogues and gardens as Wager's maple. It was formerly commoner in cultivation than now, and deserves a place in every garden where a suitable situation can be found; being perfectly hardy at least in the southern half of England, and one of the most beautiful-leaved trees we have both in spring and autumn. The seed is difficult to procure, as it ripens so early in the season—though Loudon says it ripens in England, I have never seen any,—and according to Dawson [1] should be sown directly it is ripe, and shaded and watered during the first summer. It is, however, easy to procure young trees by post from America in autumn, and such will I believe make better trees than the grafted ones which are usually sold in Europe.

Dr. Masters [2] says that this is one of the best and most elegant trees for planting in towns; but that he never saw a tree of the kind of such proportions as one on the ramparts at Ypres in Belgium.

REMARKABLE TREES

The tallest tree of this species I have seen in England is in a shrubbery at Cobham Hall, Kent, where, however, it has been too much crowded to develop its natural habit and beauty. It is about 77 feet by 9 feet, with a clean bole of nearly 30 feet. There are some trees on Ashampstead Common, Berks, of which the largest is about 65 feet by 8 feet.

At Rickmansworth Park, I measured a tree 60 feet by 12 feet which grows on the banks of the stream and spreads to a diameter of 30 paces. At Barton, Suffolk, there is a very spreading tree on a lawn, forked close to the ground, and 68 feet high, with two main stems 9 feet 3 inches and 6 feet 3 inches in girth. At Hampton Court, Herefordshire, there is an old tree in the lower park on rich alluvial soil, which, when measured by Mr. Hogg in 1881, was 65 feet by 11 feet 5 inches. When I saw it in 1905 it had lost its top and was decaying, though it had increased in girth to 12 feet 3 inches. At Woburn, in rich damp soil near a pond, there is a handsome spreading tree, with a short bole, 53 feet high by 9 feet 5 inches; and at Syon a tall slender tree, 58 feet high and only 3 feet 3 inches round, has smooth bark like that of a beech. At Arley Castle there is a fine tree 61 feet by 4 feet which, according to Mr. R. Woodward, was only planted in 1877. Smaller and younger trees are found in many gardens; but in the North and West of England we have seen none worthy of record, and Loudon mentions no trees equal to those above mentioned.

In Scotland, Mr. W. Middleton Campbell has measured a tree at Camis Eskan, near Helensburgh, which is 58 feet by 11 feet 1 inch.

In Ireland, Henry has seen no large silver maples, and one at Glasnevin, 45 feet high by $6\frac{1}{2}$ feet in girth, is badly shaped and not thriving.

[1] *Trans. Mass. Hort. Soc.* 1885, p. 153.   [2] *Gard. Chron.* xxxvi. 267 (1904).

TIMBER

All accounts agree in stating that the timber is softer, lighter, and weaker than that of the red or sugar maples, and in consequence is only used as an inferior substitute for these or other similar woods.

The sap of the silver maple is sometimes used to produce sugar in places where the sugar maple is not found, and Michaux says that though the quantity is only half as much, yet the unrefined sugar is whiter and more agreeable to the taste than common maple sugar. (H. J. E.)

ACER SACCHARUM, SUGAR MAPLE

Acer saccharum,[1] Marshall, *Arbust. Am.* 4 (1785); Sargent, *Trees N. Amer.* 632 (1905); Trelease, *Missouri Bot. Garden Report*, v. 88 (1894).

Acer saccharinum, Wangenheim, *Nordam. Holz.* 26 (1787) (Not Linnæus); Loudon, *Arb. et Frut. Brit.* i. 411 (1838).

Acer barbatum, Michaux, *Fl. Bor. Am.* ii. 252 (1803); Sargent, *Silva N. Amer.* ii. 97, t. 90 (1892).

A tree attaining in America 120 feet in height and 12 feet in girth. Bark grey and smooth on young stems, deeply furrowed on old trunks. Young branchlets glabrous, becoming brown in their first summer. Leaves (Plate 206, Fig. 12), about 5 inches long by 6 inches wide, usually five-, rarely three-lobed, cordate at the base; lobes triangular, acuminate, with one or two pairs of sinuate teeth; sinuses rounded and shallow, reaching about one-third the length of the blade; margin non-ciliate; upper surface dark green, glabrous; lower surface pale, dull, with pubescent tufts in the primary and secondary axils, elsewhere either glabrous or more or less pubescent; petiole without milky sap. Leafy stipules,[2] with bases adnate to the petiole, are occasionally developed in var. *nigra* of this species.

Flowers, monœcious or diœcious, appearing with the leaves, arising from terminal leaf-buds and from lateral leafless buds, in nearly sessile corymbs, greenish-yellow; pedicels long, thread-like, pubescent; petals absent; ovary with long scattered hairs. Fruit, ripening in autumn, glabrous; keys about an inch long; wings broad, thin, usually divergent.

This species in the form of the foliage somewhat resembles the Norway maple; but is readily distinguishable by the pale colour of the leaves beneath, and the absence of milky sap in the petioles. In winter, the buds are conical, sharp-pointed, and pubescent, showing externally 8 to 14 scales; lateral buds shortly stalked; opposite pairs of leaf-scars not united around the stem, their upper margins fringed with yellowish hairs.

---

[1] This is the name now adopted by Sargent, by Sudworth, *Check List of Forest Trees of U.S.* 91 (1898), and by other American botanists and foresters. *Acer saccharinum*, Wangenheim, is a later name, and must be dropped, especially as *Acer saccharinum*, Linnæus, is now commonly used in America for another species, the silver maple.

[2] Gray, *Amer. Naturalist*, vi. 767 (1872), and vii. 422 (1873); and Sargent, *Garden and Forest*, iv. 148 f. 27 (1891).

VARIETIES

The sugar maple is very variable in the wild state, and certain varieties of it are now recognised as distinct species by Sargent in his latest book on American trees. Two of these—*Acer floridanum*, Pax, a moderate-sized tree, growing in the Southern States and in Texas and Mexico, and *Acer leucoderme*, Small, a low tree ranging from North Carolina to Arkansas—would probably not be hardy if introduced, and need not be further mentioned by us. *Acer nigrum*, Michaux, now considered by Sargent to be a distinct species, is best treated as a variety of *A. saccharum*, and is to be carefully distinguished from var. *Rugelii*, with which it has been confused.

1. Var. *nigrum*, Britton, *Trans. N.Y. Acad. Sci.* ix. 10 (1889).

   Acer nigrum, Michaux f. *Hist. Arb. Am.* ii. 238, t. 16 (1810); Sargent, *Garden and Forest*, 1891, p. 148, f. 27, and *Trees N. Amer.* 634 (1905).

Leaves green beneath, cordate at the base, with the basal sinus closed by the approximation or overlapping of the lobes; sides of the blade drooping; lobes usually three, occasionally five; acute, entire or obtusely toothed. Bark of old trees deeply furrowed, sometimes almost black. Young branchlets orange-coloured.

This variety, according to Sargent, is widely distributed, extending from Ontario and the valley of the St. Lawrence, near Montreal, southward to Virginia and Kentucky, and westward through Michigan, Indiana, Illinois, Iowa, and Missouri to South Dakota and Kansas. It is comparatively rare near Montreal and Vermont, becoming more abundant farther west, almost replacing the type in Iowa, and the only form in South Dakota. It was first noticed by the younger Michaux on the banks of the Genesee River in New York, where it still forms a forest of considerable size.

Loudon states that the black sugar maple was introduced in 1812; but it is now very rare in cultivation. Var. *monumentale*, Temple, a tree of upright columnar habit, occasionally seen in botanic gardens, is a form of var. *nigrum*.

2. Var. *Rugelii*, Rehder, in Sargent, *Trees N. Amer.* 633 (1905).

   Acer barbatum, Michaux, *Pl. Bor. Am.* ii. 252 (1803).
   Acer barbatum, var. *nigrum*, Sargent, *Silva N. Amer.* ii. 99, t. 91 (1892).
   Acer Rugelii, Pax, in Engler, *Bot. Jahrb.* vii. 243 (1886).

Leaves pale beneath, papery in texture, three-lobed, entire or with short obtuse teeth. This is the common and frequently the only form of the sugar maple in the region from North Carolina and Georgia to Missouri; and is occasionally met with as far north as Michigan and Prince Edward's Island, leaves of this form sometimes appearing on the upper branches of trees, which bear on their lower branches typical leaves of the ordinary form of the species. This variety does not appear to be in cultivation in England.   (A. H.)

DISTRIBUTION

The sugar maple is one of the most widely and generally distributed trees in Eastern North America. The northern limit of its range on the Atlantic coast is Southern Newfoundland. It extends through Canada and the Northern States southwards along the Alleghany Mountains to Northern Georgia and West Florida, and westward along the valleys of the St. Lawrence and the Saguenay, by the shores of Lake St. John and the northern borders of the Great Lakes to the Lake of the Woods, and in the United States to Minnesota, Nebraska, Eastern Kansas, and Eastern Texas. It is common in all these regions, growing especially on rich uplands mixed with ashes and hickories, white oak, wild cherry, black birch, yellow birch, and hemlock, and often in the north forming the principal part of extensive forests. The undergrowth in some of the forests near the northern border of the United States is often composed almost entirely of young sugar maples, which grow readily under the dense shade of other trees. The type is more prevalent in the north—var. *Rugelii* and var. *nigra* in the central States, while var. *leucoderme* and var. *floridanum* appear to be the only forms found in the south.

Much of the splendour of the northern forest in early autumn is due to the abundance of the sugar maple, which is then unsurpassed by any other tree in brilliancy of colouring, the foliage turning to shades of deep red, scarlet, orange, or clear yellow.

A figure of an unusually large tree, showing the habit which it assumes when in the open, is given in *Garden and Forest*, v. 380 (1892). It grows on the farm of Mr. L. Parker, forty-five miles east of Cleveland, Ohio, and measures $13\frac{1}{2}$ feet in girth at 2 feet from the ground, with very large limbs spreading over an area 100 feet in diameter. It has been tapped annually without any apparent ill-effects, and yields each year three gallons of syrup. Another illustration in *Garden and Forest*, iii. 167 (1890), of a tree exposed on a stony hillside in New Hampshire is of a very different type, and shows the habit of an adult tree which has lost the narrow upright form of growth it usually has when young.[1]

REMARKABLE TREES

Though introduced at a very early period (the date is given by Loudon as 1735, on whose authority we know not), the sugar maple has rarely thriven in England, or, so far as we know, in Europe. The reasons for its failure to grow in this country are as mysterious as in the case of the white oak, the American beech, and other trees of the Eastern States; but it seems a short-lived tree, and seldom attains any considerable size. Loudon mentions several trees of no great age 20 to 40 feet high, and one at Purser's Cross which was 45 feet. But none of these, so far as we can learn, are now living, and some maples which have been reported under this name turn out to belong to other species. We know,

---

[1] The fastigiate tree, supposed to be of this species, is really *A. rubrum*. Cf. p. 672.

however, of several worthy of mention, of which the largest is growing at Park Place, near Henley-on-Thames, in the grounds of Mrs. Noble. This tree is in a thick wood on loamy soil overlying chalk, and if upright would be probably over 60 feet high. It leans, however, very much to one side, where the branches extend as much as twenty paces from the trunk. It has a short bole of 8 feet, which girthed 9 feet 2 inches when I measured it in 1905. There were a few fruits on it, which, however, seemed unlikely to ripen. (Plate 191.)

Sir Hugh Beevor has lately discovered a tree in the grounds of Sir Robert Dashwood at West Wycombe, which is 83 feet high by 5 feet 11 inches in girth. It is nearly dead, being probably killed by mistletoe, as many of the branches show large spindle-shaped swellings, caused by this parasite.

At Arley Castle there is a tree which measured, in 1907, 64 feet high by 4 feet 8 inches in girth. At Dropmore one in a wood is 45 feet by 4 feet 10 inches. There are also small trees at Cornbury Park and at Tortworth. There is a healthy specimen at Syon 55 feet by 5 feet 1 inch, which had some seed in 1905. At Barton, Suffolk, there is an ill-shaped tree much crowded by others, which measures about 50 feet by 4 feet 6 inches.

Mr. Bartlett reports that there are five trees of this species at Pencarrow, and that another at Tredethy, Cornwall, is 50 feet high by 3 feet 5 inches in girth.

The Hon. Vicary Gibbs informs us that at Tyntesfield, Somersetshire, a number of sugar maples are growing, which are about fifty years old. The soil being shallow they have made low heads with very stout lateral branches. He has raised some seedlings from them at Aldenham. Young trees, which I raised from seed gathered near Boston in September 1904, have grown fairly well at Colesborne, and are now about 3 feet high.

TIMBER

The wood of the sugar maple has been well known in commerce for a long period, and at one time the variety of it which is known as bird's-eye maple was very fashionable for furniture and cabinet-making, though it is now little used for first-class work in England.

The best account I know of the varieties of maple wood is in Hough's *American Woods*, i. 50-51, where he says that there are peculiar freaks in the growth of timber as yet unexplained, but of which this is one of the most important from a commercial point of view, as well as one of the most beautiful. They are known as " blister," " bird's-eye "[1] or " pin," and " curly " figures. The first two are almost peculiar to the " hard " or sugar maple. The last is found even more commonly in the red and silver or " soft " maples as they are called in the United States and Canada. The three varieties of figure are often found more or less mixed in the same tree, and it requires much experience to detect their presence in the growing tree.

The blister variety, which is much the rarest, usually has a massive trunk in

---

[1] In bird's-eye maple there is a succession of elevations and depressions in the annual layers of the wood, and Hopkins considers that this is probably due to punctures made in the bark by woodpeckers. Cf. *Garden and Forest*, vii. 373 (1894).

proportion to its top, and on removing a piece of bark the surface of the wood is found to be covered with wart-like swellings. The figure is best on the outside, falling off as the heart is approached. In order to show it, the log is cut on a lathe, which slices off a thin shaving all round, producing what is called a knife-cut veneer. The bird's-eye variety may be detected by small pits in the bark, usually inconspicuous, which correspond to small pits all over the wood, and, like the other, has a head usually small as compared with the trunk.

Maple wood also shows when cut radially a very fine silver grain, which to my eye is almost as beautiful as the other figures. The colour is normally white, but when the trees are old assumes a pink or reddish tint. It is also very much used in its plain form for shipbuilding, flooring, and all purposes where strength, durability, and close texture are required, and is largely imported in the form of prepared blocks, which, when properly fitted and laid, make a durable good floor.

Maple sugar, usually, though not always, the produce of *Acer saccharum*, is derived, by boiling, from the sap which flows from the tree in spring. Though it is in the opinion of most people one of the best kinds of sugar known, and, especially in the form of syrup eaten with buckwheat cakes, is one of the most favourite of American table delicacies; it is so little known to English readers that we do not think it necessary to describe the process of manufacture, which has been given at length by Loudon, Michaux, and other writers on American trees.[1] An article on this subject was published in *Kew Bull.*, 1895, p. 127; and an interesting letter on the domestic uses of maple sugar and maple syrup, by Miss Boyle of Maywood, New Jersey, appeared in *The Garden*, lxv. 152 (1904).   (H. J. E.)

ACER MACROPHYLLUM, Oregon Maple

Acer macrophyllum, Pursh, *Fl. Amer. Sept.* i. 267 (1814); Loudon, *Arb. et Frut. Brit.* i. 408 (1838); Sargent, *Silva N. Amer.* ii. 89, tt. 86, 87 (1892), and *Trees N. Amer.* 628 (1905).

A tree attaining in America 130 feet in height and 15 feet in girth. Bark of old trees thick, rough, deeply furrowed, and broken on the surface into small plate-like scales. Young branchlets glabrous, green, remaining green or becoming dark red in their first winter. Leaves (Plate 205, Fig. 3) very large, averaging 9 inches in breadth and length, deeply and usually narrowly cordate at the base; lobes five, with acute or acuminate apex, and large triangular lobules or teeth; sinuses deep, extending more than half-way the length of the blade, rounded at the base; margin ciliate; upper surface dark green, shining, scattered pubescent; lower surface light green, glabrescent between the nerves, with tufts of white pubescence in the axils; petiole with milky sap.

[1] The most complete account is given in *U.S. Dept. Agric.*, *Forestry Bulletin No.* 59; *The Maple Sugar Industry*, by Fox and Hubbard, Washington, 1905. Cf. also *U.S. Dept. Agric.*, *Forest Service*, *Circular* 95 (1907), which gives notes on the cultivation and economic uses of the sugar maple.

Flowers, staminate and pistillate together, in pendulous racemes, appearing when the leaves are fully grown, bright yellow, fragrant; pedicels slender, pubescent, often branched; stamens nine or ten, filaments pubescent; ovary tomentose. Fruit, ripening in autumn, brown, the carpels covered with long, pale hairs, which extend along the thickened edge of the wing; keys slightly divergent, about 2 inches long. (A. H.)

DISTRIBUTION

This species, which is the largest of the American maples, is confined to the Pacific coast, where it extends from about 55° N. in Alaska to the San Bernardino Mountains of Southern California, but never, according to Sargent, far from the coast or ascending the mountains higher than about 2000 feet. It is the largest deciduous tree in Vancouver Island except *Populus trichocarpa*, and I believe also in Washington and Oregon, though surpassed by some of the oaks in California. It attains its maximum size in the wet and mild climate of Puget Sound, especially in the Olympic Mountains, and grows with the luxuriance of a tropical tree covered with ferns, moss, and climbing plants. The beautiful photograph (Plate 193), for which I am indebted to Mrs. Browne of Tacoma, was taken near Lake Cushman in the Olympic Mountains. I cannot give exact measurements of these trees, but the height was estimated at 130 feet.[1] On Capt. Barkley's farm, north of Duncans, Vancouver Island, I measured several trees of 110 to 120 feet high, and on Swallowfield farm two trees on the banks of a river, of about the same height, one being 12 feet, the other 13 feet in girth. A gigantic spreading tree on the same farm had a swelling butt, no less than 15 paces round at the ground, but of no great height. It grows as a rule in flat meadows with Douglas fir, *Abies grandis*, and *Thuya plicata*, and likes a fairly damp soil. Farther south in the drier country of Oregon, it is smaller; Sheldon says 50 to 90 feet high by 6 to 15 feet in girth. In the dry country of Northern California about Lake Tahoe it becomes a low crooked tree only 8 to 20 inches in diameter. Its large keys are produced very abundantly, and when ripe add to the ornamental appearance of the tree.

CULTIVATION

Discovered by Menzies during Vancouver's voyage to British Columbia, it was introduced into England in 1812. Douglas sent home seeds to the Royal Horticultural Society about 1827, from which we believe the oldest trees in England have grown. But though very easy to raise and a very rapid grower when young, it does not ripen its young wood when quite young, this being often killed back by the frosts of winter, sometimes to the ground; but as the trees get older this failing decreases. Though the tree is hardy, at least as far north as East Lothian,

[1] Mr. F. R. S. Balfour of Dawick tells me that in the deep alluvial soil of the valley above Lake Cushman, this tree attains an immense size, being well sheltered by the steep mountains around. The maples grow here mixed with *Alnus oregona*, *Populus trichocarpa*, and *Thuya plicata*; and though overtopped by the last two, he estimated the maples at over 150 feet high. He also saw it of great size in the Puyallup Valley, and at the mouth of the Nisqually River in Washington; and adds that it is being extensively planted as a shade tree in the towns of the Pacific coast.

it has never become common, and is not often to be had from nurserymen. It is well worth cultivation on account of its large and beautiful foliage, and should be planted in deep moist soil in a warm aspect where it is sheltered by other trees at first. It ripens seeds in England, and I have raised plants from some sent me by the Earl of Ducie in 1900, which are now over 8 feet high.

It does fairly well as a planted tree in the lighter alkali lands of the San Joaquin valley in California, where only a few species will thrive, owing to the nature of the soil.[1]

REMARKABLE TREES

The largest specimen I know of the Oregon Maple is one at Boynton, Yorkshire, which Sir Charles Strickland to the best of his recollection planted himself about sixty years ago. It is not a well-shaped tree, as it is rather crowded, but measures 70 feet high by 6 feet in girth. There is another at Hildenley, probably of the same age, which measures 50 feet by 5 feet and bears fruit. There are several trees in Kew Gardens, the largest of which, near the entrance to the nursery, is 49 feet high by 3 feet 8 inches in girth. At Tortworth there is a tree in a rather exposed situation which is 45 feet by 5 feet 6 inches. At Syon a grafted tree is 50 feet high but only 3 feet in girth, and bore some fruit in 1905. In the Royal Avenue at Bath there is a tree 50 feet by 6 feet 2 inches, but it is not in a thriving condition, the soil being too dry to suit it. At Bicton, now the property of Lord Clinton, one of the most thriving young trees which I have seen grows near the house; and was in 1906 about 50 feet high but only 2 feet 9 inches in girth.

At Smeaton-Hepburn, East Lothian, there is a wide-spreading tree, with a bole of 6 feet, which Henry measured in 1905 as 50 feet high by 7 feet in girth.

At Glasnevin, Dublin, there is a fine tree, which in 1907 measured 51 feet high by 5 feet 4 inches in girth. There are two others about 40 feet high, with wide-spreading branches, growing in the quadrangle inside the main gate of Trinity College, Dublin.

TIMBER

This wood, though unknown in Europe, is equal in beauty, and similar in character to that of the eastern maples, and is more valued than any other native hardwood in British Columbia and Washington. In old trees it is often very well figured, though the figure is larger, bolder, and less regular than in the sugar maple, and the colour not so uniform; some parts of the heartwood being of a rich red brown. Some of the best houses in Victoria are decorated with this wood, that of the Hon. J. Dunsmuir, Lieutenant-Governor of British Columbia, being a good example. If carefully selected and well seasoned it is fit for the finest cabinet-maker's work. I had a bedstead made from it by Messrs. Weiler of Victoria which shows the beauty of the wood very well; and it could be procured in fair quantity if desired, as there are many large trees still standing in accessible places.

(H. J. E.)

[1] Hilyard, *Soils*, 481 (1906).

ACER NEGUNDO, Ash-leaved Maple, Box Elder

Acer Negundo, Linnæus, *Sp. Pl.* 1056 (1753); Sargent, *Silva N. Amer.* ii. 111, t. 96 (1892), and *Trees N. Amer.* 641 (1905).

Negundo aceroides, Moench, *Meth.* 334 (1794).

Negundo fraxinifolium, Nuttall, *Gen. Amer.* i. 253 (1818); Loudon, *Arb. et Frut. Brit.* i. 460 (1838).

A tree attaining in America 70 feet in height and 12 feet in girth; bark deeply fissured into broad rounded ridges. Young branchlets green or glaucous, glabrous. Leaves (Plate 205, Fig. 2) pinnate, turning yellow in autumn. Leaflets, three or five, stalked, ovate or oval, rounded or cuneate at the base, acuminate at the apex, serrate or toothed above the middle, often three-lobed; upper surface bright green and glabrous; lower surface pale green and with slight pubescence on the midrib and nerves; rachis glabrous.

Flowers dioecious, without petals, appearing with the leaves, from buds in the axils of the leaf-scars of the previous season, staminate flowers in fascicles, pistillate in narrow pendulous racemes. Fruit, with narrow acute nutlets, diverging at an acute angle, and thin reticulate, straight or falcate wings.

In winter, the terminal buds are about $\frac{1}{8}$ inch long, acute, with four tomentose ciliate external scales; lateral buds appressed to the twigs, with two outer visible scales; opposite leaf-scars united around the twig, narrowly crescentic, three-dotted, fringed with hairs on the margin.

Varieties

The species, extending over a vast territory, varies considerably in the wild state, the typical form described above occurring in the eastern part of its distribution. Farther west, in Colorado, Utah, and New Mexico, the branchlets and leaves become pubescent; and in California, an extreme form is met with, which is often considered to be a distinct species :—

1. Var. *californicum*, Wesmael, *Bull. Bot. Soc. Belg.* 43 (1890); Sargent, *Garden and Forest*, iv. 481 (1891), *Silva N. Amer.* ii. 112, t. 97 (1892), and *Trees N. Amer.* 643 (1905).

Acer californicum, Dietrich, *Syn.* ii. 1283 (1840).

Negundo californicum, Torrey and Gray, *Fl. N. Amer.* i. 250, 684 (1838).

This is distinguished, according to Sargent, by its darker-coloured bark; buds covered with dense tomentum; short pale persistent pubescence on the branchlets and ripe fruit; leaflets three, larger, more coarsely serrate, and more frequently lobed than in the type, and coated beneath with pale pubescence. As seen in cultivation at Kew, the leaflets are usually five and not three; and in the wild state, 5-foliolate leaves are occasionally met with, as in a specimen in the Kew herbarium collected by Lobb in California. The pubescence on the leaflets beneath is most strongly marked on the midrib and nerves, is whitish in colour, and forms prominent axil-tufts. Cf. Plate 205, Fig. 1. There are specimens of this variety at Kew; and a tree at Grayswood, near Haselmere, was about 30 feet high in 1906, and appeared to be very vigorous and thriving.

A considerable number of horticultural varieties are known :—

2. Var. *variegatum*.[1] Leaves with broad white margin. One of the most popular and most largely grown of all variegated trees. It originated as a chance branch sport in the nursery of M. Fromant at Toulouse in 1845; but remained almost unknown, till 1853, when it was awarded, at a horticultural show at Toulouse, a medal given by the Empress Eugenie.

3. Several other coloured forms are known, as var. *aureo-maculatum*, leaves spotted with yellow; var. *aureo-marginatum*, leaves with yellow margin; and var. *auratum*, leaves yellow.

4. Var. *violaceum*.[2] Young branchlets covered with a glaucous violet bloom.

5. Var. *crispum*.[3] Leaves variously cut and curled. According to Nicholson, this is not nearly so vigorous a grower as the type. (A. H.)

DISTRIBUTION

This species is the most widely distributed of North American maples, extending in its typical form from Western Vermont and Central New York, southward to Northern Florida, and westward to the Rocky Mountains. According to Sargent, it is rare east of the Alleghany Mountains, and is commonest in the basin of the Mississippi, attaining its largest size in the valley of the lower Ohio. The biggest recorded, so far as we know, is one measured by Ridgway in the Wabash Valley, which was 60 feet high by 12 feet in girth. It is one of the few eastern trees, which is quite at home in the dry prairie region, and is found on most of the rivers of the great plains, and on the foot-hills of the Rocky Mountains, where it is usually a stunted and ill-shaped tree or bush, and rarely has a clean or straight stem. Slightly modified, as regards the amount of pubescence on the leaves and branchlets, it occurs in Utah, Colorado, New Mexico, and Eastern Arizona. Var. *californicum* is met with in California in the valley of the lower Sacramento river, in the interior valleys of the coast ranges from San Francisco Bay to about lat. 35°, and in the high cañons of the San Bernardino Mountains; and has been planted, with successful results, on the alkaline lands of the San Joaquin valley.[4]

In Western America it is largely planted for shelter belts.[5] Mr. W. T. Macoun states in a recent number of the *Canadian Forestry Journal*, p. 80 (1907), that in the prairie provinces it is used largely in plantations. It grows rapidly during the first twenty years, and produces a very dense cover, which makes it a bad neighbour for slow-growing trees, but a good nurse for those which, like birch, ash, and American elm, can hold their own in its company.

---

[1] *Gard. Chron.* 1871, p. 1202, f. 275. Cf. also *Rev. Hort.* 1861, p. 268; *Gard. Chron.* 1861, p. 867; *Flore des Serres*, vii. 117 (1867). [2] *Negundo aceroides*, var. *violaceum*, Kirchner, *Arb. Musc.* 190 (1864).

[3] G. Don, in Loudon, *Arb. et Frut. Brit.* i. 460 (1838). [4] Hilyard, *Soils*, 481 (1906).

[5] Cf. *U.S. Dept. Agric. Forest Service*, Circular 86 (1907), which gives an elaborate account of the economic uses of this tree in the United States, with notes on its propagation and cultivation.

CULTIVATION

Acer Negundo was very early introduced into England, being cultivated in the garden at Fulham by Bishop Compton in 1688. According to Loudon, this tree was about 45 feet high and 7 feet 1 inch in girth in 1835.

It is by far the commonest of American maples in cultivation, the variegated form being grown in every nursery and planted extensively in shrubberies and town gardens for the sake of its colour. It grows from seed with extraordinary rapidity, attaining 5 or 6 feet high in 3 years, is apparently at home in every kind of soil, and resists all the extremes of our climate without injury. Though Loudon says that the seed must be sown in autumn, I have found that it will germinate readily when sown as late as June. Trees which came up in a bed of American ash in my nursery in June 1901, are already 15 feet high, and bearing seed freely when only seven years old.

REMARKABLE TREES

Though not often seen as a tree, yet on good soil it seems to attain almost as great size in England as in America. Loudon mentions a tree at Kenwood which was 47 feet high, 35 years after being planted. A tree at Botley, Hants, probably planted by Cobbett, was recorded[1] in 1884 as being 70 feet high by 6 feet 4 inches, but I did not find this when I visited Botley in 1906. The largest, however, that we have seen is at the Mote, near Maidstone, which I found in 1902 to be 53 feet high by 8 feet 4 inches. Henry measured one at Shiplake House, near Henley, 50 feet by 6 feet 3 inches, with a clean bole 16 feet long; and there is a large wide-spreading tree in Kew Gardens near the Director's Office which is about 40 feet high, and measures 6 feet 8 inches in girth. Another in the Oxford Botanic Garden is 4 feet in girth; and a very old tree, with a short trunk and wide-spreading branches, in Mortlock's Garden, behind the Corn Exchange at Cambridge, probably on the site of the old Botanic Garden, is about 30 feet in height and 5 feet 8 inches in girth. Miss Woolward, in 1905, measured a tree in the grounds of the Knowle Hotel at Sidmouth, 38 feet in height and 3 feet 10 inches in girth.

TIMBER

Its timber is very unlike that of other maples, for though in young trees it is whitish, the heartwood of old trees is of a most peculiar colour, purplish red with dark veins.[2] I have never seen it of sufficient size to be useful, though Michaux says it was in his time sometimes used by cabinetmakers in the west; and Sargent states that it is sometimes used for the interior finish of houses, wooden-ware, cooperage, and paper pulp. Small quantities of maple sugar are occasionally made from this tree.

(H. J. E.)

[1] *Woods and Forests*, 1884, p. 316.
[2] I made this note from specimens shown as "Box Elder" at the St. Louis Exhibition, but do not find any confirmation of this in Sargent's or Hough's works.

SEQUOIA

Sequoia, Endlicher, *Syn. Conif.* 197 (1847); Bentham et Hooker, *Gen. Pl.* iii. 429 (1880); Masters, *Journ. Linn. Soc. (Bot.)* xxx. 22 (1892).

Wellingtonia, Lindley, *Gard. Chron.* 1853, p. 823.

Washingtonia, Winslow, *Calif. Farmer*, 1854, *ex* Hooker, *Kew Journ.* vii. 29 (1855).

Gigantabies, Nelson (*Senilis*), *Pinaceæ*, 77 (1886).

Athrotaxis, Baillon, *Hist. Pl.* xii. 39 (1892).

Steinhauera, Kuntze, *Lexic. Gen. Phan.* 533 (1904).

TALL evergreen trees, belonging to the tribe Taxodineæ of the order Coniferæ. Bark thick, of two layers, the outer thick, spongy and fibrous, the inner thin, close, and firm. Branches short and stout; lateral branchlets slender, terete, and deciduous. Buds and leaves different in the two species, the leaves having an undivided fibro-vascular bundle, with a single resin canal beneath it.

Flowers monœcious, solitary, minute, appearing in early spring from buds formed in the previous autumn. Male flowers terminal or in the axils of the uppermost leaves, surrounded at the base by imbricated, ovate, acute, apiculate, involucral bracts; stamens numerous, spirally arranged on an axis; filaments short, dilated into ovate incurved sub-peltate connectives, which bear on their inner surface two to five (usually three) pendulous globose two-valved anther-cells, opening below on the back; pollen simple. Female flowers terminal, the leaves gradually passing into the bracts, which are numerous, spirally imbricated, ovate, keeled on the back, acuminate with either long or short points, and adnate to short thick rounded ovuliferous scales which bear five to seven ovules, at first erect, ultimately becoming inverted.

Cones pendulous, persistent after the fall of the seeds. Scales, formed by the enlargement of the united bracts and ovuliferous scales of the flowers, woody, with deciduous resin-glands, spirally arranged, wedge-shaped at the base, widening at the apex into oblong wrinkled discs, showing a transverse median depression, sometimes tipped by a small spine. Seeds, 5 to 7 under each scale, pendulous, oblong-ovate, compressed, with two lateral wings. Seedlings with four to six cotyledons; primary leaves linear-lanceolate, short-pointed, thin, spreading.

Several fossil species of Sequoia are known, occurring earliest in the Cretaceous period in the holarctic region, becoming very widely spread over Europe, Northern Asia, and North America in Tertiary times. Two living species, inhabiting California, are distinguished.

1. *Sequoia sempervirens*, Endlicher. Coast range of California, and crossing

the boundary line into Oregon. Buds scaly. Leaves on lateral branches linear and in two ranks in one plane. Bracts of pistillate flowers about twenty, usually with short points. Cones ripening in the first season; scales abruptly enlarged into terminal discs.

2. *Sequoia gigantea*, Decaisne. Western slopes of the Sierra Nevada in California. Buds without scales. Leaves all radially arranged, spreading or slightly appressed, ovate or lanceolate. Bracts of pistillate flowers 25 to 30, with long points. Cones ripening in the second year; scales gradually thickening from the base to the apex.

SEQUOIA SEMPERVIRENS, Redwood

Sequoia sempervirens, Endlicher, *Syn. Conif.* 198 (1847); Lawson, *Pinet. Brit.* iii. t. 52 (1884); Sargent, *Silva N. Amer.* x. 141, t. 535 (1896), and *Trees N. Amer.* 68 (1905); Masters, *Gard. Chron.* xix. 556, f. 86 (1896); Kent, Veitch's *Man. Coniferæ*, 270 (1900).

Sequoia gigantea, Endlicher, *Syn. Conif.* 198 (1847).

Sequoia religiosa, Presl, *Epimel. Bot.* 237 (1849).

Taxodium sempervirens, Lambert, *Pinus*, ii. 24 t. 7 (1824); Loudon, *Arb. et Frut. Brit.* iv. 2487 (1838).

Abies religiosa, Hooker and Arnott, *Bot. Voy. Beechey*, 160 (1841). (Not Lindley.)

Schubertia sempervirens, Spach, *Hist. Vég.* xi. 353 (1842).

A tree attaining 340 feet in height, with a slightly tapering and irregularly lobed trunk, occasionally 50 to 75 feet in girth above the enlarged and buttressed base. Bark six to twelve inches thick, divided into rounded ridges two or three feet in width, separating on the surface into long narrow fibrous scales, which on falling display the reddish-brown soft spongy fibro-cellular middle bark. Young trees pyramidal, with slender branches to near the base. Older trees in the forest with stems clean to 75 or 100 feet, the stout horizontal branches above forming an irregular narrow crown. Branchlets slender, green in the first year, gradually becoming afterwards brownish with a thin scaly bark, spreading in two ranks more or less in one plane. Buds solitary, both terminal and in the axils of two or three of the uppermost leaves, surrounded by loosely imbricated ovate acute scales, which remain persistent, dry, and brown at the base of the branchlets.

Leaves of two kinds: (1) on normal lateral branchlets, spreading in one plane in two ranks by a twist on their bases, ¼ to ¾ inch long, linear or lanceolate, ending in short cartilaginous points, slightly thickened on the revolute margins, narrowed at the base, where they become decurrent on the branchlets; upper surface dark green, with a median furrow; lower surface with a green midrib and two conspicuous whitish stomatic bands: (2) on leading branchlets, radially arranged in several ranks, appressed or spreading, about ¼ inch long, ovate or ovate-oblong, with incurved cartilaginous points; upper surface concave with a prominent green midrib and two whitish stomatic bands; lower surface rounded, indistinctly stomatiferous. Lateral branchlets with leaves of the latter kind may exceptionally occur on any part of the tree, and usually cover entire branches at the summit of large trees; while in the case of trees growing at high altitudes they are sometimes spread over the whole of the branches.

Male flowers, $\frac{1}{16}$ inch long, with rounded connectives. Female flowers with about twenty broadly ovate bracts, tipped usually with short points. Cones ellipsoidal, $\frac{3}{4}$ to 1 inch long, $\frac{1}{2}$ inch broad; scales with slender stalks, which enlarge abruptly into discs $\frac{1}{3}$ inch in breadth. Seeds light brown, $\frac{1}{16}$ inch long; wings narrower than in *S. gigantea*.

The cones ripen at the end of the first season, and are freely produced in most parts of the South of England and in Ireland, the first recorded[1] being in 1862 on a tree at Barton, Suffolk, which had been planted in 1847. Fertile seed, however, is very rare, the only instance known to us being on a large tree at Huntley Manor, Gloucestershire, from which Prof. Somerville raised seedlings in 1904. Proliferous cones[2] occur occasionally; and a cone[3] with the upper part ovulate and the lower part staminate has been observed.

The tree suckers from the root,[4] and sends up, when cut, numerous shoots[5] from the stool. Fasciation[6] has been observed in the suckers in the redwood forest in California. Dr. Masters[7] described and figured the peculiar woody excrescences, which are sometimes formed at the base of the stem of young trees, which have been raised from cuttings.

VARIETIES

In the wild state there is some variation, as noticed above, in the occasional occurrence of lateral branches with foliage like that of the leading shoots. Several varieties have been obtained in cultivation :—

1. Var. *albospica* (var. *adpressa*). Tips of young shoots creamy white in colour. Leaves small and dense upon the twigs, resembling those of *Taxus baccata adpressa*.

2. Var. *glauca*. Leaves linear, acute, $\frac{1}{4}$ inch long, glaucous, loosely imbricated, appressed or spreading.

3. Var. *taxifolia*. Leaves broader than in the type.

A tree, pendulous in habit, is growing at Dropmore.

DISTRIBUTION

The redwood occurs on the western slopes, valleys, and alluvial flats of the coast range, from the Chetco river in Oregon to Salmon Creek Cañon, twelve miles south of Punta Gorda in Monterey county, California, and ascends from sea-level to 2000 or rarely 3000 feet. It occupies a narrow strip of country along the sea coast, about 500 miles in length from north to south, and is not found inland beyond

[1] Bunbury, *Arboretum Notes*, 166 (1889).   [2] *Proc. Calif. Acad. Sc.* v. 170, t. 16, f. 3 (1895).

[3] *Bot. Gazette*, xxxviii. 2 (1904).

[4] Two suckers are growing beside a tree, 60 feet high, at Shiplake House, near Henley.

[5] In *Journ. Roy. Hort. Soc.* xix. 432 (1896), it is stated that redwood coppice shoots are believed to have been used for producing hop-poles in Kent; but this must have been an experiment on a small scale, and without any practical value. At Arley, shoots from the stool of a tree, which was felled, made a growth of 4 feet in their first year.

[6] Pierce, in *Proc. Calif. Acad. Sc.* ii. 83 (1901), who also gives an account of peculiar white-coloured suckers, which are often seen in California.   [7] *Gard. Chron.* xi. 372, fig. 53 (1879).

the influence of the sea fogs. A large portion of the area, originally covered by the tree, has of late years been destroyed by fires and by felling for lumber. In ancient times the redwood grew considerably to the southward of its present limit, as is proved by logs being found by well-borers in various parts of the coast range, where it does not now exist, as far south as Los Angeles and San Diego.

In Oregon there are only about 2000 acres of redwood, in two small forests, on the Chetco river, six miles from its mouth, and on the Winchuck river. The redwood belt, properly so called, which is a continuous forest of the species, begins on the northern boundary of California, and ends in Mendocino county, where it attains its maximum width, about thirty-five miles. Farther north the belt narrows, being only ten miles broad in Del Norte county. South of Mendocino county the redwood is only met with in small isolated forests.

In Monterey county[1] the groves are small, the most southerly forest of any importance being in the Santa Cruz Mountains, where the tree is common. The State of California appropriated in 1901 $250,000 for the purchase of the redwood forest of the Big Basin in Santa Cruz county, and this is now known as the California Redwood Park.[2] Prof. Jepson tells me that the area of the park is about 3800 acres, 2500 acres of which are covered with timber, consisting of Redwood mixed with Tan Oak, Madroña, and Douglas Fir. A fine grove, known as the Santa Cruz " Big Trees," is famous. A small grove, now practically destroyed, existed fifty years ago on the east side of the bay of San Francisco in Alameda county. At present the tree grows in the Mount Diabolo range in only one limited locality, Redwood Peak, in the Oakland Hills, directly opposite the Golden Gate. In Napa valley the tree is rather common ; and crossing over the summit of Howell Mountain it descends the slope towards Pope valley. This is the point where the redwood grows farthest from the ocean, and the only locality where it is found to the east of the divide of the coast range. In Marin county there are only a few isolated groves, mainly used as picnic grounds ; and in Sonoma county a few scattered claims still remain uncut.

The redwood belt, which I visited in 1906, near its northern limit at Crescent City, is the most impressive of all forests, being remarkable not only for the immense size[3] of the trees, but also for their extraordinary density upon the ground. A single acre has yielded 100,000 cubic feet of merchantable timber. The favourable conditions of the soil and climate account in great measure for this extreme productiveness ; but I am inclined to think that the mode of reproduction by suckers and by coppice shoots, explains in part the density with which the trunks stand upon the ground. A large proportion of the old trees are sprouts from ancient

[1] Cf. Jepson, *Flora W. Mid. California*, 24 (1901) ; and C. H. Shinn, *Cycl. Am. Hort.* iv. 1660 (1902).

[2] Prof. Jepson in a recent letter says that there is practically no Redwood in the National Forest Reserves ; but a few groves in private hands are as safe as if under State or National control, namely :—Redwood Cañon by Mount Tamalpais, near San Francisco ; Bohemian Club Grove in Sonoma county ; and Armstrong Grove in the same county. These comprise 500 to 1000 acres each.

[3] Mr. J. H. Maiden, Director of the Sydney Botanic Garden, in an article in the *Sydney Morning Herald*, quoted in the *London Pharmaceutical Journal*, April 30, 1904, states that the excessive heights claimed for eucalyptus trees in Australia are unreliable, and considers that the redwood, accurately measured by Sargent as 340 feet, is the tallest tree in the world. Von Mueller, in *Eucalyptographia*, Decade 5 (1879-1884), gives, on the authority of Mr. D. Boyle, the measurement of a fallen *Eucalyptus amygdalina* as 420 feet, and states that Mr. G. Robinson, a competent surveyor, measured another tree of this species as 471 feet ; and it is unknown to me on what grounds Mr. Maiden has questioned these measurements.

trees, as is readily seen by the way in which one-sided stems are often grouped around a hollow, from which the old stump has rotted away. At the present day reproduction is mainly effected by suckers, the proportion of these to seedlings being as 100 to 1. Seeds do not germinate except in open places, and young seedlings, requiring plenty of light to grow, are usually suppressed by the shade of the suckers, which, being well nourished by the roots of the parent tree, grow fast in dense shade.

The habit of the tree perpetuating itself by suckers seems to have impaired the vitality of the seed, as only 15 to 25 per cent of it proved fertile in experiments made by Mr. P. Rock of Golden Gate Park.

The topography of the redwood belt is uneven, and the character of the forest in consequence is very varied. The mountains of the coast range rise to altitudes of 1000 to 2000 feet, and consist of two or three ridges parallel to the coast, through which rivers and streams have cut deep valleys in some places, and formed wide alluvial flats in others. On the steep slopes and at the higher elevations, where the soil is shallow and dry, the redwood is always mixed with Douglas, hemlock, *Abies grandis*, and two or three other species, and is comparatively small in size and less dense upon the ground. It is only at low altitudes, in the deep soil of alluvial flats and in ravines, where the water-supply is great, that the redwood grows as practically pure forest, and attains a great size and density ; but even here a few trees of Sitka spruce and hemlock are usually associated with it. Absolutely pure stands, however, occur on flat tracts near streams, and in these the shade is so great that nothing grows upon the ground but *Oxalis* and a few tufts of *Aspidium munitum*. I saw a stand of this kind close to the Smith River, where the trees were of enormous size and of incredible density upon the ground. One tree measured 51 feet in girth. The river bank was fringed with *Alnus oregona* 50 to 60 feet high, behind which were two or three rows of taller *Umbellularia* ; and a single Lawson cypress, 200 feet high, had taken refuge on the river bank. Behind this screen there were only redwoods towering far above the other trees. On the slopes the ground cover was dense and impenetrable, consisting mainly of *Aspidium* attaining an immense size, *Acer circinatum*, *Rhamnus Purshiana*, *Gaultheria Shallon*, *Rubus*, etc. According to R. T. Fisher, of the U.S. Forestry Service, of whose paper[1] I have made use in this account, the redwood slopes, where the tree is mixed in varying proportions, cover fifty times as large an area as the redwood flats, where the tree is pure or nearly so.

Near Crescent City the flat which extends for about three miles in width from the ocean to the first hill of the coast range was originally covered with a mixture of redwood, Sitka spruce, and hemlock, most of which is now cut away. On the bluffs of the sea-shore a few small trees of *Pinus contorta* take refuge, while behind them and inland there are scattered groves of second-growth spruce, about 50 feet high. The first slope, exposed to the south-west and rising to 500 feet, is a dense stand of virgin spruce and hemlock, the trees attaining 200 feet high by 15 feet in girth. Crossing the hill to the north-east slope the first redwoods are seen, and from here inland for about eight miles over rolling country the redwood is the dominant tree, enormous in size and thick upon the ground. Afterwards, ascending the gorge of

[1] "The Redwood": *U.S. Forestry Bulletin*, No. 38 (1903).

the Smith River, Douglas fir begins to prevail, the redwoods becoming gradually fewer and smaller; and the last ones were seen twelve miles inland at about 1000 feet elevation.

The prevailing formation in the redwood belt is sandstone; and the tree attains its maximum either on deep sandy loam or on gravel full of moisture. The climate is remarkably even and moderate, with warm days, cool nights, and scarcely any frost even in winter; while the air is charged with humidity, and the annual rainfall amounts to from 60 to 80 inches. The following observations, taken in 1900 at Crescent City, show the nature of the climate in which the redwood thrives :—

	Rainfall in Inches.	Temperature, Fahrenheit. Maximum.	Minimum.
January	11	64°	32°
February	10	61°	33°
March	6	63°	36°
April.	6	70°	32°
May	5	63°	39°
June	2	67°	41°
July		67°	40°
August	0·3	71°	40°
September	0.6	76°	41°
October	11	70°	37°
November	6	72°	33°
December	8	61°	27°
Total rainfall . . .	65.9		

The tree is not found in the interior valleys to the east of the coast range, where the summer is comparatively hot and dry, and only a moderate amount of rain falls in winter.

Dr. Mayr,[1] reproduces a sketch of the largest redwood he saw in December 1885 near Santa Cruz. The mean of three measurements made it 308 feet high by 46 feet in girth at 6½ feet from the ground, above the swollen base. The first large green branches were at 230 feet up. This tree was still standing in 1903. He also gives an excellent illustration[2] of the appearance of a redwood forest after lumbering and fire have devastated it, which reminds me strongly of similar scenes in the Douglas fir forests of Oregon and Washington.

Fisher gives several tables showing the composition of the species and the size of the trees in the redwood belt. At Scotia, on an alluvial flat, there are 100 redwoods to the acre, no other species being present, and of these thirty-six were over 20 inches, and averaged 76 inches in diameter. Mayr[2] gives the following figures for the best pure stand which he measured :—57 trees to the acre, averaging 275 feet in timber height and 23 feet in girth; total cubic contents, exclusive of branches, 199,000 cubic feet per acre. The tallest redwood recorded[3] was measured in 1896 by Professor Sargent. This tree grew on the Eel River, and was 662

[1] *Fremdländ. Wald- u. Parkbäume*, tt. 19, 20 (1906); cf. also *Waldungen*, p. 268 and frontispiece.
[2] *Waldungen von Nordamerika*, 267 (1900).   [3] *Garden and Forest*, 1897, p. 42.

years old, 340 feet in total height, 230 feet to the first branch, and 10 feet 5 inches in diameter at 6 feet from the ground. I found it impossible to obtain in the dense forests accurate heights; but I saw lying on the ground a tree blown down many years ago which measured 240 feet, the top having been broken off at a point where the stem was still 3 feet in diameter. In the logging camp near Crescent City a tree which had just been felled measured 45 feet in girth at 6 feet from the base, and 24 feet in girth at 144 feet up; the top, knotty and full of branches, had smashed in the fall, and was rejected as being useless as timber.

Most of the trees cut are from 400 to 800 years old. After 500 years it usually begins to die at the top and to fall off in growth. The oldest redwood found by Fisher was 1373 years of age.

Isolated trees are occasionally blown down by the wind; but no considerable tracts are ever overthrown. There is no tap-root; but the lateral roots, numerous and stout, strike downwards, usually at a sharp angle, and form a compact mass, in shape like an inverted funnel. Such roots both anchor the tree securely and provide it with a large supply of moisture.

Fires are of rare occurrence in the damp northern part of the redwood belt; but farther south, where the climate is drier, they are frequent in August and September. Usually however, only young trees and undergrowth are burned, the larger trees remaining unharmed. Injuries by fire or by the fracture of the branches by the wind, which involve the sapwood, are supposed to be the cause of the curious burrs and protuberances which are found on many trees.

Where the forests have been cut down, the better land has been permanently put under fruit, grain, or pasture; but on the worse lands the tree will survive and is in no danger of extinction, owing to its astonishing power of reproduction by sprouts. There are many fine stands of second-growth redwood; as in Mendocino county, where young trees, only 45 years old, are nearly 100 feet high and 20 to 30 inches in diameter. In Sonoma county, second-growth timber is being cut commercially, and though sappy, makes good boards for boxes.

The Spaniards, near San Francisco Bay, were the first to cut the redwood forests; but their operations were on a very small scale. Regular felling only began in 1850, and at first, as redwood timber was little valued, only Douglas fir and Sitka spruce were taken out of the redwood belt. Of late years, redwood timber has greatly increased in value, and the introduction of machinery has made lumbering more easy and profitable. Only big companies, however, can work with profit, as the outfit is very expensive, consisting of sawmills, many miles of railroad in the forest, locomotives of a special type for ascending steep gradients, waggons, donkey-engines, and logging camps, with a large staff of workmen.

No cutting is being done at the present time in the southern counties. The largest sawmills are in Mendocino county; and they had cleared, by the year 1900, 150,000 acres, or a quarter of the total acreage, including the largest and best stands. In the same year the mills had cleared in Humboldt county 65,000 acres and in Del Norte county 2000 acres. Since that year cutting has been going on at an accelerated pace, and the quantity now felled annually is enormous.   (A. H.)

INTRODUCTION

It is stated by Kent in Veitch's *Coniferæ* that Hartweg, who collected in California and Mexico for the Royal Horticultural Society in 1846 and 1847, was the introducer, but there is no evidence to confirm this statement, either in Hartweg's letters, which are printed in the Journal of the Society, or in Gordon's account of the conifers of which he sent home seeds.

The first mention I can find of the tree in the *Gardeners' Chronicle* is on March 17, 1849, when James Duncan, gardener at Basing Park, wrote that he had planted out in July 1847 a plant 9 inches high, which had stood two winters without protection. In the same journal for 1851, p. 246, it is stated that in 1845 there was a plant 2 feet 7 inches high at Holker, near Ulverston ; [1] and on the authority of Mr. Frost that a grafted tree planted at Dropmore in 1845 was 18 feet high in 1851. It is evident that all of these must have been raised earlier than 1848, when Hartweg returned to England, and in his letter, received by the Society on November 4, 1846,[2] though he speaks of having seen the tree on the mountains of Santa Cruz, he says nothing about having collected or sent home seeds.

In *A Synopsis of the Coniferous Plants grown in Great Britain and Sold by Knight and Perry, at Chelsea*, published by Longmans, London, without date, but probably about 1850, it is stated on pp. 45, 46, that the redwood was introduced in 1843, when plants were sent to Knight and Perry by Dr. Fischer of St. Petersburg, who received seeds of it from America. I have inquired of M. Fischer de Waldheim, Director of the Botanic Gardens at St. Petersburg, whether he knew who was the actual collector, but he replies that there is nothing in the archives which will give this information.

CULTIVATION

I have never seen a plant raised from seed grown in this country, though I believe it will ripen in the south-west. Imported seed, so far as I have tried it, germinates badly, and the seedlings are tender at first, and should not be planted without protection till they are two or three years old, as the young growth will usually suffer from frost.

I am inclined to think that many of the plants sold by nurserymen are raised from cuttings, and purchased plants certainly seem hardier than the seedlings I have raised, most of which were killed to the ground in 1905 and 1907, though they shot up many suckers the following year. I have no evidence to show whether trees raised from cuttings will grow into tall, straight trees, as in the case of Cryptomeria.

According to a note[3] by Mr. Frost in 1851, the first plants sent out by Knight and Perry were grafted, but this seems very unlikely, as there is no stock except that of the Wellingtonia which would seem at all suitable.

---

[1] Mr. Fenner tells me that this tree is now only 65 feet high by 6 feet in girth, and has been damaged by wind at the top. Suckers are growing from roots two feet from the main stem.

[2] *Journ. Hort. Soc.* ii. 124 (1847).   [3] *Gard. Chron.* 1851, p. 246.

Though in this country the tree has proved fairly hardy in most localities, it is certainly more susceptible to frost in spring and autumn, especially when young, than Wellingtonia, and cannot be looked on as a really hardy tree except in well-drained soils, and in situations where it is well sheltered from dry frosty winds. All the really fine specimens I have seen are in unusually favourable places, and I should not recommend the tree for planting largely except in the south-west and west of England and Scotland, though in Ireland it seems to be more generally flourishing.

It does not dislike lime in the soil, and though the top is always killed back in hard winters, grows fast even at Colesborne, a tree planted on the site of the old house here about 1855, being now no less than 11 feet in girth though only about 55 feet high. Another planted later on thin dry soil is only about 35 feet by 4 feet.

Its long feathering branches, which droop to the ground and sometimes take root, make it a very ornamental tree, but as a timber tree its value remains doubtful. Whilst young it is very liable to be barked at the ground by mice, which have destroyed more than half of those which I have planted, and Sir W. Thiselton-Dyer tells me that squirrels[1] attack the bark at Kew.

Specimens of its timber grown in England are very inferior to the imported wood, on account of the rapid increase in girth which the tree makes unless crowded; and the only places I have seen where it seemed at all likely to be profitable are at Whitfield and Penllergare, though in Ireland there may be better hopes of its economic value, in favourable situations.

The following opinions, taken from the reports[2] published by the Conifer Conference, express very well the condition of this tree in twenty selected places :—

ENGLAND

Pampisford, Cambridge	Often injured by frost.
Golden Grove, Caermarthen	Leader sometimes frosted.
Scorrier, Cornwall	Requires shelter from wind.
Tortworth, Gloucester	In a shady place.
Linton, Kent	Lost 3 feet of top last winter.
Howick, Northumberland	Not thriving well.

SCOTLAND

Inveraray, Argyllshire	Often loses its leader.
Whittinghame, East Lothian	Vigorous; well sheltered.
Fordell, Fife	Very fine specimens.
Murthly, Perth	Fine in damp places.
Scone, Perth	In fine health; grows well.
The Cairnies, Perth	Doing well in shelter.
Castle Leod, Ross	Fine specimen.

[1] Squirrels are fond of making their nests out of the bark, but do no injury to trees, which have attained a considerable size. Cf. *Gard. Chron.* 1866, p. 413.

[2] *Journ. Roy. Hort. Soc.* xiv. pp. 483 *seq.* (1892). Cf. also *Gard. Chron.* 1866, p. 1043, where an abstract is given of Mr. Palmer's statistics of the effects of the severe winter of 1860-1861 on this tree, planted in 113 different places in England, Scotland, and Ireland.

IRELAND

Shane's Castle, Antrim . . .	Grows rapidly here.
Fota, Cork	Free-growing, fine tree.
Woodstock, Kilkenny . . .	A handsome tree.
Birr Castle, King's County . .	Thriving well.
Adare Manor, Limerick . . .	Thriving; gales break leader.
Baron's Court, Tyrone . . .	Fine rapid grower.
Coollattin, Wicklow . . .	Very fine specimen.

REMARKABLE TREES

Among the great number of large redwoods I have seen at various places in England, I think the finest is one at Claremont, growing near the borders of the lake in a very sheltered position (Plate 194), which in 1903 measured 95 feet by 12 feet and in 1907 98 feet by 12 feet 9 inches. At Melbury there is a tree not so tall but thicker, which in 1906 was 85 feet by 15 feet 1 inch. At Fonthill Abbey there is a remarkable twin tree which grows in a damp hollow, dividing at the ground into two trunks which are 98 to 100 feet high by 10 feet and 9 feet 3 inches in girth respectively. At Boconnoc in Cornwall there is a tree which in 1851 was already 16 feet high, and in 1891 was reported as measuring 75 feet by 13 feet, but when I measured it in 1905 it had lost its top, and was then only 68 feet by 14½ feet. At Dropmore, Buckinghamshire, a tree,[1] remarkable for its pendulous branches and branchlets, is 94 feet high and 11 feet in girth. Three years ago, according to Mr. Page, the gardener, it was 10 feet 6 inches in girth, so that it is still making rapid growth. This tree was planted in 1845, when it was a foot high, having been bought for five guineas at Knight and Perry's nursery. It is bearing this year an immense number of cones; but no attempt has ever been made to raise seedlings.

In a sheltered dell known as the Wilderness, at Cuffnells, near Lyndhurst, the seat of R. Hargreaves, Esq., are three splendid redwoods, which were planted about the year 1855 by his father. These measure 102 feet by 10 feet 8 inches, 98 feet by 15 feet, and 105 feet by 10 feet 10 inches respectively, the last being equal or superior to the one at Claremont, and growing close to a magnificent tree of *Pinus insignis*, which will be figured in our next volume.

At Beauport, Sussex, there is a tree with very pendulous branches bearing cones on the ends of the twigs, 73 feet by 9 feet 6 inches, and a larger one of the ordinary form, 85 feet by 11 feet 5 inches; and at Hemstead in Kent there is a tree not quite so tall as the Cryptomeria growing by its side (see Plate 42), and of about the same age. In the eastern counties the best we have seen are at Hardwicke House, Suffolk, where a tree in 1905 was 74 feet by 11 feet 10 inches, and at Barton, where in an exposed situation on the lawn there is one of 71 feet by 8 feet. This seems to be the only survivor of four which Sir Charles J. F. Bunbury[2]

---

[1] Erroneously reported to have been 114 feet in height in 1903 in *Journ. Board of Agriculture*, x. 345 (1904).
[2] *Arboretum Notes*, 166 (1889).

planted in 1847 and 1848, one of which bore cones in 1862. In Fulmodestone Wood, Norfolk, a tree planted in 1855 was 67 feet by 8½ feet when Henry measured it in 1904. At The Coppice, Henley, at 300 feet elevation, a tree planted in 1864 was 73 feet high by 9 feet 5 inches in girth in 1905. At Bayfordbury, Herts, the best specimen is 74 feet by 10 feet 2 inches. At St. George's Hill, near Byfleet, a tree growing on Bagshot sand was in 1904 75 feet by 8 feet 3 inches.

In Gloucestershire there are good trees at Tortworth, at Highnam, at William-strip Park, and at Huntley Manor, where Prof. Somerville measured one in 1904 76 feet by 13 feet.

At Whitfield, Herefordshire, there is a group of eleven very fine trees on an area of only 25 yards square, of which the tallest is about 95 feet by 12 feet, and the others would average 80 to 90 feet in height, by about 9 feet in girth, and contain from 100 to 120 cubic feet per tree. An acre of such trees, as thickly grown as this, would produce from 8000 to 10,000 cubic feet, which, considering that their age is not more than about fifty years, is a very remarkable yield of timber.

At Penllergare, near Swansea, there is another somewhat similar group of twelve trees on a triangle of which the sides are only about twenty yards, in which the best tree was about 75 feet by 10 feet, and the others from 4 feet to 8 feet in girth ; only one of these was a really bad tree, and two of them were forked.

At Penrhyn Castle there are several very fine trees, of which the best that I measured in 1906 was about 90 feet by 12 feet 3 inches, but others may be larger. At Coed Coch, Abergele, there is a splendid tree which Mr. A. Hunter, the gardener at this place, measured in 1905, when it was 90 feet in height and 12½ feet in girth.

In Scotland the redwood seems to grow best in Perthshire, where all the largest that I have seen are found, namely, one at Castle Menzies, which in 1892 was 74 feet by 4 feet 6 inches ; and at Moncrieffe, where a tree, mentioned by Hunter as 42 feet by 4 feet 11 inches in 1883, was in 1907 about 65 feet by 9 feet, and though healthy looks as though it had lost its top more than once.

A tree at Falkland Palace, Fifeshire, said to be then one of the finest in Scotland, was in 1892 65 to 70 feet by 9 feet. At Smeaton - Hepburn, East Lothian, in 1902, there was a tree 57 feet by 9 feet, planted in 1844, which had lost its top on several occasions. At Castle Kennedy the wind is evidently too severe for it, as the largest measured by Henry in 1904 was only 39 feet by 7 feet.

In Ireland the redwood has attained large dimensions in those districts which have a mild climate and a heavy rainfall. In Queen's County, where the rainfall is only moderate and sharp frosts occur, it has done badly, and is in marked contrast to the splendid Wellingtonias which are growing beside it in several places.

At Castlemartyr, Cork, there are several large specimens, one, about 70 feet in height, being in 1907 16½ feet in girth at 5 feet from the ground. At Fota, in the same county, a tree in 1903 measured 90 feet high by 10 feet in girth. This was reported [1] in 1891 to be 75 feet by 7½ feet. At Coollattin, in Wicklow, there

---

[1] *Journ. Roy. Hort. Soc.* xiv. 549 (1892).

are three trees, nearly equal in size, one of which measured in 1906 77 feet by 11 feet 5 inches. This[1] was 55 feet by 8½ feet in 1891. At Hamwood, Co. Meath, a tree, said to have been planted in 1847, was 59 feet high by 11 feet in girth in 1905. At Woodstock, in Kilkenny, there is a tree, which in 1904 was 91 feet by 13 feet 3 inches. This[1] was 68 feet by 10 feet 4 inches in 1891. At Churchill, Armagh, a tree,[2] planted in a bog in 1862, was 60 feet high by 6½ feet in 1884; but had lost its leader several times. Henry did not see this tree on his visit to Churchill in 1904. At the Conifer Conference of 1891 good trees were also reported to be growing at Shane's Castle in Antrim, Clonbrock in Galway, Courtown in Wexford, and Powerscourt in Wicklow.

The largest tree of this species that I have seen in Europe is a well-shaped one, branching to the ground, on the Isola Madre in Lake Maggiore, which in 1906 was no less than 104 feet high by 14 feet 8 inches in girth.

Pardé[3] says that the trees at Les Barres were mostly killed to the ground in the winter of 1879-80, but have thrown up vigorous shoots which produce cones every year; and at Segrez I saw a tree which was killed in the winter of 1870-71 and afterwards threw up six or seven straight stems from the stool, which are now over 50 feet high.

TIMBER

The best timber, as a rule, is produced by trees growing on alluvial flats, that from trees growing on the slopes being hard and flinty. The sapwood, which is of no service, is whitish in colour and 1 or 2 inches in thickness. The heartwood varies in colour from light pink to dark mahogany, and is esteemed for many purposes. It is light in weight, soft, straight-grained, is easily worked; and although it requires much filling, is capable of taking a fine polish. Its durability is attested by the fact that trees which have lain for centuries in the forest have been taken to the sawmill and converted into useful lumber. Mr. D. N. M'Chesney says[4] that in Manila the wood has been found proof against the attacks of the white ant, together with that of *Tsuga Albertiana* and *Libocedrus decurrens*, while the timber of *Pinus ponderosa*, *Picea Engelmanni*, and the Douglas fir suffered badly from these destructive insects.

In Europe the Californian redwood has established a market for itself, but supplies have been steadily falling off for some years. It is, however, still shipped in considerable quantities to China, Japan, Honolulu, and Australia. In its native country it is employed for both exterior and interior fittings of houses, sleepers, electric light and telephone poles, shingles, tanks, and vats. Very frequently the grain is bold, wavy, and very handsome; and is in demand for ceilings and large panels. Some splendid examples were used for the decoration of the Californian Court at the St. Louis Exhibition of 1904. The large burrs which are not uncommon in some districts, when cut into slabs, make very fine table tops showing a mass of close small eyes of a deep red colour. A very large plank of the wood was exhibited at the World's Fair, Chicago, 1903, which measured

---

[1] *Journ. Roy. Hort. Soc.* xiv. 556, 565 (1892).   [2] *Woods and Forests*, 1884, p. 624.

[3] *Arbor. Nat. des Barres*, 52 (1906).

[4] *Bull. Nos. 30 and 33, New Series, Division of Entomology; U.S. Department of Agriculture* (1901), p. 95.

$16\frac{1}{2}$ feet wide, $12\frac{3}{4}$ feet long, and 5 inches thick. The tree from which this extraordinary plank was cut was felled in Humboldt County, and was said to be 300 feet in height by 35 feet in diameter. Planks of 5 feet in width are imported, and I have myself purchased boards 4 feet in width absolutely free from knots and defects. In England this wood is chiefly used for the inside linings of furniture, and it is said to be one of the finest woods in the world for large signboards, as it maintains a remarkable consistency of shape under the most trying conditions of climate and exposure. The value in Liverpool in 1907 was from 2s. 2d. to 2s. 6d. per cubic foot, a price which cannot be said to encourage shipments that have to pay the cost of so long a sea carriage. The wood is usually of slow growth, and the annual rings are from thirty to fifty to the inch. The cells are so large that they can be seen by unaided vision. Resin ducts are almost entirely absent in both species of Sequoia but have been found by Jeffrey[1] in the flowering shoot and in the first annual ring of vigorous branches of adult *S. gigantea*, and in the wood of the shoot and root of redwood as the result of injury. (H. J. E.)

SEQUOIA GIGANTEA, WELLINGTONIA, BIG TREE

Sequoia gigantea,[2] Decaisne, *Bull. Bot. Soc. France*, i. 70 (1854), and *Rev. Hort.* 1855, p. 9, f. 1 (not Endlicher); Masters, *Gard. Chron.* xix. 556, f. 85 (1896); Sargent, *Bot. Gazette*, xliv. 226 (1907).

Sequoia Wellingtonia, Seemann, *Bonplandia*, iii. 27 (1855); Lawson, *Pinet. Brit.* iii. 299, tt. 37, 51, 53 (1884); Sargent, *Silva N. Amer.* x. 145, t. 536 (1896), and *Trees N. Amer.* 69 (1905); Kent, Veitch's *Man. Coniferæ*, 274 (1900).

Sequoia Washingtoniana, Sudworth, *Check List Forest Trees, U.S.* 28 (1898).

Wellingtonia gigantea, Lindley, *Gard. Chron.* 1853, pp. 820, 823; W. J. Hooker, *Bot. Mag.* 4777, 4778 (1854).

Taxodium Washingtonianum, Winslow, *Calif. Farmer*, 1854, ex Hooker, *Kew Journ.* vii. 29 (1855).

Taxodium giganteum, Kellogg and Behr, *Proc. Cal. Acad.* i. 151 (1855).

Washingtonia Californica, Winslow, *loc. cit.*

A tree attaining 320 feet in height, with a tapering stem, occasionally 90 feet in girth above the much enlarged and buttressed base. Young trees narrowly pyramidal. Old trees free of branches to 100 or 150 feet, with an irregular crown of short thickened branches. Trunk fluted with broad, low, rounded ridges; bark 1 to 2 feet thick divided into lobes 4 to 5 feet wide, corresponding to those of the trunk, separating into loose reddish fibrous scales, which expose the spongy middle bark. Branchlets pendulous, not distichously arranged but in dense masses, green in the first year, afterwards gradually becoming brownish with a thin scaly bark. Buds minute, without scales.

Leaves persistent for four years, arranged on the branchlets in approximately three ranks; on the main axes ovate acuminate, up to $\frac{1}{2}$ inch long; on the lateral axes lanceolate, acute, $\frac{1}{8}$ to $\frac{1}{4}$ inch long; appressed and decurrent at the base, free

---

[1] *Mem. Boston Soc. Nat. Hist.* v. 441 (1903).

[2] The tree was first described by Lindley, who called it *Wellingtonia gigantea*. Wellingtonia cannot be maintained as a distinct genus. *Sequoia gigantea* is the correct name, according to the rules of botanical nomenclature promulgated by the Vienna Congress of 1905, and is now adopted by Sargent.

and spreading from beyond the middle, rigid, ending in sharp cartilaginous points; lower surface green, rounded or keeled; upper surface with green midrib and two inconspicuous bands of stomata.

Staminate flowers, $\frac{1}{6}$ to $\frac{1}{8}$ inch long, with acute or acuminate connectives. Pistillate flowers with twenty-five to forty pale yellow bracts gradually narrowed into long slender points. Cones ripening in the second year, ovoid-oblong, 2 to 3 inches long by $1\frac{1}{2}$ to 2 inches wide, brownish; scales gradually thickening from the base to the dilated disc, which is $\frac{3}{4}$ to 1 inch broad, and often bears a reflexed spine in the centre of the transverse depression. Seeds, $\frac{1}{8}$ to $\frac{1}{4}$ inch long, light-brown, apiculate at the apex, surrounded by laterally united, often unequal wings, which are broader than the body of the seed. Proliferous cones have been observed.[1]

Wellingtonia produces cones freely in many parts of the British Isles, but these are smaller in size as a rule than those of the wild tree and rarely contain mature seed. Mr. Richards informed me that he had sown a large quantity of seed, produced by a tree growing at Penrhyn on the lawn in an isolated sunny position, and only obtained eight seedlings. Barnes[2] raised young plants from seeds produced by a tree at Bicton. At Orton Longueville, Mr. Harding[3] succeeded in raising six seedlings out of 100 seeds. The tree cones well at Dropmore, but has seldom if ever produced fertile seed there.

Wellingtonia differs markedly from the redwood, in not reproducing itself either by suckers from the root or by coppice shoots. In its native forests, seed is produced in great abundance, and numerous seedlings occur everywhere in the southern part of the area of distribution of the species; but in the northern groves seedlings are said to be totally wanting.

VARIETIES

None have been noticed in the wild state. Several have appeared in cultivation in Europe.

1. Var. *pendula.* Branches bent downwards at the base, and hanging for their whole length close to the stem, forming in young plants a slender pyramid and in older examples a tall narrow column. This remarkable variety was obtained out of the seed-bed by Lalande of Nantes in 1863, and was put upon the market in 1873 by Paillet of Châtenay-les-Sceaux, near Paris. The best tree[4] of this kind is growing at M. Allard's arboretum at Angers, in France, and when measured by Elwes in 1907 was 44 feet high by 3 feet in girth, but only 13 feet round the branches. At Bicton,[5] this variety is represented by a tree which in 1902 was 33 feet high by 26 inches in girth at 2 feet from the ground. At Brettargh Holt,[5] Kendal, the residence of Charles Walker, Esq., a weeping Wellingtonia was reported to be 22 feet high in the same year. Another example, aged 26 years, growing at Dalkeith Palace and reported to be $19\frac{1}{2}$ feet high in 1902, was figured in the *Gardeners' Chronicle.*[5] A specimen

---

[1] Carrière, *Rev. Hort.* 1887, p. 509, f. 103. Cf. *Gard. Chron.* ii. 649 (1887).
[2] *Gard. Chron.* 1868, p. 872.   [3] *Ibid.* xxix. 55 (1901).
[4] Described and figured by Rehder in Möller's *Deutsche Gärtner Zeitung*, March 22, 1902.
[5] *Gard. Chron.* xxxi. p. 388, fig. 136, and p. 435; and xxxii. p. 23 (1902).

30 feet in height at Berkhampstead, Herts, is mentioned by Webster.[1] In the *Revue Horticole*, 1906, p. 395, f. 157, a curious weeping Wellingtonia, growing at the Trianon, is figured. The stem, which is 42 feet in length, bends over and is supported on one side by a prop. Barron also obtained a weeping form, which was sold as *S. gigantea Barroni pendula*.[2]

2. Var. *aurea* (var. *aureo-variegata*). The young shoots are amber-coloured at first, but speedily become deep yellow, the colour being pretty uniform over the whole tree. The original plant was a seedling, which Hartland[3] of the Lough Nurseries, Cork, received in 1856. It began to show colour when it was about a foot high, and after it had attained 8 feet, a large number of golden Wellingtonias were propagated from it by grafting. A specimen 20 feet high was growing[4] in the public garden at Denbigh in 1887. We have seen no trees of this variety of a considerable size.

3. Several other varieties, which I have not seen, are mentioned by Beissner[5] as *glauca*, *argentea*, *Holmsii*, and *pygmæa*.   (A. H.)

DISTRIBUTION

Wellingtonia has a restricted distribution, being confined to the western slopes of the Sierra Nevada of California, in an interrupted belt at elevations of from 5000 to 8400 feet above sea-level, extending from the middle fork of the American River (lat. 39°) to the head of Deer Creek, just south of lat. 36°.

I am indebted to Mr. Gifford Pinchot for the most recent account which has been officially published of the big trees of California,[6] illustrated by some excellent photographs ; from which it appears that John Bidwell in 1841 was really the first to discover this tree in the Calaveras Grove, Prof. Brewer of Yale having been the first scientific visitor in 1864 to this and the Mariposa Grove. Mr. Whitney, in the *Yosemite Guide Book* (1870) described eight of the then known groves, namely :—

1. The North Grove in Placer county is on a tributary of the middle fork of the

[1] *Hardy Coniferous Trees*, 113 (1896).

[2] Another weeping form is said to have originated in Little and Ballantyne's nursery at Carlisle ; but the original tree died in 1877. Cf. *Journal of Forestry*, iii. 260 (1879).   [3] Letter to Kew.

[4] *Gard. Chron.* ii. 276 (1887).   [5] *Nadelholzkunde*, 165 (1891).

[6] A " Report on the Stanislaus and Lake Tahoe Forest Reserves, by G. B. Sudworth," *Bulletin No. 28* ; *U.S. Dept. of Agriculture, Division of Forestry*, published at Washington in 1900, which gives the following table of measurements of thirty of the big trees in the Calaveras Grove :—

Tree No.	Diameter 6 feet above Ground.	Height.	Tree No.	Diameter 6 feet above Ground.	Height.	Tree No.	Diameter 6 feet above Ground.	Height.
	Feet.	Feet.		Feet.	Feet.		Feet.	Feet.
1	9.0	235	11	12.5	250	21	15.0	325
2	9.0	251	12	12.5	266	22	15.5	268
3	9.5	260	13	13.0	286	23	15.5	272
4	10.0	237	14	13.5	320	24	15.5	289
5	10.0	243	15	14.0	259	25	16.0	262
6	10.0	261	16	14.0	265	26	16.0	275
7	10.5	248	17	14.0	269	27	16.5	266
8	11.0	255	18	14.5	278	28	16.5	268
9	11.0	260	19	15.0	285	29	16.5	288
10	12.0	248	20	15.0	307	30	19.5	315

American river, at an elevation of 5100 feet, and 70 miles north of the Calaveras Grove. It now contains only six trees, of which the two largest are 240 and 220 feet high.

2. The Calaveras Grove[1] is at an elevation of 4750 feet, occupying a belt only 3200 by 700 feet in extent, and containing 90 to 100 trees of large size, besides a considerable number of small ones. The largest tree standing here was barked up to 116 feet high, and the bark set up in the Crystal Palace at Sydenham, where it was afterwards destroyed by fire. This tree was 302 feet high and 96 feet in girth at the ground. It was felled by boring holes all round the trunk with pump augers, which occupied five men for twenty-two days. An even larger tree in this grove was the "Father of the Forest," cut down in 1853, and now lying on the ground; it has been hollowed out by fire so that a man can ride through it on horseback for a distance of 82 feet. Its extreme length, so far as could be judged from the remains, was 365 feet, and its circumference at the base is said to have been 110 feet.

The largest living trees in this grove were as follows, only nine being, according to Whitney, over 300 feet :—

"The two Sentinels," over 300 feet high by 23 feet in diameter.
"The Pride of the Forest," 300   „   23   „
"Abraham Lincoln,"   320   „   18   „
"Starr King,"   360   „   girth not stated.
"General Scott,"   325   „   „
"Keystone State,"   325   „   diameter at 6 feet, 14 feet 3 inches.
"General Jackson,"   319   „   „   „   12 „ 7 „
"Mother of the Forest,"   315   „   „   „   19 „ 4 „
                                                    without bark.
"Daniel Webster,"   307   „   „   „   15 feet

3. The Stanislaus or South Calaveras Grove is about six miles south-east of the last, and is said to have contained, when Mr. Sterry owned it, 1380 trees from 1 foot to 34 feet in diameter, but the number now existing is much less. The largest standing trees mentioned were as follows :—

"Columbus."
"New York,"   over 300 feet high by 104 feet in girth.
"Ohio,"   „   311   „   103   „
"Massachusetts," „   307   „   98   „

Besides these, "Smith's Cabin," a hollow tree in which a hunter is said to have lived for three years, 21 feet by 16 feet in size inside, and "Old Goliath," a fallen tree said to be 100 feet or more in circumference, were remarkable for their size.

About 25 miles south-east of this is a grove called the Crane Flat Grove, most of the trees in which are said to be rather smaller than those in the Calaveras Groves.

[1] Cf. note 6 on p. 701.

4. The Mariposa Grove is the one best known to English travellers, and is usually visited from Clark's ranch on the road to the Yosemite Valley. It consists of two nearly distinct groves, the upper one being compact, on an area of 3700 by 2300 feet and containing 365 trees over 1 foot in diameter, besides a great number of small ones. The southern division or lower grove is said to contain only half as many Sequoias, and these are more mixed with other trees, such as Douglas fir, sugar pine (*Pinus Lambertiana*), *Abies concolor*, and *Libocedrus decurrens*. Many of the trees in both of them have been much injured by fire, which has destroyed many of the younger ones within the groves; but there are on the outskirts several small natural groups of young trees up to 6 or 8 inches in diameter. The largest tree here is the " Grizzly Giant," whose photograph is well known in many English houses, and which is 93 feet in girth at the ground, and 64 feet at eleven feet up; some of its branches are fully 6 feet in diameter. The tallest tree in this grove, according to Whitney, is 272 feet, and another is 270 feet by 26 feet in diameter at the base; only six in all are over 250 feet high.

5. The Fresno Grove is about 14 miles south-east of Clark's ranch, and is about $2\frac{1}{2}$ miles long by 1 to 2 miles wide. It contains 500 to 600 trees, of which the largest is 81 feet in girth at 3 feet from the ground.

6. About 50 miles south-east of the Fresno Grove, along the slope of the sierra between the King's and Kaweah rivers, is by far the most extensive forest that has been found. It is about 30 miles north-east of Visalia, and is scattered over an area 8 to 10 miles long and 4 to 5 wide, at an elevation of about 4500 to 7000 feet. The average size of these trees is much smaller, only 10 to 12 feet in diameter, the largest measured, near Thomas's Mill, being 106 feet in girth near the ground, where a considerable portion has been burnt off. At 12 feet from the ground this tree was 75 feet in girth, its height being 276 feet, though the top was dead.

7. There are two other groves on the Tule river, of which the most northerly is 30 miles from the King's River Grove. These were discovered in 1867 by Mr. D'Heureuse when exploring for the Geological Survey. They extend over an area of several square miles and contain a considerable number of trees, of which no measurements are given.

8. Besides these there are small and little known groves on Dinkey Creek, a tributary of King's river, and on the headwaters of the Merced river, the last said to contain less than 100 trees.

Prof. W. L. Jepson of the University of California has been good enough to send us the following enumeration[1] of the existing groves of Wellingtonia, most of which he has visited this year :—

---

[1] This list will be published in Prof. Jepson's forthcoming work on the Trees of California.

[TABLE

TABLE OF BIG TREE GROVES AND FORESTS. BY WILLIS L. JEPSON.

Groves.	Locality.	Area in Acres.	Altitude.	No. of Trees.
NORTH GROVES.				
1. North Grove	Middle Fork American River, Placer County	...	4300–5000	6
2. Calaveras Grove	Stanislaus River, Calaveras County	51	4800	101
3. Stanislaus Grove	Stanislaus River, Tuolumne County	1000	5000	1380
4. Tuolumne Grove	Merced-Tuolumne Divide, Tuolumne County	...	5800	30
5. Merced Grove	Merced-Tuolumne Divide, Mariposa County	...	5500	60
6. Mariposa Grove	South Fork Merced River, Mariposa County	...	6000	{ 365 / 182
7. Fresno Grove	Headwaters Fresno River, Madera County	2500	5000	500
SOUTH GROVES.				
8. M'Kinley Grove	Dinkey Creek, Fresno County	...	5000	75
9. Converse Basin Forest	South Fork King's River, Fresno County	5000	6000–6500	...*
10. Boulder Creek Forest	Boulder Creek, South Fork King's River, Fresno County	1500	6500–7000	...
11. General Grant Forest	Gen. Grant Nat. Park, Fresno and Tulare Counties	{ 2000 to 3000	6500–7000	...
12. Redwood Cañon Forest	Redwood and Eshom Creeks, Tulare County	3000	5500–6500	...
13. North Kaweah Forest	North Fork Kaweah River, Tulare County	1500	6000–7000	...
14. Giant Forest	Marble Fork Kaweah River, Tulare County	2300	6500–7000	...
15. Cliff Creek Grove	Middle Fork Kaweah River, Tulare County	...	7000	...
16. Harmon Meadow Grove	Middle Fork Kaweah River, Tulare County	...	7000 ?	...
17. Mineral King Forest	East Fork Kaweah River, Tulare County	3000	5500–6500	...
18. Lake Cañon Grove	East Fork Kaweah River, Tulare County	...	...	...
19. Mule Gulch Grove	East Fork Kaweah River, Tulare County	...	...	...
20. Homer's Peak Forest	East Fork Kaweah River, Tulare County	...	5500–7000	...
21. South Kaweah Forest	South Fork Kaweah River, Tulare County	{ 3000 to 5000	5000–7000	...
22. Dillon Forest	North Fork Tule River, Tulare County	1000	5000–7000	...
23. Tule River Forest	Middle Fork Tule River, Tulare County	3500	5500–7000	...
24. Pixley Grove	Middle Fork Tule River, Tulare County	...	6500–7000	...
25. Fleitz Forest	Middle Fork Tule River, Tulare County	...	5000–6500	...
26. Putnam Mill Forest	Middle Fork Tule River, Tulare County	4000	5500–6000	...
27. Kessing Groves	South Fork Tule River, Tulare County	...	5500–7000	...
28. Indian Reservation Forest	South Fork Tule River, Tulare County	1500	6000	...
29. Deer Creek Grove	South Fork Deer Creek	300	7000	150
30. Freeman Valley Forest	Kern River, Tulare County	1000	5500–6500	...
31. Kern River Groves	Kern River, Tulare County	700	6500–7000	...

* Blanks in this column indicate that the trees are so numerous that they have not been counted.

One of the best accounts of the Wellingtonias and their surroundings is in Muir's *Mountains of California.* He states that the young trees have slender branches growing with great regularity down to the ground, as we see them on an English lawn; but when the tree attains 500 or 600 years old, the spiry, feathery, juvenile habit merges into the firm rounded dome-like habit of middle age, which in its turn takes on the eccentric picturesqueness of old age. The foliage of the saplings is dark bluish-green, while that of the older trees ripens to a warm brownish-yellow tint like that of Libocedrus. The bark is rich cinnamon brown, purplish in young trees and in shaded portions of the old ones. In winter the trees break out into bloom, myriads of small four-sided staminate flowers crowding the ends of the smaller sprays, colouring the whole tree, and when ripe dusting the air and the ground with golden pollen. The fertile cones are bright grass green, about 2 inches long by $1\frac{1}{2}$ inch wide, and are composed of about forty firm scales densely packed, with three to eight seeds at the base of each, a single cone thus containing 200 to 300 seeds, which are about $\frac{1}{4}$ inch long by $\frac{3}{16}$ inch wide, with a thin flat margin. The cones are very freely produced; and on two branches, $1\frac{1}{2}$ to 2 feet in diameter, Mr. Muir counted no less than 480. But of the millions of seeds produced, very few germinate; and of these not one in ten thousand lives through the vicissitudes of storm, drought, fire, and crushing by snow, to which they are exposed in youth.

Natural reproduction in the groves, when they have been protected from fire and grazing, is said to be at a standstill, owing to the dry humus beneath the trees forming an unsuitable seed bed; and it is only in the forests on the south fork of the Kaweah, and on the Tule river, where young trees of all ages can be found in abundance.

The damage, waste, and loss which has occurred in those groves which have been partially cut for timber is said to be enormous. When a large tree is felled its immense weight breaks a great part of the top into useless fragments, and crushes many other trees in its fall; whilst the usual means adopted to break up the logs into pieces which can be handled is by blasting; and this destroys another large part of the timber. When the best is removed, a mass of broken branches, timber, and bark, often 5 or 6 feet in depth, is left on the ground, which is later destroyed by fire; leaving complete devastation in place of the most beautiful forest; and it is said that owing to various causes, the lumbering of these forests has often been quite unprofitable to their owners.

Mayr[1] estimates the age of the largest tree which he measured, 33 feet in diameter at 13 feet above the ground, to be 4250 years. Sir Joseph Hooker[2] told Bunbury that, as the Wellingtonia makes repeated growths in the year, it is more difficult than is the case in other conifers to distinguish the shoot of one year from that of the preceding year; and he suspected that more than one ring of growth is formed in each year, and that in consequence the estimates of enormous age of this species are probably fallacious.

---

[1] *Waldungen Nordamerika*, 343 (1890).　　　　[2] Lyell, *Life of Sir C. J. F. Bunbury*, ii. 227 (1906).

Introduction

The Wellingtonia was introduced by Mr. J. D. Matthew, who visited the Calaveras Grove in July 1853, and sent home seeds immediately afterwards. In *Pinetum Britannicum*, p. 318, eleven trees of this origin are traced, namely, two each at Gourdiehill, Megginch Castle, and Ballendean, near Inchture, and one each at the Kinnoul Nursery near Perth, Newburgh, Balbirnie, Inchry House, and Eglinton Castle; but none of these were so large when he wrote as those raised from later consignments. Lobb visited the Calaveras Grove in the autumn of the same year, and returned[1] to England in December 1853, bringing with him a large quantity of seed and two living plants. The latter were planted out in Veitch's nursery at Exeter, but only survived three or four years.

Cultivation [2]

The culture of the Wellingtonia presents no difficulty if care is taken to have the roots thoroughly spread out when planted out. They are often kept too long in pots, which causes their main root to curl round, and when this has assumed a corkscrew shape it never loses it. If the tree is transplanted every year or so while young, it may safely be removed when 4 or 5 feet high. As regards soil and situation it is more accommodating than the redwood, and even in heavy soil is rarely injured by spring frosts. It grows very fast in most places up to 40 or 50 feet high, and then, unless the soil is deep and well drained, often becomes stunted and increases more in girth than height. If planted in a park or field pastured by stock, it must be very carefully fenced, as horses and cattle will gnaw its bark persistently and do it much injury. I noticed a good instance of this in the park at Mark's Hall, Essex, where some Wellingtonias had been so much bitten by cattle that they resembled the trees in a toy Noah's Ark, one about 35 years old being only 12 feet high by 3 feet in girth. When surrounded by other trees, where it cannot extend its branches laterally, the girth is much less in proportion. A tree that I saw in a plantation at Powis Castle, which was growing extremely well, was 75 feet high and only 7 feet 3 inches in girth, whilst one in the lower park at the same place was 81 feet by 16 feet.

Remarkable Trees

When first introduced this tree made such a sensation in the horticultural world, that it was planted almost everywhere, and there are specimens at every place of importance in the United Kingdom, many of which are very nearly equal in size. The tallest at Windsor Castle was already 21 feet high in 1865, and is now, as I am

---

[1] *Hortus Veitchii*, 39, 346 (1906).

[2] An interesting article on the causes of success or failure in plantations on a large scale of this tree in South Hampshire appeared in *Gard. Chron.* ix. 794 (1878).

told by Mr. A. MacKellar, 85½ feet high and 12½ feet in girth at 3 feet from the ground.

The largest and finest tree I have measured myself is in an open but well sheltered glade near the lake at Fonthill Abbey, and was, in November 1906, certainly over 100 feet and probably 105 feet high, by 17 feet in girth (Plate 106). This tree was raised at Eaton Hall from seed sent to Lord Stalbridge in 1861, and is not so old by seven years as many others in this country.

There are two very fine trees at Poltimore Park, Devonshire, the seat of Lord Poltimore, one of which I measured in August 1906, and found to be 98 feet by 16 feet 9 inches, but the gardener, Mr. Slade,[1] thinks it is taller. Near the Temple, at Highclere, there is a tree which I saw in 1903, of which the size is given by Mr. Storie, the forester, as 97 feet by 13 feet, but I could not verify this measurement myself.

One of the largest in girth that I know, is an ugly tree at Powderham whose top has long been broken, and which has formed immense branches, so thickly crowded that I could only get the tape round it with assistance. In 1906 it was no less than 17 feet 8 inches; but I must observe that the exact girth of such trees is of little importance, as it depends very much on the height at which the measurement is taken.

At Beauport, Sussex, there are two fine trees, one of which, planted in 1856, was, in 1904, 86 feet high by 13 feet 8 inches in girth. The other, a younger tree, measured 83 feet by 11 feet 11 inches. Sir Hugh Beevor measured in 1904 a tree at Hardwicke, Suffolk, 80 feet by 12 feet 4 inches, and another at Wooton, 85 feet by 14 feet 6 inches. Mr. R. Woodward, jun., reports in 1906 two trees at Wexham Place, Stoke Pogis, the residence of E. H. Wilding, Esq., which measured 79 feet by 9 feet 6 inches and 78 feet by 10 feet 8 inches. Canon Ellacombe[2] planted a tree at Bitton in 1855, which had attained 70 feet in height in 1888. A tree[3] at Wrest Park, Bedfordshire, planted in 1856, was, in 1900, 74 feet high by 15 feet 3 inches in girth.

At Strathfieldsaye there is a fine large tree,[4] planted in 1857, which I measured in 1903, and found to be 85 feet by about 12 feet. In 1907 it was 90 feet high, but the branches were so thick that I could not get the girth with sufficient accuracy to say how much increase it had made. A number of branches have become layered—a not uncommon occurrence in damp and sheltered situations—but when they have been allowed to remain long after taking root, it is perhaps better not to take them off, as this disfigures the tree for some time. There is also a fine avenue of this tree at Strathfieldsaye, which is more regular and satisfactory in growth than some others which I have seen. Another at Orton Longueville is no less than 700 yards from east to west, but some of the trees when I saw them seemed to be suffering from the wetness of the subsoil, the tops of many being stunted. There is also a fine

[1] Cf. Mr. Slade's remarks on this tree in *Gard. Chron.* xxvii. 406 (1900).
[2] *Gard. Chron.* iii. 801 (1888). [3] *Ibid.* xxvii. 373, fig. 121 (1900).
[4] In 1868, this tree was 24½ feet high; in 1872, 30 feet high; in 1895, 71 feet high; and in 1899, 79 feet. A cutting from it was struck, and planted out in 1875, when about 2 feet high, and had attained, in 1896, 30 feet in height and 6 feet in girth, at 4 feet from the ground. Cf. *Gard. Chron.* xix. 8 (1896) and xxvi. 162 (1899).

avenue at Luton Park, in which most of the trees are good specimens, but unless the soil is good and uniform throughout, and the trees are selected with great care, I should not recommend this tree for avenues.

By far the best avenue of this tree that I have seen is one near Wellington College, which was planted in the year 1869 by the late John Walter, Esq., of Bear Wood, Berks, on a light, sandy soil which, however, seems to have suited the trees remarkably well, and on which the symmetry of their tops and uniformity of the growth is remarkable. It is about 1200 yards long, running about north and south, and is 25 yards in width. The trees are planted at 18 yards apart, which is about the right distance for this tree. The average height of the trees is 75 to 80 feet, and the largest that I measured on the west side near the top was 87 feet by 21 feet. Plate 198, taken specially for this work, gives a very good impression of its appearance. Mr. C. E. Salmon tells me that there is another avenue of this tree which was planted in 1871 by the late Mr. J. Walter in front of Bear Wood House.

At Aston Clinton, Bucks, the residence of Lady A. de Rothschild, there is a group of closely planted Wellingtonias with tall clean stems, tapering only slightly, and carrying timber size to 50 or 60 feet. The ground, which is a circular area 120 feet in diameter, is covered with decayed leaves and is free from herbaceous vegetation. On it there are seventy-two trees in all, ranging in height from 60 to 75 feet, and in girth from 4 feet 4 inches to 8 feet 2 inches at 6 feet from the ground, above the swollen base. This clump was planted in 1869, according to Mr. W. H. Warren, who has kindly sent us a photograph, reproduced in Plate 197. This beautiful grove, which was seen by Henry in 1906, shows how well the tree succeeds when planted densely on good land.

At Brickendon Grange, Hertford, the property of John Trotter, Esq., in a wood, composed of a mixture of common spruce, Wellingtonia, *Cupressus macrocarpa*, *Abies Lowiana*, and *Pinus ponderosa*, all planted at the same time, in 1861, the comparative girths of trees, measured by Mr. H. Clinton Baker, in December 1907, without selection, at 5 feet from the ground, are as follows:—

Wellingtonia: 3 feet, 5 feet, 5 feet, 5 feet 2 inches, 5 feet 3 inches, 5 feet 10 inches.

Cupressus macrocarpa: 4 feet, 4 feet 2 inches, 4 feet 8 inches, 4 feet 9 inches, 5 feet 3 inches, 5 feet 6 inches, 5 feet 7 inches, 5 feet 10 inches, 5 feet 10 inches.

Abies Lowiana: 4 feet, 4 feet 10 inches.

Pinus ponderosa: 5 feet 2 inches, 5 feet 3 inches.

Common Spruce: 2 feet 2 inches, 2 feet 5 inches, 2 feet 6 inches, 2 feet 7 inches, 2 feet 8 inches, 2 feet 8 inches, 2 feet 10 inches, 3 feet 1 inch, 3 feet 3 inches.

Sir John Stirling Maxwell sends me the following measurements of some Wellingtonias planted in 1864 or 1865 at Cloverley Hall, Shropshire, the seat of Capt. Heywood Lonsdale, in mixture with spruce and larch.

No.	Height.	Quarter-girth at 5 Feet.	No.	Height.	Quarter-girth at 5 Feet.
1. . .	64 feet	26 inches	12. . .	68 feet	24 inches
2. . .	66 ,,	24 ,,	13. . .	61 ,,	19 ,,
3. . .	67 ,,	30 ,,	14. . .	65 ,,	23 ,,
4. . .	66 ,,	24 ,,	15. . .	71 ,,	24 ,,
5. . .	61 ,,	24 ,,	16. . .	69 ,,	27 ,,
6. . .	73 ,,	34 ,,	17. . .	66 ,,	29 ,,
7. . .	62 ,,	23 ,,	18. . .	72 ,,	27 ,,
8. . .	60 ,,	19½ ,,	19. . .	68 ,,	27½ ,,
9. . .	63 ,,	20 ,,	20. . .	62 ,,	21 ,,
10. . .	65 ,,	18½ ,,	21. . .	70 ,,	23½ ,,
11. . .	58 ,,	23 ,,			

He also measured some at 3 and 6 feet to show their rate of growth as compared with other trees, as follows :—

	Girth at 3 Feet. Feet. Inches.		Girth at 6 Feet. Feet. Inches.		
Wellingtonia, 1 . . .	8	0½	6	3	These trees
,, 2 . . .	8	7	7	2	were not se-
,, 3 . . .	9	2	7	2½	lected with
,, 4 . . .	9	5½	7	9	care, but were
Common spruce, 1 . .	5	7	4	5	all good spe-
,, 2 . .	3	9	3	4½	cimens of
Larch	5	6½	4	10	their kind.
Austrian pine . . .	6	0½	5	8½	

A tree planted in the pleasure ground at Cloverley by the late Mr. W. E. Gladstone in 1872, now measures 56 by 8½ feet; another planted Jan. 1, 1864, is now 65 by 11 feet.

In Scotland, the Wellingtonia has not attained as great a height as in England, but seems to grow well in many places. The finest I have seen is at Murthly, planted in 1857; this in 1891 measured 66½ feet by 9 feet 3 inches, and when I measured it in September 1906, had increased to 86 feet by 12 feet 5 inches. There is one at Castle Menzies which Hunter says was planted out of a pot in 1858, when it cost three guineas, and in 1883 measured 44 feet by 9 feet 3 inches. This has not grown much taller, though it had attained the immense girth of 21 feet when I last saw it in 1907. At Smeaton-Hepburn, East Lothian, a tree, planted in 1855, was measured in 1905 by Henry as 78 feet by 12 feet 9 inches. At Keir, Perthshire, there are several trees, the tallest of which measured, in the same year, 71 feet by 9 feet. At Haddo House, Aberdeenshire, a tree planted in 1857 and reported [1] to be 50 feet by 8 feet 4 inches in 1891, was, in 1904, 68 feet by 11 feet.

The largest trees, reported [2] by Renwick and M'Kay, are one at Buchanan Castle, Stirlingshire, which was 71 feet high by 9 feet 3 inches in girth in 1900,

[1] *Journ. Roy. Hort. Soc.* xiv. 501 (1892).   [2] *Brit. Assoc. Glasgow*, 1901, *Fauna, Flora, and Geology*, 144.

and another at Glendoune, Ayrshire, which was 9 feet 4 inches in girth in 1898.

In Ireland there are many fine trees, the tallest in the British Isles in 1891, of those reported at the Conifer Conference,[1] being one at Shanbally in Tipperary, which was then 80 feet in height by 8½ feet in girth.

The finest seen by Henry is growing at Ballykilcavan, Queen's County, the seat of Sir Hunt H. A. Johnson-Walsh, Bart., and measures 95 feet in height and 10 feet 10 inches in girth.   At Brockley Park, a few miles distant, the seat of W. Young, Esq., a tree measures 73 feet by 10 feet 9 inches.   At Emo Park, in the same county, there is a fine avenue, though the trees are growing on poor shallow limestone soil.   They are planted about 35 yards apart, and average 70 feet high by 10 feet in girth.   On the lawn at this place there is a finer tree, 81 feet high by 10 feet 4 inches ; and beside it, a redwood, planted at the same time, is only 50 feet high and doing badly.

At Coollattin, Wicklow, a tree measured, in 1906, 78 feet by 12 feet.   It produces fruit freely, but the seed does not mature and when sown has never produced seedlings.   At Churchill, Armagh, a tree, planted in a bog, was in 1905 67 feet by 12 feet, and looked very healthy.   Two trees, 77 feet and 73 feet high, were reported[2] to be growing in 1897 at Fassaroe, near Bray, in Wicklow.

The largest Wellingtonia of which I have heard on the Continent is a tree near the Hotel Bonnemaison at Bagnères de Luchon (Haute Garonne), of which a large photograph has been kindly sent me by the Hon. W. Rothschild.   This splendid tree measures rather over 91 high by 25 feet in girth at the ground.   A tree at Locarno,[3] on the northern end of Lake Maggiore, has attained, in 17 years after planting, a height of 72 feet and a girth of 9 feet 2 inches.   The species appears to be quite hardy in the severe climate of Munich, as Dr. Mayr says that at Grafrath one only 10 years old was nearly 19 feet high, and had endured a frost of $-25°$ Cent. without any injury ; though in the winter of 1902-1903, when the thermometer fell to $-28°$ Cent., the branches on the sunny side of the tree were somewhat browned.   Trees, however, at Berlin,[4] which had attained 30 to 40 feet in height, succumbed to the severe cold of the winter 1893-1894.

TIMBER

The timber is very light, soft, weak, and brittle, varying in colour from pale yellowish-brown to rich red-brown, with whitish sapwood, which occupies one to two hundred rings.   In native specimens these are extremely close, fifteen or twenty to the inch in some cases.

The wood is said to be very durable in contact with the ground, and is largely used for making vine stakes and for shingles, also to some extent for building, fencing, and box-making.   So far as I know it is never imported to Europe, and has no commercial value except locally.

[1] *Journ. Roy. Hort. Soc.* xiv. 571 (1892).   [2] *Gard. Chron.* xxii. 385 (1897).
[3] Christ, *Flore de la Suisse*, 77 (1907).   [4] Bolle, in *Garden and Forest*, 1894, p. 95.

In the British Museum of Natural History there is a section taken at about 18 feet from the ground, of a tree felled in 1892 at Fresno. The annual rings show that it was 1335 years old.

In *Garden and Forest*, v. 541-547, there is an account, with two excellent photographs, of the felling of a large Sequoia from which a specimen was taken for the Jesup Timber Collection in the Natural History Museum at New York. One of the pictures shows the tree in the act of falling; the other shows the stump, which at ground level was 90 feet in girth, and on its bark and outer edge fifty men are standing, and there is room for a hundred more. Mr. Moore, the Superintendent of the King's River Lumber Company, on whose land the tree was cut, estimated its contents at 400,000 feet, board measure, equal to about 40,000 cubic feet. (H. J. E.)

Printed by R. & R. CLARK, LIMITED, *Edinburgh.*

PLATE 120.

GYMNOCLADUS AT CLAREMONT

PLATE 127.

CEDAR ON MOUNT LEBANON

PLATE 128.

LEBANON CEDAR AT PAINSHILL

LEBANON CEDAR AT GOODWOOD

PLATE 129.

PLATE 130.

LEBANON CEDAR AT STRATHFIELDSAYE

PLATE 131.

LEBANON CEDAR AT PETWORTH

PLATE 132.

PLATE 133.

LEBANON CEDAR AT STRATTON STRAWLESS

Plate 134.

LEBANON CEDAR AT BIRCHANGER

PLATE 135.

CEDAR AVENUE AT DROPMORE

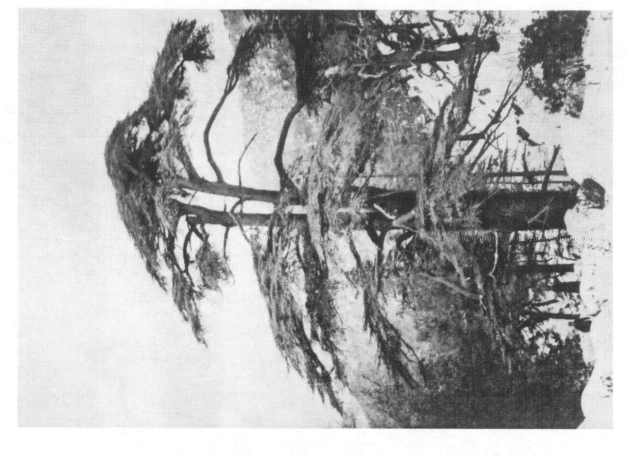

PLATE 136.

A

B

ALGERIAN CEDARS AT TÉNIET EL HÂAD

PLATE 137.

ALGERIAN CEDAR AT ASHAMPSTEAD

ALGERIAN CEDAR AT FOTA

PLATE 138.

PLATE 139.

DEODARS IN THE HIMALAYA

PLATE 140.

PLATE 141.

LIBOCEDRUS CHILENSIS IN CHILE

PLATE 142.

LIBOCEDRUS DECURRENS AT FROGMORE

PLATE 143.

CUNNINGHAMIA SINENSIS AT BAGSHOT PARK

PLATE 144.

B

NVSSA SYLVATICA IN AMERICA

A

LIQUIDAMBAR IN AMERICA

NYSSA AT STRATHFIELDSAYE

PLATE 145.

PLATE 146.

SASSAFRAS AT CLAREMONT

PLATE 147.

CORYLUS COLURNA AT WOLLATON PARK

PLATE 148.

HORNBEAM AT CORNBURY PARK

POLLARD HORNBEAMS AT BAYFORDBURY

PLATE 149.

PLATE 150.

HORNBEAM AT EASTON LODGE

PLATE 151.

HORNBEAMS AT WEALD PARK

PLATE 152.

HORNBEAM AT GORDON CASTLE

PLATE 153.

HOP-HORNBEAM AT LANGLEY PARK, NORFOLK

PLATE 154.

TASMANIAN BEECH AT FOTA

PLATE 155.

EVERGREEN BEECH AT BICTON

BEECH FOREST IN CHILE

PLATE 156.

PLATE 157.

ARBUTUS AT KILLARNEY

PLATE 158.

ARBUTUS HYBRIDA AT SEDBURY PARK

B

SCIADOPITYS AT HEMSTED

PLATE 159.

A

SCIADOPITYS IN JAPAN

PLATE 160.

SCOTS PINE AVENUE AT CARCLEW

PLATE 161.

SCOTS PINE AT BRAMSHILL

PLATE 162.

SCOTS PINE AT INVERARAY CASTLE

PLATE 163.

SCOTS PINE AT DUNKELD

SCOTS PINE AT GORDON CASTLE

PLATE 164.

PLATE 165.

SCOTS PINE AT LOCH MORLICH

SCOTS PINE AT ABERNETHY

PLATE 166.

PLATE 167.

SCOTS PINE AT ABERNETHY

PLATE 168.

SCOTS PINE IN BALLOCHBUIE FOREST

PLATE 169.

SCOTS PINE AT BALLOCHBUIE

PLATE 170.

HICKORY AT BUTE HOUSE

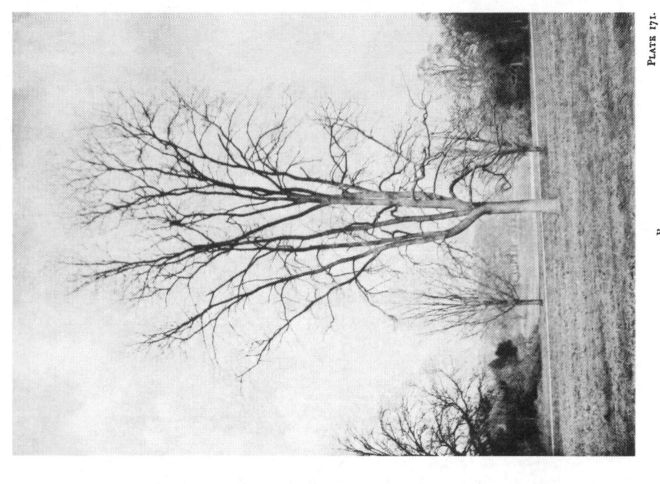

B

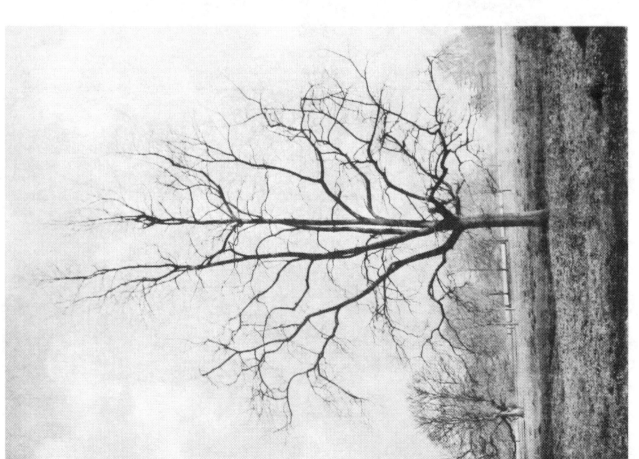

A

HICKORIES IN SYSTON PARK

PLATE 171.

HICKORY AT KEW

PLATE 172.

PLATE 173.

CARYA ALBA AT BROCKLESBY PARK

ORIENTAL PLANE AT ELY

PLATE 174.

PLATE 175.

ORIENTAL PLANE AT CORSHAM COURT

B

ORIENTAL PLANE IN SYRIA

A

WESTERN PLANE IN AMERICA

Plate 176.

PLATE 177.

RED MAPLE AT BAGSHOT PARK

LONDON PLANE AT ALBURY

PLATE 178.

SYCAMORE AT COLESBORNE

PLATE 179.

SYCAMORE AT NEWBATTLE

PLATE 180.

SYCAMORE AT DRUMLANRIG

PLATE 181.

SYCAMORE AT CASTLE MENZIES

PLATE 182.

SYCAMORE IN SWITZERLAND

PLATE 183.

PLATE 184.

COMMON MAPLE AT CASSIOBURY

COMMON MAPLE AT LANGLEY PARK, NORFOLK

PLATE 185.

PLATE 186.

NORWAY MAPLE AT EMSWORTH.

PLATE 187.

NORWAY MAPLE AT PARK PLACE

NORWAY MAPLE AT COLESBORNE

PLATE 188.

PLATE 189.

ITALIAN MAPLE AT HARGHAM

MONTPELLIER MAPLE AT RICKMANSWORTH

PLATE 190.

PLATE 191.

SUGAR MAPLE AT PARK PLACE

PLATE 192.

RED MAPLE AT WHITTON

WESTERN MAPLE IN AMERICA

PLATE 193.

REDWOOD AT CLAREMONT

PLATE 194.

PLATE 195.

REDWOOD FOREST IN CALIFORNIA

WELLINGTONIA AT FONTHILL

WELLINGTONIA AT ASTON CLINTON

PLATE 197.

PLATE 198.

WELLINGTONIA AVENUE NEAR WELLINGTON COLLEGE

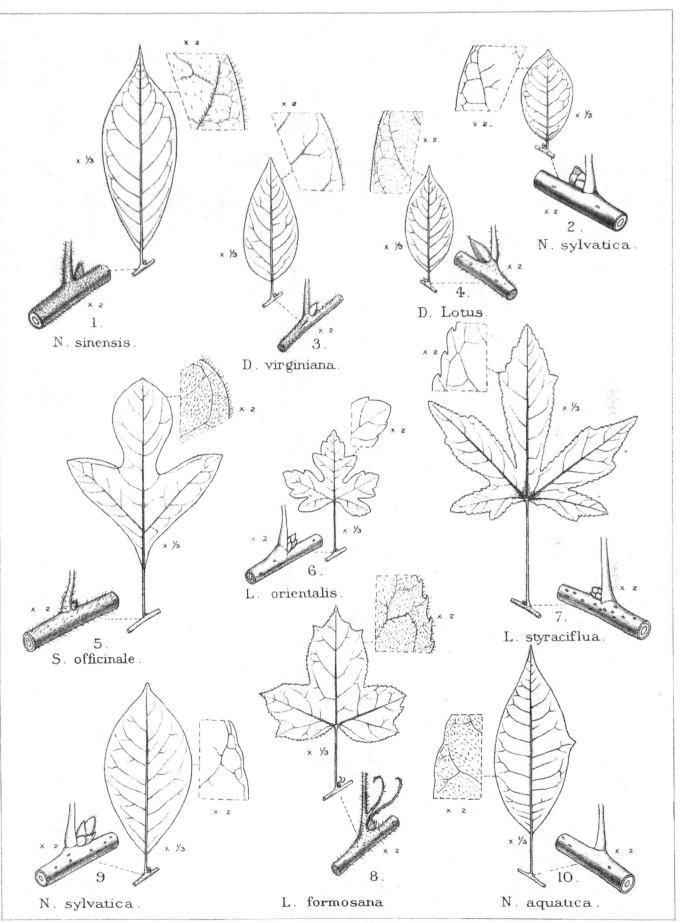

1.
N. sinensis.

2.
N. sylvatica.

3.
D. virginiana.

4.
D. Lotus.

5.
S. officinale.

6.
L. orientalis.

7.
L. styraciflua.

8.
L. formosana.

9.
N. sylvatica.

10.
N. aquatica.

Huitt, del. Huth, lith.

PLATE 199.

NYSSA, DIOSPYROS, SASSAFRAS, AND LIQUIDAMBAR.

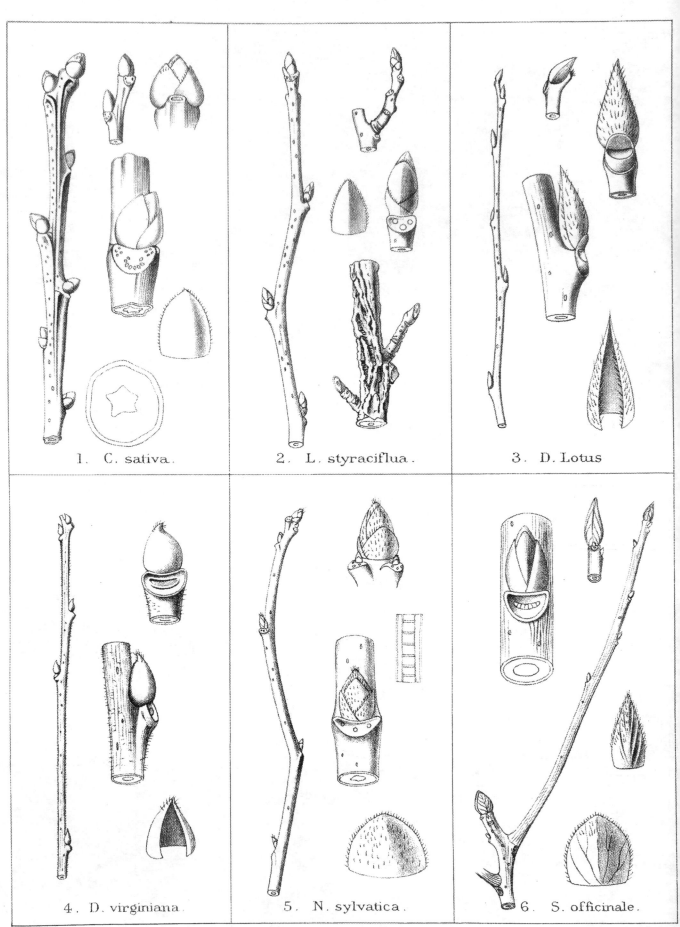

1. C. sativa.

2. L. styraciflua.

3. D. Lotus

4. D. virginiana.

5. N. sylvatica.

6. S. officinale.

PLATE 200.

CASTANEA, LIQUIDAMBAR, DIOSPYROS, NYSSA, AND SASSAFRAS.

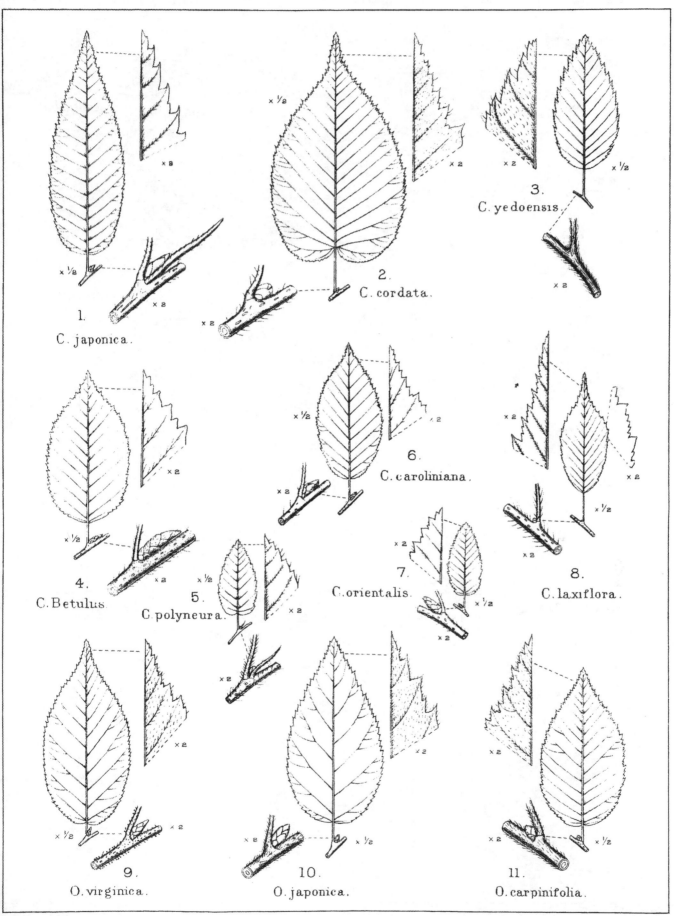

1. C. japonica.

2. C. cordata.

3. C. yedoensis.

4. C. Betulus.

5. C. polyneura.

6. C. caroliniana.

7. C. orientalis.

8. C. laxiflora.

9. O. virginica.

10. O. japonica.

11. O. carpinifolia.

Huitt del, Huth lith

PLATE 201.

CARPINUS AND OSTRYA.

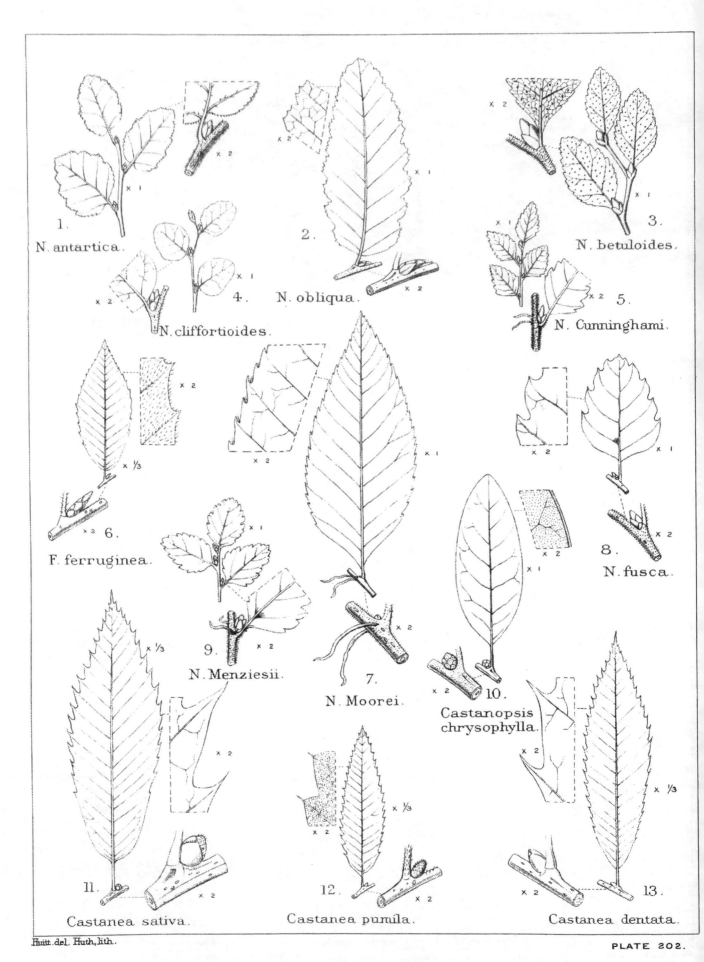

1. N. antartica.

2. N. obliqua.

3. N. betuloides.

4. N. cliffortioides.

5. N. Cunninghami.

6. F. ferruginea.

7. N. Moorei.

8. N. fusca.

9. N. Menziesii.

10. Castanopsis chrysophylla.

11. Castanea sativa.

12. Castanea pumila.

13. Castanea dentata.

Fuitt.del. Huth,lith.

PLATE 202.

NOTHOFAGUS, FAGUS, CASTANOPSIS, AND CASTANEA.

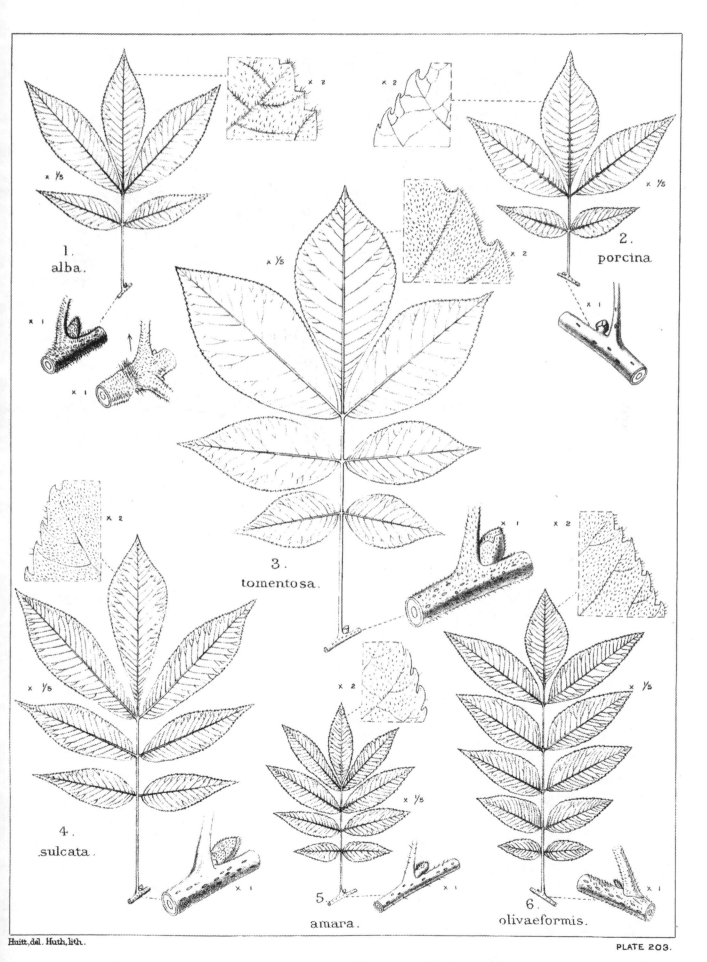

1.
alba.

2.
porcina

3.
tomentosa.

4.
sulcata.

5.
amara.

6.
olivaeformis.

PLATE 203.

CARYA.

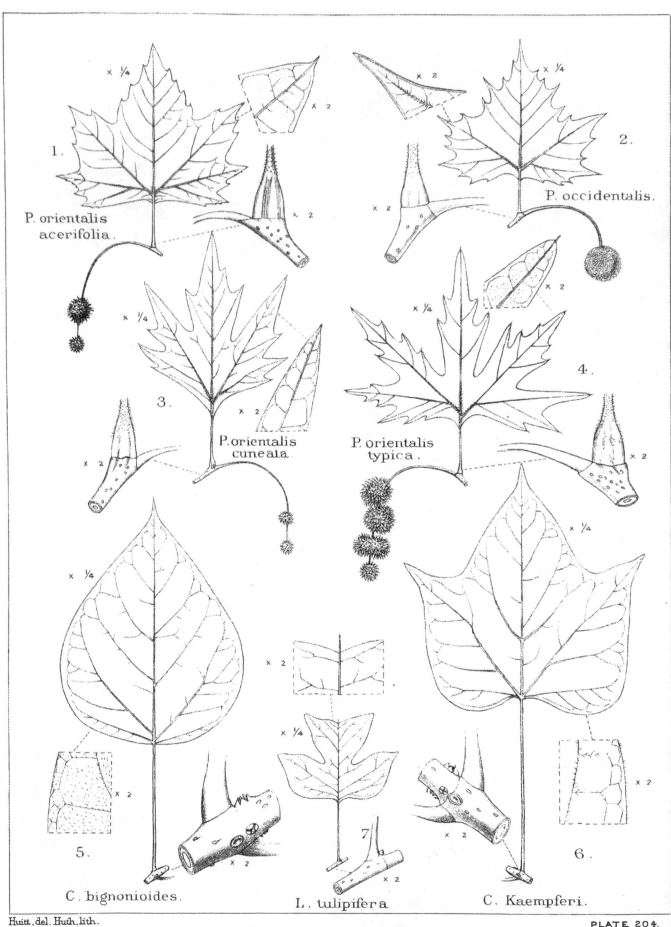

1.

× ¼

P. orientalis
acerifolia.

× 2

× 2

2.

× 2

× ¼

P. occidentalis.

× 2

× 2

3.

× ¼

× 2

× 2

P. orientalis
cuneata.

× ¼

× 2

P. orientalis
typica.

× 2

4.

× ¼

× 2

× 2

5.

× ¼

× 2

× 2

C. bignonioides.

× 2

L. tulipifera

× ¼

7.

× 2

× ¼

× 2

C. Kaempferi.

6.

× 2

PLATE 204.

PLATANUS, CATALPA, AND LIRIODENDRON.

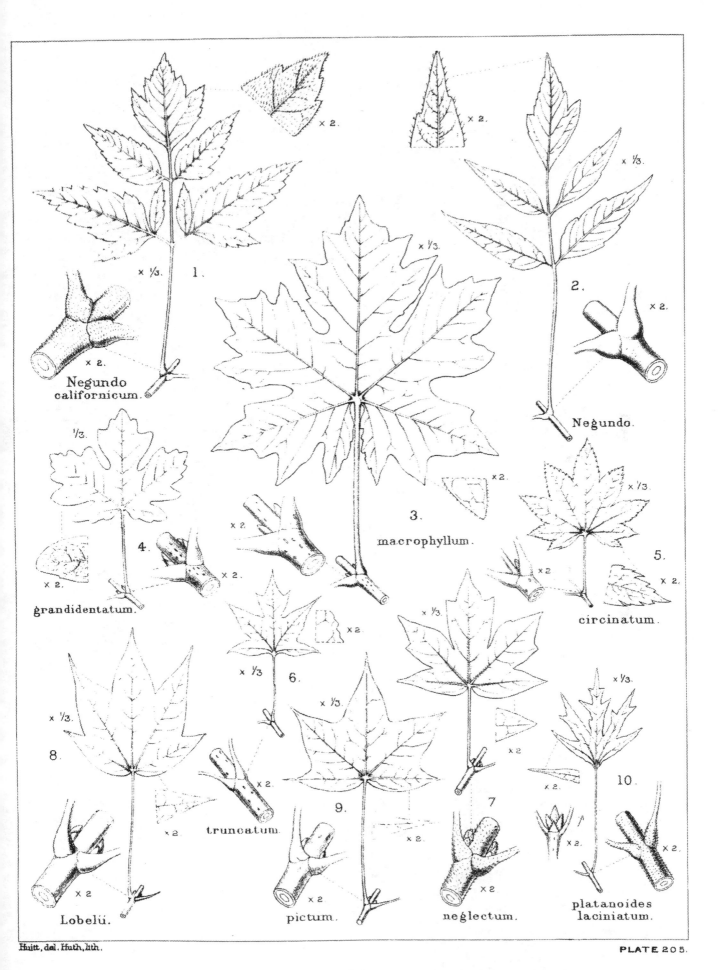

1. Negundo californicum.

2. Negundo.

3. macrophyllum.

4. grandidentatum.

5. circinatum.

6.

7. neglectum.

8. Lobelii.

9. pictum.

10. platanoides laciniatum.

truncatum.

Huitt, del. Huth, lith.

PLATE 205.

ACER.

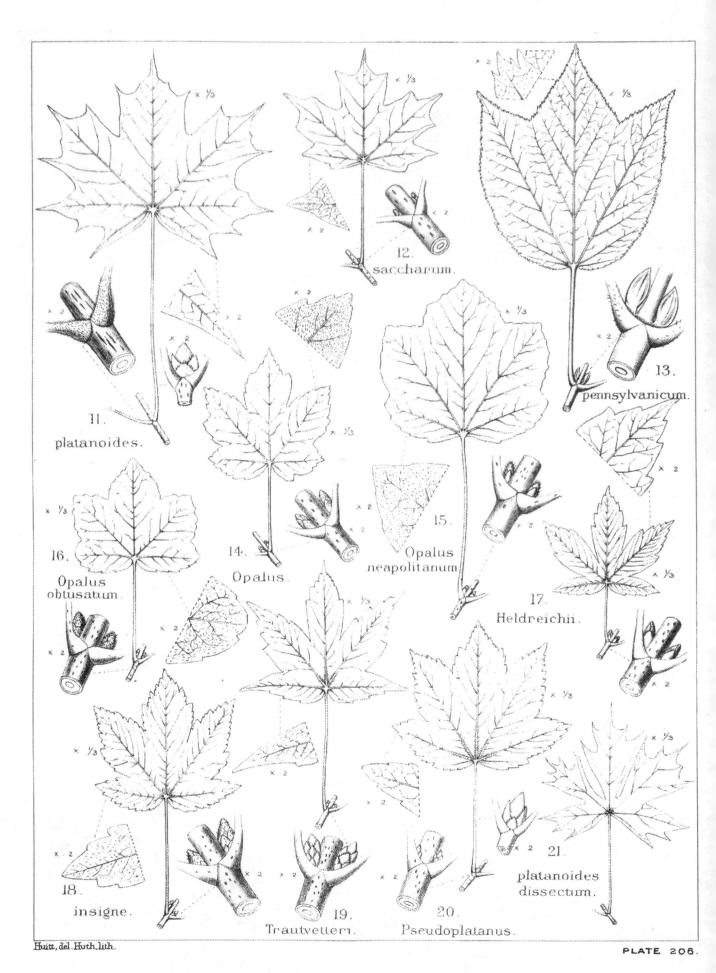

11. platanoides.

12. saccharum.

13. pennsylvanicum.

14. Opalus.

15. Opalus neapolitanum.

16. Opalus obtusatum.

17. Heldreichii.

18. insigne.

19. Trautvetteri.

20. Pseudoplatanus.

21. platanoides dissectum.

Huitt, del. Huth, lith.

PLATE 206.

ACER.

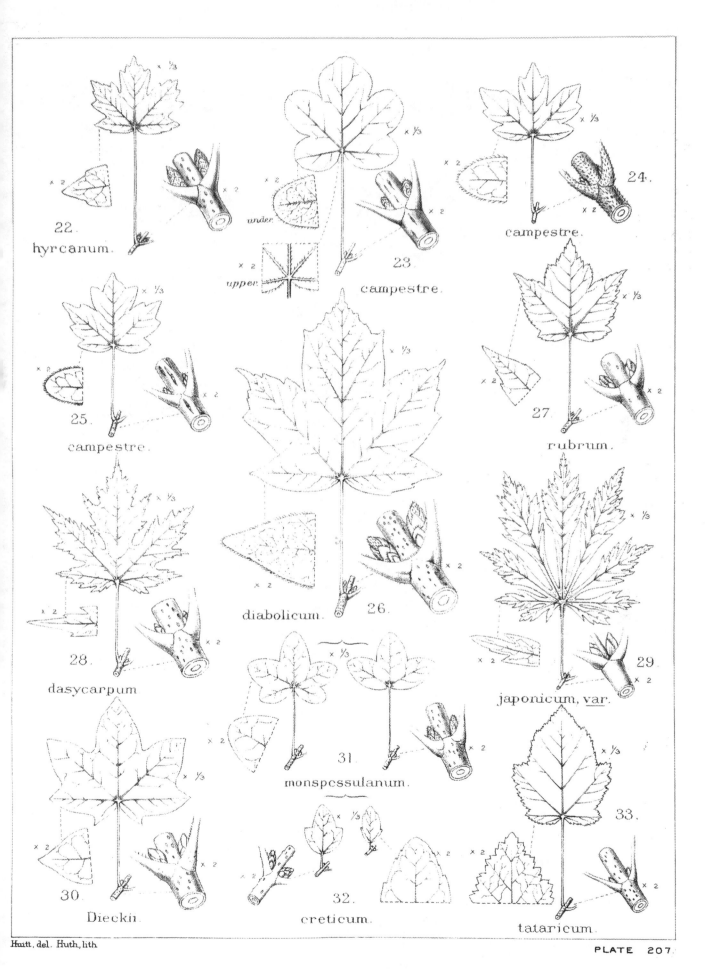

22. hyrcanum.

23. campestre.

under

upper

24. campestre.

25. campestre.

26. diabolicum.

27. rubrum.

28. dasycarpum

29. japonicum, var.

30. Dieckii.

31. monspessulanum.

32. creticum.

33. tataricum.

Hutt, del. Huth, lith.

PLATE 207.

ACER.

Printed in the United States
By Bookmasters